BY CHARLES MURRAY

Losing Ground: American Social Policy 1950–1980
In Pursuit: Of Happiness and Good Government

APOLLO

THE RACE TO THE MOON

Charles Murray

AND

Catherine Bly Cox

Simon and Schuster

New York   London   Toronto   Sydney   Tokyo

SIMON AND SCHUSTER
Simon & Schuster Building
Rockefeller Center
1230 Avenue of the Americas
New York, New York 10020

SIMON AND SCHUSTER and colophon are registered trademarks
of Simon & Schuster Inc.

Designed by Irving Perkins Associates
Manufactured in the United States of America

1 3 5 7 9 10 8 6 4 2

Library of Congress Cataloging in Publication Data
Murray, Charles A.
Apollo, the race to the moon / Charles Murray and Catherine Bly Cox

p.   cm.
Includes index
1. Project Apollo (U.S.)   I. Cox, Catherine Bly.   II. Title.
TL789.8.U6A558   1989
629.45′4--dc20   89-6333
                                              CIP
ISBN 0-671-61101-1

For all the people who gave their best to Apollo—
and for their families, who did too.

Contents

Acknowledgments

The idea for *Apollo* came from Jack Trombka, who told us fascinating stories about life in Building 30. Then, after we had written a précis but decided we didn't have time to do the book, *Apollo* survived because Amanda Urban, our agent, refused to take us seriously—and worked out a way that gave us time after all. To Jack and to Binky go our lasting gratitude for making it possible to live in the world of Apollo for the past four years.

At Simon and Schuster, Alice Mayhew first helped us to shape this ungainly project and then provided wise guidance the rest of the way, while David Shipley subjected each chapter, paragraph, and word to an astonishingly accurate editorial judgment.

Robert Sherrod generously opened to us his unique and monumental archive of material on the Apollo Program. We still wish we could read the book that only he could write, but in the absence of that, let it be understood that this book is partly his.

We received so much help from so many people in NASA that we will simply thank en masse the people in the public affairs offices, history offices, and photographic offices of Johnson Space Center, Kennedy Space Center, Marshall Space Flight Center, Langley Research Center, and NASA headquarters—plus, and especially, the audio office at J.S.C. Out of all those people, we must single out for special thanks Janet Kovacevich,

ACKNOWLEDGMENTS

Mike Gentry, Bob Lessells, Dionna Ormsbee, Lee Saegesser, Terry White, and Dick Young. Special thanks go as well to Sylvia Kennick and John Dojka of the George M. Low Papers at Rensselaer Polytechnic Institute.

Apollo depended ultimately on our interviews with the people who lived the experience. The interviews ranged from ten-minute telephone conversations to multiple sessions lasting for hours. The name of each contributor is listed in the discussion "*Apollo* as History" at the back of the book. We are grateful to them all.

Among the contributors, some went far beyond any ordinary standard of helpfulness. People rummaged through basements for flight control manuals, copied folders of old correspondence, showed us diaries, loaned us pictures, took repeated phone calls asking for clarification of one more detail, and occasionally fed us. Many of them patiently reviewed draft, often through two or three versions. We tried to pick out a few for special thanks, but it was too hard to know where to stop.

At least, however, many of our contributors have gotten in return what we hope is the pleasure of seeing their personal experiences described in the text. We would like to take this occasion to thank those who didn't, the unsung contributors who added so richly to our understanding of Apollo: George Abbey, Lee Belew, Woody Bethay, Michelle Brekke, Forrest Burns, Jay Campbell, Charles Clary, Pete Clements, Jim Cooper, John Cox, Chris Critzos, Connie Critzos, Paul Donnelly, Michael Duke, Lyn Dunseith, Robert Farquhar, Bob Fricke, Billie Gibson, Walter Haeussermann, Jack Heberlig, Bill Hess, Clay Hicks, Dick Hoover, John Humphrey, Carl Huss, Fletcher Kurtz, Jack Lee, Larry Lettow, Ed Lineberry, Mary R. Low, Bryce Lowry, Al Martin, Alexander McCool, Dave McKay, Joe Mechelay, Harold Miller, John Miller, Sonny Morea, Warren North, George Page, Henry Pearson, Bill Phinney, Andy Pickett, Don Puddy, Paul Purser, Ernie Reyes, Jack Riley, Glover Robinson, Ted Sasseen, Norman Sears, Jack Sleith, George Smith, Bill Sneed, Buddy Sparkman, John Stonesifer, Chet Wasileski, Walt Weisman, Charlie Welly, Terry White, Don Whiting, John Wood, Gary Woods, DeMarquis Wyatt, Ken Young, Raymond Zedekar, and Ed Zirnfus.

To all the men and women of Apollo who helped us, sung and unsung, please know that we are facing now the question many of you faced after the landing of Seventeen: What can ever be as much fun again?

Catherine Bly Cox & Charles Murray
Washington, D.C.
16 February 1989

Apollo Abbreviations

We have tried to make this book read the way the people of Apollo sound when they talk about the race to the moon. This has meant being faithful to a variety of idiosyncrasies. Take, for example, "Space Task Group," the organization that began the manned space program. People who were part of the Space Task Group never speak of it as "S.T.G.," so neither do we. The Kennedy Space Center may be called either "K.S.C." or "Kennedy," whereas the Marshall Space Flight Center is usually called "Marshall," not "M.S.F.C." We unquestioningly follow suit.

The big problem comes with acronyms, some of which are pronounced, some of which are spelled out. For example, people on the inside always refer to the National Advisory Committee for Aeronautics as "N.A.C.A.," pronounced "en ay see ay." They never say "nacka." But (don't ask why) they pronounce NASA "nassa." The same inexplicable distinction applies to lunar-orbit rendezvous, L.O.R. (always "el oh ar," never "lore") and the descent propulsion system, DPS (always "dips," never "dee pee ess").

Confronted with this arbitrary mix, we use an arbitrary rule: If an acronym is spelled out, we put periods after the letters—hence N.A.C.A. and L.O.R. If an acronym is pronounced as a word, no periods—hence NASA and DPS.

Prologue: May 25, 1961

On Thursday, May 25, 1961, President John F. Kennedy awoke to a glorious early-summer day, the kind that compensates Washingtonians for the oppressive summer to follow. The pleasant chill of the night was giving way to temperatures in the seventies, the skies were clear, the gardens on the Ellipse were in full flower. It was the right setting for a new beginning, and a new beginning was what the President intended to propose.

In the first year of John Kennedy's administration, the country was in transition between Eisenhower's fifties and the years that would be remembered as The Sixties. In movie theaters, the hits still tended to be big stories with big stars playing good guys—*The Guns of Navarone* with Gregory Peck and *Spartacus* were drawing crowds that spring. Popular music was still Elvis Presley and Connie Francis, the Marcels and the Shirelles—it would be more than two years before the Beatles had their first hit. Prices were still fifties-style too: *The Washington Post* that Thursday morning informed its readers that they could buy porterhouse steak for seventy-nine cents a pound at the A&P or a four-bedroom house in fashionable Chevy Chase for a price in the "mid-20s." As for unrest on the campuses, there was indeed a loud, disruptive demonstration at Harvard that spring—but it was to protest the substitution of English for the traditional Latin on diplomas. Timothy Leary was at Harvard too, still an anonymous teaching assistant, just beginning to experiment with a new drug called lysergic acid diethylamide, abbreviated L.S.D.

But the *Post* that morning also had harbingers of things to come. "MISSISSIPPI JAILS 27 RIDERS," read the banner headline, referring to two busloads of Freedom Riders who had been arrested in Jackson,

13

Mississippi. Their offenses consisted of trying to use the whites-only restroom and eat in the whites-only cafeteria. In other front-page news, Secretary of State Dean Rusk had gone out to Andrews Air Force Base to welcome Lyndon B. Johnson back from his first overseas trip as Vice-President. Johnson had been to Southeast Asia. There had been rumors that the United States might send troops to the troubled region, but Johnson firmly denied them. "Nowhere in Asia was there a call for American troops," he told reporters, although he of course "would not want to forever foreclose the possibility."

A military adventure was the last thing Kennedy wanted, in Asia or anywhere else. For one thing, he didn't want to spend any more money than he had to. In 1961, the federal government was planning to spend almost $95 billion, a huge sum, and Kennedy's economic advisers were warning him that the budget deficit might exceed $3 billion. Also, military adventures had not been working out very well recently. Five weeks earlier, a C.I.A.-sponsored brigade of Cuban refugees had landed at the Bay of Pigs. They had been stopped on the beach and killed or imprisoned, and the United States stood condemned for being at once reckless and irresolute. The very phrase "Bay of Pigs" was already entering the lexicon as a synonym for debacle.

The failed Cuban invasion was not unrelated to another of the *Post*'s front-page stories, regarding a speech the President was about to make. "President Kennedy will address a joint session of Congress at 12:30 today on 'urgent national needs' in what will amount to a second State of the Union message," read the *Post*'s lead. The story speculated about what the President might say and then turned to his motives: "Ever since the Cuban invasion fiasco the bloom has been off the bright rose of the early days of the new Administration," and today's speech, the *Post* explained, was part of an effort to recover the élan of the early spring. That Kennedy was going to Capitol Hill to deliver the speech personally—a step usually reserved for the State of the Union Address and the most momentous of occasions—indicated how seriously the new administration viewed the situation. It was a recent decision. Kennedy had originally planned to send Congress a written message; not until the day before had he sent word to House Speaker Sam Rayburn to call the joint session.

Eight hundred miles south of Washington, at Cape Canaveral, Florida, the morning was typically warm, humid, and sunny. But by the time the President began speaking, the Cape's volatile weather had changed. Dark clouds gathered in the eastern sky and lightning was beginning to flash offshore.

Rocco Petrone was driving to lunch at the cafeteria near Hangar S. Petrone, a thirty-five-year-old Army major, had just put a lid on his military career. Too absorbed by rockets and space travel to leave the Cape, he had turned down a chance to go to the Command and General Staff School. He was now head of the Heavy Space Vehicle Systems Office in Kurt Debus's Launch Operations Directorate. On May 25, Petrone was preoccupied with working out last-minute changes to the launch complex at Pad 34 that would be used for the new Saturn rockets.

Sitting beside Petrone in the car was Albert Zeiler, a member of Wernher von Braun's German rocket team and now one of Petrone's colleagues at the Cape. Petrone suggested to Zeiler that he turn on the radio; President Kennedy was scheduled to address Congress, and rumor had it that Kennedy would say something about the space program.

Kennedy was already speaking. When they reached the cafeteria, Petrone and Zeiler stayed in the car, rain beating against the roof as they listened to the rest of the speech.

It was a long speech, and the part they had been waiting for didn't come until near the end. By that time there was so much static from the storm that Petrone and Zeiler could make out only part of what the President was saying—something about "achievements in space," "longer strides." But Rocco Petrone did hear the important part, the passage that he remembered forever after as words that moved a nation. What he heard John F. Kennedy say was this:

> I believe that this nation should commit itself to achieving the goal, before this decade is out, of landing a man on the moon and returning him safely to earth. No single space project in this period will be more exciting, or more impressive to mankind, or more important for the long-range exploration of space; and none will be so difficult or expensive to accomplish.

Rocco Petrone turned to Albert Zeiler with a grin on his face and said, "Al, we've got our work cut out for us." Petrone, a man of theatrical flair, loved the drama of a nation undertaking this enormous challenge in full public view. He thought of it as saying to the world, "Here's the line we're going to cross."

Over at Hangar S, members of the Space Task Group that was in charge of America's manned space program were preparing a Mercury capsule for Gus Grissom's mission. Just three weeks earlier, Al Shepard's suborbital flight had given the Space Task Group their first fifteen minutes of experience in manned space flight. Merritt Preston, chief of the Group's contingent at the Cape, walked into the cramped office portion of the hangar and announced, "Well, we got it. We got the moon flight." Scott Simpkinson, who was in no mood for jokes after months of sixteen-hour days wrestling with the capricious Mercuries, growled something like, "Shut up, I ain't got time to be jokin' with you," and continued with his work. Simpkinson hadn't heard anything about this moon business. And Kennedy wanted them to do it by the end of the decade? He didn't take it seriously—"Hell, no."

Sitting at a nearby desk, Sam Beddingfield, the engineer responsible for the pyrotechnics and the recovery systems on the Mercury capsule, looked up at Preston. Beddingfield took Preston seriously, but he didn't get excited. He thought of himself and his fellow engineers at the Cape as ditchdiggers: They would be glad to dig any ditch anybody wanted when someone told them how wide, how deep, in which direction, and by when. Beddingfield figured that Kennedy had given them a good by-when. He didn't know whether they could actually get to the moon before 1970, but it was useful to have a fixed schedule to work to. It helped organize things.

Dick Koos, a youngster of twenty-four, was out on the floor of Hangar S, trying to get the procedures trainer to work. The procedures trainer drove the dials and switches in a mock-up of the Mercury capsule that the astronauts used for practice. By now it was supposed to be simulating orbital flights, but the trainer was a balky creature, full of bugs and glitches, and it always broke down while the astronaut was still putatively over Madagascar or some such place. When he heard the news, all Koos could think of were the television newsreels of rocket launches he used to watch in the enlisted men's day room at Fort Bliss back when he was an Army Pfc. Every time, it had seemed to Koos, the rockets would dance around on the pad and then blow up. He was still getting used to the idea of putting a man on top of one of those things—and now came word about the President's speech. It was more confusing than exciting. Here they were trying to get the procedures trainer to run for three hours at a time, and Kennedy was saying "Go to the moon"? Somebody, Koos thought, must know something he didn't.

A man who knew a lot that Dick Koos didn't, Robert Gilruth—director of the Space Task Group, chief operational officer of the American manned space program—was flying over the farms of the Mississippi Valley on his way to Tulsa when the speech began. He had asked the pilot to set up a radio patch so he could listen. Gilruth had known for a month that the President wanted to do something dramatic and he was looking forward to hearing the President ask for an enlarged and accelerated space program.

Unlike Rocco Petrone at the Cape, Robert Gilruth could understand the words easily enough. The reception was quite clear. The problem lay in what he was hearing. An accelerated program, yes; Gilruth had wanted that. A lunar landing, yes, in an orderly fashion, with time to work through all the difficulties that such an enterprise was bound to encounter. But not this. As the words sank in, as he realized that the President of the United States was asking an enthusiastic Congress to commit the nation's honor to getting a man to the moon and back before the end of the decade, Robert Gilruth, the man who was supposed to make good on it, was—he could think of just the one word for it—aghast.

Eight years and eight weeks later, on July 20, 1969, Neil Armstrong stepped off the ladder of the lunar module Eagle *onto the Sea of Tranquillity. This is the story of what happened in between, and something of what went before and came after. It is not a history, but the tale of a few of the people of Apollo. Some of them held high positions; some worked in the trenches. A few were in the public eye; the rest were not. They have in common that they remained on the ground (for this is not a story about the astronauts) and that each played a part, large or small, in putting men on the moon and returning them safely to earth. These few must stand for all the others, for Apollo was an epic, and epics must be captured in miniature.*

BOOK I

GATHERING

I talk to people who say, "Gosh, John, all we gotta do is think back twenty-five years ago and we can go to Mars the same way." I say, "No, you can't. It was a unique set of circumstances that lined up all those dominoes."
                                                            —John Aaron

1

"That famous Space Task Group is akin to the Mayflower"

On an April Saturday in 1959, a young Canadian engineer named Owen Maynard and his family crossed the United States border at Niagara Falls and made their way south. Five of them were crowded into their new green Plymouth with yellow trim—Maynard, his wife, Helen, and three small children in the back seat—plus enough luggage to set up house-keeping. Maynard, a trim man who claimed to be five foot seven "on a good day," had a perpetually curious, boyish look that made him seem younger than his thirty-four years. He was also a man with an adventurous streak—in World War II, still in his teens, he had been the youngest pilot to fly the high-performance Mosquito—and this was one of the qualities that had drawn him to this migration.

The Maynards drove all day on the narrow two-lane highways, out of a wintry Canadian April, across New York and Pennsylvania into Maryland, around Washington, D.C., and over the Potomac River into a full-blown Virginia spring. They stopped for the night south of Washington; the next morning, they continued south. In Richmond, they turned off busy Highway 1 onto an even narrower two-lane road, State Highway 60, heading southeast. The towns became fewer now, and smaller. Fifty miles outside Richmond they came to the only sizable town on the route, Williamsburg, and after that it seemed there was nothing but forests and marshland and an occasional farmhouse.

This was the "Peninsula," as it was known locally, a stretch of land bounded by the James River to the south, the Chesapeake Bay to the east, and the York River to the north. The rivers still had few bridges in 1959, and the sense of moving away from the rest of the world, even back in time, was unsettling. "You felt as if you were going to the end of the world," said one of the other Canadians who came south during that period. "Of course, it *was* like the end of the world to everybody who went there, because you couldn't get anywhere. You got down onto the Peninsula and that was it."

In mid-afternoon the Maynards arrived at the tip of the Peninsula and at their destination, a town called Hampton. It was still a backwater fishing town in those years—crab boats floating in a brackish tidal river, sea gulls shrieking overhead, ramshackle boatyards. The dilapidated Langley Hotel was the only place to stay in town, so Maynard's new employer had arranged for them to put up at the Hotel Chamberlain at Old Port Comfort, a few miles east. It was an improvement over downtown Hampton, but still a far cry from the tidy residential neighborhood in Bloordale, Ontario, that the Maynards had left behind.

On Monday morning, April 20, 1959, Owen Maynard drove a few miles northward from Hampton to report to work. His new employer was situated on a large government reservation divided into an East Area and a West Area. Between them lay a large airfield operated by the Air Force. Maynard was directed to the eastern and older half of the facility. The drive hugged the curve of Back River, opening out onto the Bay, with open water off to his left and broad empty fields on the right. On the outskirts of the East Area, Maynard came to his first indication that this was an unusual sort of place, a long wooden shed coming in off the water. The shed was a quarter of a mile in length—the hydrodynamic testing tank.

Past the shed, he came to a cluster of brick buildings that looked almost like a college campus, with neatly trimmed lawns and fields and tree-shaded streets. But interspersed among the dignified old buildings with their broad porches and elaborate brickwork were exotic ones of inexplicable shapes, huge spheres and structures with U-shaped appendages made of what looked like vacuum-cleaner hose—vacuum-cleaner hose twenty or thirty feet in diameter.

These latter buildings were the wind tunnels that Maynard had seen so often in his aeronautical engineering journals. Because he had heard and read so much about all this over the years, it felt familiar to him, as if this were a dimly remembered home to which he was finally returning. And in a sense he was coming home, as, in a sense, was any man in his profession who came to this place. For this was Langley, North America's oldest aeronautical research center and mother to all the rest. Owen Maynard, aeronautical engineer, had come to be a part of Langley's latest offspring, the U.S. manned space program.

1

The reason an American space agency was recruiting Canadian engineers went back to November of the preceding year, when a new agency called the "National Aeronautics and Space Administration," instantly shortened to "NASA," was cobbled together by Congress and President Eisenhower. NASA was in turn a reaction to the panic caused by the Soviet Union's launch of Sputnik I in October of 1957.

A small globe of metal with a few crude instruments weighing 184 pounds, Sputnik I was not much of a satellite by later standards. But it was the planet's first artificial satellite, and it was not America's. Two weeks later, the Russians compounded their triumph by orbiting Sputnik II, a much larger satellite with a dog named Laika aboard. Not only did Sputnik II carry the dog, which suggested that the Soviets were thinking about putting human beings into space, the final stage of the rocket had remained attached to the satellite—which meant, incredibly and ominously, that the Soviet rocket had managed to put a six-ton weight into earth orbit.* The United States, on the other hand, was working on a grapefruit-sized satellite weighing three and a half pounds. Two months later, the Department of the Navy tried to launch this puny competitor to Sputnik in front of television and newspaper cameras from around the world. The Vanguard rocket being used for the launch rose four feet into

---

* In ordinary use, "rocket," "booster," and "launch vehicle" refer to the same thing—the lifting portion of the assembly, of which the other portion is the payload. "Booster" is sometimes used more specifically to refer to the first stage. A "missile" is a launch vehicle with a warhead.

the air, fell back, and crumpled onto the pad in a spectacular explosion. A few days later, the Soviet delegate to the United Nations inquired solicitously whether the United States was interested in receiving aid earmarked for underdeveloped countries.

The public furor surrounding these events had been immense. Sputnik was the first time the Soviets had demonstrated superiority to the United States in any technological endeavor. It was especially galling to see them do it in a field as visible, as exotic, and as potentially dangerous as rockets and space exploration. In their humiliation, Americans lashed out at a variety of targets—the educational system, the military, Eisenhower's golf, American consumerism. There was also a disquieting edge to the furor. The *Dallas News,* for example, wondered whether there weren't after all "some advantages of tight, totalitarian control." It was a thought heard elsewhere: Maybe a little more regimentation and discipline in this country wouldn't be such a bad thing. *Life* joined the clamor. In an article entitled "Arguing the Case for Being Panicky," the author likened the American of the 1950s to "some fat Roman lolling in the baths of Caracalla shortly before the Visigoths sacked his city."

To President Eisenhower, this was all a lot of hysterical poppycock; but he also decided that he couldn't ignore it. The United States was going to have a space program whether or not he wanted it, that much was clear. All he could do was keep it within bounds.

One obvious place for the space program was Huntsville, Alabama, where the United States Army had stashed away the finest rocket scientists in the world: Wernher von Braun's Germans, "the rocket team," as it had been known in Germany, the men who had built the V-2. Von Braun's team could easily have beaten the Russians to the first satellite—their Jupiter had been capable of putting up a satellite for more than a year. But the Germans had been held back by a combination of jurisdictional disputes (only the Air Force was supposed to build missiles with intercontinental range) and the administration's lukewarm interest in rocket technology. These barriers fell quickly after Sputnik. Given the go-ahead, von Braun's people put America's first satellite, Explorer I, into orbit on January 31, 1958.

But Eisenhower was not willing to put the space program under von Braun. There were many reasons for this—among others, it would have been embarrassing for the American space program to be so blatantly a German enterprise. More important, von Braun was tied to the Army, and Ike was adamant that the American space program not be run by the military. This former General of the Army and Supreme Commander of the largest single military force in the history of the world was already harboring the fears of the military-industrial complex that he would voice at the end of his term.

Eisenhower found the perfect agency for his purpose in a venerable organization called the National Advisory Committee for Aeronautics, known to its friends as the N.A.C.A. The N.A.C.A. had been around for forty years, quietly going about its business of aeronautical research. It wasn't even an "agency" in the usual sense of the term but rather a collection of aeronautical research facilities. Ike chose to hand over the satellite business to this small, unheralded group of civilians. He submitted his proposal to Congress in April, the legislation was passed in June, and on October 1, 1958, a transmogrified N.A.C.A. became NASA, responsible for America's space program.

NASA in the beginning consisted of four "centers"—the semi-autonomous, competitive, insubordinate duchies that had formerly made up the N.A.C.A. The Lewis Research Center was in Cleveland, Ohio, a blue-collar center for a blue-collar town, a place where airplane propulsion systems were developed and tested. The Ames Research Center was just south of San Francisco, created during World War II to service the West Coast's growing aircraft industry. It now boasted a huge wind tunnel capable of testing full-size models of new aircraft. Also on the West Coast, down south in the Mojave Desert, was Edwards High Speed Flight Station, famous for Chuck Yeager and Scott Crossfield, where the X-1 had broken the sound barrier and the X-15 was about to fly.

Ames and Edwards would play little role in the manned space program. Lewis would contribute some critically important men to the program, but would do little of the actual work. It was left to the mother center, Langley, established in 1919, to serve as the birthplace of the manned space program. To understand how men came to walk on the moon, it is first of all necessary to understand something about the eccentric and beguiling and sometimes exasperating little world that was Langley.

In 1917, the United States government realized to its dismay that the United States, home of the Wright brothers, couldn't build a plane to compete with the ones that England and France and Germany were flying in the Great War. So the government created the N.A.C.A. and provided funds for a research center which, for reasons having to do with land prices and flying weather, was located near Hampton. The facility's full name was the Samuel P. Langley Memorial Aeronautical Laboratory, but everyone called it "Langley."

The people who decided to put it there failed to consider whether young engineers graduating from America's best schools in places like New Haven and Princeton would want to live on the tip of a rural Virginia peninsula. As small and isolated as it may have seemed to the men arriving in 1959, it was infinitely more so in earlier years. In 1917, when Langley began, Hampton was a place where people still referred to the War Between the States as "the late war."

This isolation meant that a great many bright young aeronautical engineers did not want to come to Langley, and instead chose more cosmopolitan places. It also meant that the ones who did make the trek down the peninsula tended to be highly motivated and a little odd. That was the first of the things that made Langley a world of its own.

The citizens of Hampton, Virginians to the marrow, didn't think much of the arriving Yankee engineers with their college educations. The Hamptonians called the newcomers "Nacka nuts," or sometimes the "brain busters," and had as little to do with them as possible.[*] So the engineers fraternized with one another almost all the time, on and off the job.

Another circumstance isolating Langley was the idiosyncrasy of a fellow named John Victory, the N.A.C.A.'s executive secretary in Washington from its founding in 1917 until its demise in 1958. Under Victory, the ultimate fussy bureaucrat, Langley developed the wackiest administrative system in the history of the United States government. Because the executive secretary could make life unbearable for the engineer-in-charge down at Langley who failed to meet his demands, the engineers-in-charge from the 1920s onward completely centralized Langley's communications with the outside world. Every outgoing letter was reviewed by the individual branch chiefs and then sent out over the signature of the engineer-in-charge himself. Incoming letters were treated in the same way, opened and read by the engineer-in-charge's office before being routed to the people to whom they were addressed. Telephone calls were subject to the same kind of regulation. And because Victory insisted on conformity to the most minutely prescribed conventions of writing style and usage, Langley soon developed a common and idiosyncratic language into which new employees were quickly indoctrinated ("We don't say it that way here at Langley").

The system was almost comically restrictive, but it created a peaceful cocoon within which the engineers of Langley lived and worked, buffered

---

[*] The blame for the estrangement was not all on one side. Soon after arriving in Hampton, the first engineer-in-charge was invited to speak at the Rotary Club. He took for his theme the great benefits Hampton was going to derive from having Langley nearby. Finally, he told the Rotarians, Hampton would get a little culture.

from the politics of Washington, buffered from the exigencies of a competitive aircraft industry. Buffered as well by geography, it is not surprising that the engineers out on the Peninsula soon developed their own comprehensive, seamless way of working—the Langley Way.

As the decades passed, the N.A.C.A. remained small. It wanted to remain small. "There was no one there with big ambitions of making a billion-dollar agency out of it," one old Langley hand remembered. "The fact is, we avoided big projects." And they went about their research at their own Langley pace. A Langley veteran recalled one night when he stayed late—not very late, perhaps six o'clock—and got locked within the building. He called Security, and someone came over and let him out. "Let that be a lesson to you," he was told. "Don't stay after working hours." Another one-time Langley engineer remembered nostalgically that when you worked at Langley, "time was not really a factor. The question was the elegance of what you were going to create." That, however, was a major proviso: If you were a Langley engineer, you didn't have to work long hours, but you did have to be perfect—or, at least, never publish a mistake.

To that end, every research report that Langley published went through an agonizing review process, "almost like writing a Ph.D. thesis," in the words of one who suffered through it many times. When an engineer had written his report, it would be reviewed by his branch chief, his division chief, specialists selected by the office of the engineer-in-charge, and, finally, an editorial committee chosen specifically for each paper. At each level of review, the author would be told to make revisions. That was for the technical content. At the end, the Editorial Office would vet the paper for syntax and grammar (so the prose would conform to "the way we say things at Langley"). It could take months to get what Langley called a "Technical Note" through the process.

But if the pace was slow, the work was indeed elegant. Langley built superb facilities to do all its own development and testing. Langley never contracted any of its work to outsiders—a point of great pride. It had the best wind tunnels, the best model-builders, the best technicians, the most rigorous standards. An Air Force general, having listened to one of N.A.C.A.'s day-long technical briefings, once approached John Victory. The presentations had been superb, he said. Could Victory possibly send some people to the general's command, to teach them N.A.C.A.'s methods? No point in it, Victory sniffed. The general's people wouldn't be able to use them. The Air Force didn't have the necessary discipline.

Langley also sustained a uniquely collegial atmosphere. "What was so marvelous about that N.A.C.A. group,'' said a senior NASA official who himself came out of industry, ''was the opportunity to do real research and to really move things forward. I've heard those stories from the fellows—'I went to work for such-and-such company and found myself on a drawing board, said, ''Aw, to hell with it,'' and then went down to Langley and a little later I'm working with Tommy Thompson. . . .' '' Engineers could do that at Langley—walk in as bright young kids fresh out of school and start working side by side with some of the best aeronautical minds in the country—men like Floyd (Tommy) Thompson, Hartley Soulé, Bob Jones, Abe Silverstein.

In the Langley world there was a thin line between work and play. Some Langley engineers flew their own airplanes. Others designed their own houses, leading to some of the strangest-looking homes on the Eastern seaboard. But most commonly, the engineers of Langley would go home and build model airplanes—not from kits, but from their own designs, carved with their own tools, using their own hands. They had a club called the Brain Busters, after the Hampton label for Langley engineers. On Sunday afternoons they would go down to the open meadows adjoining the airfield, a senior branch chief like William Hewitt Phillips alongside a brash junior engineer like Max Faget or a local Hampton boy like Caldwell Johnson or a gangly shop technician like Jack Kinzler. Each would bring his latest creation, a marvelously crafted machine with design touches that might be years ahead of their time. They paid no attention to rank, only to how the airplane performed and how it could be improved.

From this strange mixture of ambling Southern pace and obsessive perfectionism, family-like closeness and devotion to vocation, Langley turned out gems of aeronautical research. ''It wasn't like NASA,'' reminisced Caldwell Johnson, who would become one of the lead designers of the Apollo spacecraft. ''The press didn't care about it—to them, it was a dull bunch of gray buildings with gray people who worked with slide rules and wrote long equations on boards.'' But to the engineers within, it was a unique and wonderful place. ''Just a splendid organization,'' Johnson remembered fondly, and that's the way all the old Langley hands talked. Langley in those years became a kind of Elysium for young engineers in love with aeronautical research.

On the same day that NASA officially replaced the N.A.C.A., Keith Glennan, NASA's first administrator, announced that the United States would put a man into space. The Russians were openly working toward this end; the United States didn't want to be humiliated again; and Eisenhower (once again, reluctantly) went along with a small program to try to beat them. Looking for people to do it, Glennan turned to Langley and, more specifically, to an engineer named Robert R. Gilruth.

Gilruth, forty-five, avuncular and balding, was the prototypical Langley engineer. As a boy in Duluth, Minnesota, Gilruth had made model airplanes out of balsa with rubber-band engines. By the time he was in his teens, he was writing to the N.A.C.A. for technical papers on wing sections. At the University of Minnesota he helped design and build the Laird Watt, the fastest airplane in the world. He went to work at Langley in 1937 and was assigned to Flight Test Engineering. By the time World War II began, he ran the division.

Unlike Wernher von Braun, who had been fascinated by space flight from the time he was a teenager, Gilruth got into the space business by accident. It was the middle of World War II, and Gilruth was receiving urgent requests for data on drag at transonic speeds. The Langley wind tunnels couldn't provide it, because wind tunnels in those days were all closed-throat, and closed-throat tunnels choked at about Mach .85. Since Gilruth couldn't get data on transonic speeds in the wind tunnel, he had to create something that actually went faster than the speed of sound. Gilruth calculated that a properly streamlined shape dropped from a sufficient height would break the sound barrier. He and his engineers created such a shape, added some instrumentation, and dropped it from a B-29 at 30,000 feet. The idea worked. The bomblike shape achieved a velocity of Mach 1.2 and Robert Gilruth provided the American aviation industry with its first good look at the transonic flight environment.

Then it occurred to Gilruth and his engineers that if they mounted their supersonic shape on a rocket, they could drive the Mach numbers much higher. That was an attractive prospect to the N.A.C.A. and the Air Force, because the X-1, the first plane intended to fly faster than the speed of sound, was already on the drawing boards. In early 1945 Gilruth was given a small supplementary appropriation to open a testing facility. He chose a site at Wallops Island, Virginia, a deserted stretch of Atlantic coast across the Chesapeake Bay seventy miles northeast of Langley. Space flight still had nothing to do with it. Gilruth, now running a unit called the Pilotless Aircraft Research Division (P.A.R.D.), was just trying to get data about heat transfer at high Mach numbers.

But there wasn't much to do at Wallops in the off-hours, and Max Faget had a collection of *Astounding Science Fiction* that everybody kept swapping around, Gilruth as enthusiastically as anyone. Besides, it was impossible to fire rockets into the sky for long without beginning to think about space flight. During the late 1940s and early 1950s, as P.A.R.D. began working with multi-stage rockets, Gilruth and his engineers became the American civilians who knew more than anybody else about how to build a craft that could be put on top of a rocket and survive the dynamic stresses of launch and then the heat and deceleration of entry.

When Eisenhower decreed that military experts on these topics were out of bounds, Gilruth became the natural choice for the person to lead an effort to send a man into space. He was appointed director of a small new entity created on November 5, 1958, and called the "Space Task Group."

"That famous Space Task Group is akin to the *Mayflower*," one man who wasn't a member pointed out, "considering how many people tell you they were in it." As a matter of historical record, there were just forty-five in the first boat. From Langley, there were twenty-seven men (all engineers) and eight women (all secretaries and "computers," as the women who worked the calculating machines were called). Another ten engineers were assigned to the Space Task Group from Lewis Lab in Cleveland.

From the beginning, the Space Task Group had a catch-as-catch-can quality. Half of the men assigned to the Space Task Group from Langley were handpicked by Gilruth. But Tommy Thompson, the director at Langley, didn't want his laboratories stripped of their best men. Gilruth recalled later that Thompson "was all for me, because he knew that if we didn't succeed, NASA wouldn't succeed. But when I gave him my memo of all the people I wanted he said, 'Okay, Bob, but for every one that you want to take from Langley, I want to add one that *I* want you to take from Langley.' " And so that's how it worked out. "All of them to this day wonder which half they were in," Caldwell Johnson wryly observed.

Practically everyone was young. At forty-five, Gilruth was the old man. His deputy, Charles Donlan, was forty-two. The head of engineering, Max Faget, was thirty-seven, as was Chuck Mathews, head of flight operations. The rest were mostly in their early thirties or twenties, all the way down to the baby of the original forty-five, Glynn Lunney at twenty-one.

They were uniformly young partly because so few of the senior staff from the old N.A.C.A. wanted to be part of the Space Task Group. In 1958 and 1959, manned space flight did not look like the kind of venture on which to bet one's career. So far, Congress and the President had authorized only Project Mercury, in which a man riding in one of the Space Task Group's capsules atop one of von Braun's Redstones would be lobbed into the upper reaches of the atmosphere; on subsequent Mercury missions, he would be launched into earth orbit on the Air Force's Atlas. No one knew what was supposed to happen after this initial series of launches. No one even knew whether men could function in space. No one knew whether there was anything useful for them to do even if they could function. And the whole thing was going to be unspeakably expensive, with no obvious return on the investment. To most of the senior engineers at Langley, manned space flight had the earmarks of a fad.

Nor was it just that manned space flight was a risky career path. To many of the Langley engineers, it was an unattractive way to spend one's time. "You must remember," said Chris Kraft, the man who would later become famous as the manned space program's director of Flight Operations, "that 'space' was a word that wasn't even allowed in the N.A.C.A. library. If it was anything that had to do with space, that didn't have anything to do with airplanes, so why were we working on it?" To one young engineer in P.A.R.D., it seemed as if the people who went to the Space Task Group were "outcasts." A senior Langley engineer came to him and told him not to go with the Space Task Group. "I have a job for you in my division," he said to the young man. "You don't want to ruin your career. There's nothing going to come of this, and you're going to be hurt by it."

It just wasn't the Langley Way, this business of "implementing a manned satellite project," as the order creating the Space Task Group charter had phrased their mission. Langley didn't "implement" things; Langley did research, and it did everything itself. This manned space project was going to contract work to the outside. It would turn its engineers into bureaucrats. Gilruth remembered colleagues coming up to him during the first years and asking him, "Well, have you let any good contracts today?" Far from envying him his new position, most of his peers thought that he had chosen "a horrible fate," and none of them took the same road. Not one of the other engineers of Gilruth's age or older transferred into the program. "They wanted to support us," said one engineer who went to the Space Task Group, "but through the traditional research avenues." Thus, in the years that followed, old-line Langley personnel would decisively affect the decision on how to get to the moon and would help to develop the lunar-landing trainer at Langley. But they did not join up.

The old Langley staff were right to be suspicious. The Space Task Group *was* going to cut corners and ignore protocol in ways that pained those trained in the Langley Way. It *was* going to contract out much of its work. And it was going to abandon airplanes, the only proper flying machine for a Langley engineer to work on. The Space Task Group was going to do something that had never been done before.

Glynn Lunney later realized that the ignorance of these early days was bliss. "A lot of the senior engineers thought the project was crazy, and they were knowledgeable enough in the ways of the world to ask whether they wanted to risk their reputations," Lunney recalled. "I fortunately was unknowledgeable in the ways of the world and said, 'Gee, that looks like it would be a hell of a lot of fun to me—let's go do that!' "

Jack Kinzler reacted the same way. In the year before the Space Task Group was formed, Kinzler, one of the master craftsmen in whom Langley took so much pride, had begun to get excited about space. Wernher von Braun had come up from Huntsville to Langley and given a talk about space flight. P.A.R.D. had given some in-house seminars. "I got so consumed with space I was just waiting for somebody to initiate an actual program," Kinzler said later. "So when I was asked to join the Task Group with Bob, I said I would drop anything to do that." As the word spread, Kinzler was overrun with applicants. "I'd have fellows come up to me and say, 'I want to go with you, Jack, I've been reading about it, I've been hearing about it.' So I got the dedicated guys who read all the space magazines and cared about the program."

Sometimes, it wasn't such a calculated decision. The workload facing the Space Task Group was so overwhelming that it exerted a kind of gravitational pull. John Mayer, for example, one of the original forty-five, had formerly been part of the Aircraft Loads Branch, along with Carl Huss and Ted Kukunsky, who weren't. But no more than a week after the Space Task Group had been formed, Mayer was back visiting the Aircraft Loads Lab in the West Area, asking his friends to do some computer runs for him. The branch chief of Aircraft Loads said okay, Carl and Ted could bootleg a little work to help out Johnny. And within another two weeks, Huss and Kukunsky were doing more work for Mayer than they were for Aircraft Loads. "Ted and I sat across from each other at the same desk," Huss recalled. "One day we looked at each other and asked why we didn't transfer over to the Space Task Group. So we did."

The first group was young and enthusiastic, but it was also tiny for such an ambitious project. By February 1959, the original forty-five had grown to only a little more than a hundred people, and yet they were supposed to put a man into space. Moreover, "there was this gap," as Lunney put it. "We had these super generals and these super privates who were learning how to be corporals. But we didn't have a hell of a lot of guys in between."

And that's where Owen Maynard came in. In the spring of 1959, as the Space Task Group's burden was threatening to overwhelm it, the Canadian government unintentionally gave the American space program its luckiest break since Wernher von Braun had surrendered to the Americans.

2

For several years, the Canadian aircraft corporation A. V. Roe, known as AVRO, had been designing and building what was expected to be the most advanced interceptor in the NATO inventory, a long-range, all-weather, fly-by-wire aircraft with a combat speed of Mach 1.5, expected eventually to increase to Mach 3.0. The plane was called the Arrow, known to the engineers as the C.F.-105. A prototype had flown its first test flights in October 1958, and by February 1959 the Arrow had reached a speed of nearly Mach 2.1 and was within a few more test flights of setting several world's records. But then the Conservative Party under John Diefenbaker won a parliamentary majority and replaced the Liberal government. The new government decided to cancel the Arrow. The decision was announced in Parliament on February 20, 1959, at eleven in the morning. Four hours later, AVRO, in a move that was intended to embarrass the government and force the issue to a head, announced over its P.A. system that 14,000 employees—virtually everyone associated with the Arrow—were being laid off. The government responded by forbidding AVRO to spend another cent on the project. Within a few days, crews with cutting torches were in the Malton hangars, slicing up the Arrows for scrap.

The cancellation turned into a political scandal of huge proportions and brought anguish and economic hardship to many of the 14,000 employees at AVRO. But there was a bright side, if you happened to be Bob Gilruth and desperately short of talent. Four thousand engineers who had designed the most advanced supersonic airplane of its day were suddenly looking for jobs.

Jim Chamberlin, chief of design for the Arrow, had visited Langley a few years earlier. It had been during the flying saucer craze, and Chamberlin, who had designed a real flying saucer, a pancake-shaped thing with jets, came down to Langley to talk about it. When the Arrow was canceled, Chamberlin thought of the Space Task Group and talked to his chief engineer, who got in touch with the Canadian government. The Canadian government got in touch with NASA, and within a few weeks it had been concluded that some of the AVRO engineers would spend two years working for the American manned space program before returning to AVRO (when, presumably, AVRO would have new work for them). This would be good for the United States, which needed the talent, and good for Canada, whose engineers would return with experience in a new technology.[*]

AVRO let Chamberlin put together a book with background information on approximately 150 of the best people in his design team and take it down to Langley. Gilruth, Donlan, and two other Space Task Group managers flew up to Ontario and interviewed seventy-five of the candidates. They offered jobs to thirty-five of them.

One of the thirty-five was Owen Maynard. Maynard was interviewed on a Saturday morning. On Sunday morning, he got a call from Gilruth offering him a job and asking if he could please let them know whether he would be accepting the job by, say, one o'clock that afternoon? It didn't take Maynard that long—not only did he want to help build the Mercury capsule, he wanted to fly it, and he enthusiastically touted his experience with the Mosquito. No, Gilruth said hastily, they weren't recruiting flight crews on this trip. But they were glad to sign him on as an engineer.

Others hesitated. "You have to understand," Rod Rose, one of several Englishmen in the group, pointed out, "there's considerable prejudice north of the border about coming south of the border. South of the border is a big ogre." So at first Rose refused the offer, but Jim Chamberlin gave him a two-hour pep talk and persuaded him that his future lay at Langley.

Twenty-five of the thirty-five accepted jobs right away and another five came along later—in all, thirty men were added to the Space Task Group, which at that time still numbered not many more than a hundred people. Gilruth and his interviewers returned from Canada elated. However improbably, Project Mercury had skimmed off the top layer of talent at AVRO. Tecwyn Roberts, a Welshman, remembered Gilruth laughing

---

[*] This two-year arrangement never took hold. By the time the AVRO engineers were heading south, they had been given green cards, good for an indefinite stay in the States, and few of them were thinking in terms of returning.

about it later. "We thought of taking more of your crowd from AVRO," Gilruth said to Roberts, "but we figured twenty-five percent aliens in the American space program was sufficient."

The Canadians (as they were known, despite the mix of Scotsmen, Welshmen, Irishmen, and even a Frenchman) never gained much public recognition for their contribution to the manned space program, but to the people within the program, their contribution was incalculable. "They had it all over us in some areas—just brilliant guys," one of the original Space Task Group engineers remembered. "They were more mature and dignified and they were bright as hell and talented and cordial and professional. To a man." Glynn Lunney, like many other Americans who were with the Space Task Group, marveled at the serendipity of it all, with the Canadians "washing up on the shores of Langley Field" at the critical point. They were, thought Lunney, "the leavening of the bread."

But when Owen Maynard drove into Langley for the first time, he was just a young engineer trying to find his way in a strange place. Finally he found his destination, Building 58, tucked in beside the nineteen-foot wind tunnel on Dodd Street. It was a two-story red brick building with a little portico in front, built back in the 1920s as Langley's first headquarters and still called the "Administration Building." By 1959 it had long since been second-rate space, used for the East Area's cafeteria during World War II and subsequently for miscellaneous storage and office space. Behind it was Building 104, called the Technical Services Building, said to be the oldest structure anywhere at Langley. Also made of red brick, it was so obscured by its newer neighbors that it was only partially visible from the street. These two buildings constituted the sum total of the facilities of Project Mercury.

Maynard parked his car and went into the Administration Building. The little lobby inside the door was charmingly old-fashioned, a small rotunda with painted murals of the history of flight—Icarus, the Wright brothers, and Professor Samuel P. Langley himself. The offices radiated from it, and they too had a certain shabby elegance, with floors of oiled oak and ceilings twelve feet high. But the building was pathetically humble compared to the AVRO facilities that Maynard had just left.

The Technical Services Building in back turned out to be humbler yet. It felt like a worn-out junior high school building, with antique lighting, a roof that leaked during rainstorms, and creaking risers on the old wooden stairs. The nineteen-foot wind tunnel next door made a terrific racket. There wasn't any air conditioning, and Maynard would soon find that in the heat and humidity of the Virginia summer, he had to take care lest the perspiration from his arms ruin his drawings.

As Maynard looked about and made inquiries and listened to the shop talk, he found that it came down to this: About 140 engineers (including the Canadian contingent), most of them youngsters, with borrowed quarters and a strained budget, were supposed to put a man into space and redeem America's technological prestige in the eyes of the world. He was still adjusting to this realization when, a few weeks later, he met a strange fellow named Max Faget. Faget, he discovered, was not content with just getting a man safely into orbit in the Mercury capsule. He was thinking about putting a man on the moon.

2

"I could picture the astronauts looking down at it with binoculars"

Immediately to the north of the White House stands Lafayette Square, a formal park of grass and trees crisscrossed by brick paths. During the first half of the nineteenth century, when wealthy Washingtonians built townhouses along the three sides of the square facing the White House, Lafayette Square became the most fashionable address in the city. One of these houses still stands at the northeast corner of the square. It is one of the larger houses, washed in a pale lemon, with a handsome bay window overlooking the park. Built in 1820, it was given to the widow of the fourth president of the United States, in payment of a debt owed to her husband, and it has been called by her name ever since: Dolley Madison House.

In the spring of 1959, Dolley Madison House was the headquarters of NASA, an organization still so compact that its entire headquarters staff could be housed in that one townhouse and a small adjoining office building on H Street. One spring day not long after Owen Maynard's introduction to Langley, Max Faget walked up the front steps on H Street and was ushered into the Federalist drawing room that now served as the office of the NASA administrator.

Faget had come up from Langley at Keith Glennan's invitation to make a presentation to an informal seminar that the administrator periodically convened on an ad hoc basis. "It was almost like a symposium," Faget recalled, "not to deal with management issues, but more or less to blue-sky things." Wernher von Braun was attending that day, up from Huntsville. He wasn't yet part of NASA, but Glennan liked to call on him for advice, and von Braun, who longed to be part of the space program, was glad to oblige. William Pickering, director of the Jet Propulsion Laboratory, had flown in from Pasadena. Representing headquarters, besides Glennan himself, were Abe Silverstein, director of Space Flight Programs, and Donald Ostrander, director of Launch Vehicle Programs. It was as august a group as little NASA could assemble.

In 1959, Max Faget was chief of the Flight Systems Division of the Space Task Group—a top position within the Space Task Group, but one that left him far junior to the others in the meeting. Yet Glennan had called him up from Langley, because Glennan had an exotic problem for the group to blue-sky. Already, Max Faget was becoming known in the flight engineering world as a man who could do that like nobody else.

1

"Maxime Faget" is a wonderfully felicitous name for a spacecraft designer, calling up an image of some quirky artisan genius at Cartier or Dior—not an inaccurate image for describing the real Maxime Faget. Faget would never be able to compete with von Braun or the astronauts for public recognition, but within the space-flight fraternity he would acquire a unique reputation. A story is told about a time in the 1970s when a famous author sought out some old-time Langley engineers to ask for their help with a novel about the space program. The famous author had envisioned a central character who was an engineering eminence. The engineer would have graduated with highest honors from Princeton, the nation's preeminent aeronautical engineering school, and have been a Rhodes scholar at Oxford. Then he would have returned to Princeton, this time to the Institute of Advanced Studies, from which he would become the anonymous genius behind the American space program. Could the Langley engineers think of a person whom the author might use as a model for this character? One of them said yes, we used to have a fellow like that around here. But he hesitated. Would it be okay if he was born in British Honduras and got his degree from Louisiana State?

From an old Louisiana family of French extraction, Max Faget was born in British Honduras because his father, an eminent physician in the Public Health Service, was conducting research there on tropical diseases. The young Faget got his engineering degree from Louisiana State in 1943 and went from there to combat duty on submarines in the Pacific. After he was discharged, he and a friend from L.S.U. stopped by to interview at Langley on a lark. Faget didn't know a thing about aviation, Bob Gilruth remembered later, but "he was bright and he was interested," and Gilruth was impressed by someone who would volunteer for submarine duty in wartime. So Max Faget was hired, the man who went on to become the principal creative force behind the development of American manned spacecraft from Mercury through the shuttle.

It was not immediately apparent to people who first met Faget that he was a budding legend. "I was being interviewed by a guy about where I'd like to work," recalled Tom Markley, who went to Langley in 1956. "It was just my first week out of college, and so I said, 'I don't know.' He said, 'You've got a physics background. I think maybe you ought to work for Max Faget.' I said okay. So he called Max down and pretty soon this little guy in a pair of jeans and a white T-shirt walks in. I thought, What have I gotten myself into?"

At that time, Faget was first making his name in the world beyond Langley as the man who developed the "blunt-body" shape for the Mercury capsule.[*] It was a considerable triumph. Others, notably at Ames Research Center, had been arguing for a lifting body, a shape with some of the characteristics of an aerodynamic wing. Faget demonstrated that the blunt ballistic shape not only would be simpler and faster to implement (an exceedingly important consideration after Sputnik), but would also have enough lift to be maneuverable.

Markley first realized what Faget was up to sometime in 1957, before the Space Task Group was formed. Markley was walking through the P.A.R.D. shop at Langley when he looked up and saw Max, all by himself, standing on a balcony overlooking the shop. He was throwing what looked like paper plates out over the balcony. "I thought he was

[*] H. Julian Allen and Alfred J. Eggers of the Ames Research Center originated the blunt-body concept in the early 1950s for use in ballistic missile nose cones. Faget developed the use of the blunt-body shape for manned spacecraft, wherein the spacecraft entered the atmosphere with the broad end of the cone absorbing the shock and heat.

crazy at first," said Markley. "I just stood there and looked at him." He finally walked up the steps to the balcony. As he got closer he could see that Faget had taped pairs of paper plates together, back to front.

"Max, what are you doing?" Markley asked.

"I think these things will really fly. We have some lift over drag in this thing," Faget answered, as he sailed another pair of plates out over the shop floor. Down on the floor, the shop technicians continued to work unperturbed. They were used to this sort of thing from Max.

Standing only five feet six inches tall, slight, natty in his bow tie when he got dressed up for outsiders, Faget looked deceptively boyish and harmless. This impression usually lasted only briefly, however. "He was opinionated, completely outspoken—he wouldn't pull back an iota," said John Disher, who had to deal with him from headquarters. "If you had a dumb idea or he thought you were dumb, he'd tell you to your face." Along with that bluntness went supreme self-confidence. "It never occurs to him that he can be wrong," said his closest collaborator, Caldwell Johnson. "A lot of people have got a great deal of self-doubt, they hesitate to do things, because they're not all that damn sure they're right. Max is never in any doubt in his mind whether he's right or wrong. And that's good. Goddamn, if you don't have a few people like that, you're not going to get anywhere."

And yet, oddly for a man of such strong opinions and large ego, Faget seems to have been liked by just about everyone who worked with him. "A sweetheart," said Owen Maynard. "He was always laughing and joking," said Markley, "a great guy to grow up under." "Max was a very likable guy," Disher confirmed. "He just detested higher authority."

Perhaps the best way to describe Faget's style is cheerful ruthlessness. His associates recalled knock-down, drag-out technical arguments with him. Faget's voice would rise, his face would flush—and then it would pass as quickly as a summer storm and Max would be off on something else. The only thing you had to remember when you were around Faget was that once he got absorbed in something, you were well advised to keep a safe distance. One engineer who used to play squash with him in the astronauts' gym at Houston recalled that Faget would "beat the hell out of you with the racket if you didn't watch out." The two men were the best of friends, had been for years—but, nonetheless, "you had to play defensively, because Max didn't make any effort whatsoever to avoid hitting you. And you'd lose a lot of points that way."

On the other hand, once Faget's absorption in something was finished, that was it. He was curiously detached about his spacecraft once they had left his hands. He didn't go to see them launched. He could have—he wasn't required in Mission Control the way Operations people were—but he always managed to be too busy with something else. In later years he relented and went to see one of the Apollo launches, but only one.* Even when he was in the Control Center, Faget seemed indifferent to the flights. It was an outgrowth of the traditional cleavage of design people and operations people in the world of aviation. "Max completely hated Operations," one said. Caldwell Johnson explained: "Max always pissed and moaned about the Operations guys—about how you design a good spacecraft, and then they come along and screw it all up. And that's not so. That's not so at all. But it's kind of a built-in professional disagreement."

Faget seldom bothered with a drafting table. He worked from ordinary coordinated paper, the kind sold at the corner drugstore. And while he knew what all the alphas and epsilons of the engineering world meant, he didn't start with them when confronting a problem. Once when a visitor asked him whether he was an "intuitive" engineer, he answered,

> I think I'm an intuitive engineer, yes. There is such a thing as a talent for things. I think I have a talent for it. . . . You know, it depends on how you want to use your tools, and how you want to do your analysis. I can almost go in a trance. . . . I can get fully absorbed in a problem, to the point of almost—well, you really have to distract me to get my attention.

How long does this last? he was asked.

> I don't know. I really don't know. Oh—sometimes days and weeks. You can't stand it but so long. I won't say a trance—you know, you just stare at the wall. Your brain is working, I know that. It takes energy: You're tired after you've come out of it.

To his colleagues, the products of Faget's trances sometimes seemed more like imaginative leaps than solid, precise engineering—one of them used to claim that Faget stood for "Flat-Ass Guess Every Time." But the consensus of the engineers who worked closely with Faget was that his batting average was astonishingly high. "Max had brilliant ideas, and he was usually right," said John Disher, and that was the way that he came to be perceived throughout NASA. "This country has owed a debt of

---

* Nor did he change in later years, even after he retired from NASA. Max Faget, the lead designer of the shuttle, as of the end of 1988 had never gone to Cape Canaveral to watch a shuttle launch.

gratitude to Max for a long time,'' maintained Glynn Lunney—an Operations man. Forty years after he was first hired at Langley, a colleague since N.A.C.A. days would describe Max Faget by saying, ''The United States could run for the next hundred years on the ideas Max had while he was shaving this morning.''

2

On this sunny spring morning in 1959, Faget was still just a creative young engineer, and on that basis he had been chosen by Glennan to make a presentation on how a manned lunar landing might be accomplished. ''It was the first time I can remember actually considering it,'' Faget said later, ''and I really hadn't thought about it very much before the meeting, to tell you the truth.'' But the problem seemed fairly straightforward to him, at least in its broad outlines. What you'd want to do, he told his audience, was to take it in stages. ''I thought of first flying out and looping around the moon to get a look at it, at least get out of earth orbit and get a look at the moon a little bit closer. I could picture the astronauts looking down at it with binoculars,'' Faget remembered, laughing. ''Just gonna go whipping on by and come back. The next time, after we got that under our belts, we ought to try to orbit the moon. The nice thing about that would be that you'd be able to fly over possible landing sites and pick one out from your reconnaissance photographs.''

Finally, on a third mission, Faget told the group, the spacecraft would go into lunar orbit and check out the landing site ''just like anybody that flies over a new field before they land on it. And then you would go down and land, one or two orbits later.'' At that point, Faget remembered, Pickering and von Braun began objecting. What was all this business about going into orbit around the moon? ''Max, you're completely overlooking all that we're going to learn in the Surveyor Program,'' von Braun said. Surveyor, an unmanned lunar probe, wasn't going to orbit, it would just go straight in and make a soft landing. And after that had been done a number of times, and the risks were understood, they would do the same thing with the first lunar landing.

Faget didn't know quite what to say. Here were Pickering and von Braun—respectively, director of NASA's most advanced engine laboratory and the most celebrated rocket engineer in the world—assuming that they would go straight in, and here was Faget, an engineer who worked for Bob Gilruth on the Space Task Group, saying to himself that this didn't make any sense at all. He was trying to picture it in his mind. If you went straight in, you would be coming in at almost 10,000 feet per second. You'd have to conserve propellant, so you wouldn't want to retrofire until about a hundred miles above the lunar surface. At 10,000 feet per second, a hundred miles takes about fifty-three seconds. That wasn't a whole lot of time. If the engine didn't fire right away . . .

"It would be a pretty bad day if when you lit up the rockets, they didn't light," Faget observed. Pickering brushed it aside. "We'll have done this a number of times!" he said. "We're already working on the techniques!" And he went into a discourse on the kind of rocket they would use to slow down and the kind they would use to land. It would all be worked out. Faget, not ordinarily bashful, confined himself to murmuring, "Well, gee whiz, you know . . ."

"The thinking at that time was very primitive," Faget said later. But it was interesting as hell, and Faget began to think about it more and more—especially what the spacecraft would have to do, and how it would have to be configured.

3

As the spring of 1959 wore on, Glennan decided that it was time to start thinking about a lunar mission in a more organized fashion. He set up a formal panel called the "Research Steering Committee on Manned Space Flight." To history, it is known as the "Goett Committee," after its chairman, a NASA engineer named Harry Goett. Each center was told to nominate a representative. The Space Task Group got to send a representative, too, and Bob Gilruth chose Faget. At its first meeting at Dolley Madison House in late May 1959, the Goett Committee wrote down a first cut at the list of priorities. They listed nine steps in manned space flight, in order. A lunar landing was far down the list, at number seven, for this was a sober, realistic committee, not a group of dreamers. But two members of the committee were less sober and sensible and argued tenaciously for a lunar landing as NASA's next objective. One of them was Max Faget. He was joined by another young engineer who had come out of Lewis rather than Langley, not a member of the Space Task

Group, but the man who was destined to be the guiding hand for the lunar program as a whole as Faget was for the spacecraft design. His name was George Low.

In those unbuttoned days, it was hard to tell who was running what. George Low was putatively NASA's top man for manned space flight. Any organization chart would make that clear: NASA's administrator was in the top box, and then below that was a row of boxes representing "Offices," one of which was Abe Silverstein's Office of Space Flight Programs. Under that was George Low, who was "Program Chief, Manned Space Flight"—a title which would seem to indicate that Bob Gilruth, head of the Space Task Group, worked for him.

In reality, Low was in Washington with his fancy title only because he had been dragged away from what he really wanted to be, which was an assistant division chief under Max Faget. Low had held that job for a few weeks earlier in the year, commuting happily from Cleveland to Langley, until Abe Silverstein, Low's old mentor at Lewis, called Gilruth and told him that he needed Low up in Washington. "I'm not the Washington type," Low said to Silverstein. "Yes," said Silverstein, "but I need you," and so Low reluctantly agreed, exacting from Silverstein a promise that the assignment would be temporary.

Then Low turned around and did the same thing to his old friend and colleague from Lewis, John Disher. In what Disher remembered as "one of the few slightly underhanded things I ever saw George do," Disher was also reassigned from the Space Task Group to Washington against his will. There they were, sitting in their offices in the Dolley Madison House, supposedly managing NASA's manned space program. But what it really amounted to, Disher remembered, was three men—George Low ("Program Chief, Manned Space Flight"), Disher himself ("Chief, Advanced Manned Space Flight"), Warren North ("Chief, Manned Satellites," meaning Mercury)—and a couple of secretaries.

Austrian born, thirty-four years old, to all appearances the most mild-mannered and professorial of men, George Low wanted NASA to go to the moon, the sooner the better. Not only did he want to go to the moon, he wanted to land on it. After the first meeting, he lobbied his colleagues on the Goett Committee, reasoning with them in his quiet, precise voice still tinged with an Austrian accent. Manned space flight needed a goal to sustain it, he argued, a dramatic focal point. A lunar landing was that goal.

Low made converts. In its second meeting in June 1959, the Goett

Committee went on record as saying that a manned lunar landing should be NASA's next objective after Mercury. But it was one thing for a research committee to say such things, another for NASA to make it official policy, and 1959 was too soon. Low and Disher found that out in July when they tried to get Abe Silverstein to go along. They were working on a draft of a paper entitled "Space Flight Development, Advanced Technology, Manned Space Flight: Long Range Plans," which would eventually be their official statement of plans and priorities. It included eight pages describing a manned lunar landing program, with a funding schedule that would land men on the moon by the late 1960s.

Abe Silverstein, who was politically more savvy than his enthusiastic young assistants, read it warily. Silverstein himself had nothing against going to the moon. But the Eisenhower administration was cool to manned space flight and frigid on large spending programs. There was no point in jeopardizing the programs that NASA could reasonably hope to see approved. Besides, as the latest progress reports from Cape Canaveral were showing, NASA was still a long, long way from being credible. As of the time Silverstein was being asked to put a lunar mission on NASA's agenda, the Space Task Group team down at the Cape couldn't even get its first rocket off the ground. Silverstein bounced the draft back to Low and Disher with instructions to take out the part about the moon.

3

"Those days were out of the Dark Ages"

On June 9, 1959, the first contingent of Space Task Group engineers disembarked from a chartered D.C.-3 at the Cape. They were thirty-five strong—six engineers, a draftsman, twenty-seven technicians, and a secretary, the only woman, Emily Ertle. The little party was led by Scott Simpkinson, who, like almost all the rest of his crew, had come to the Space Task Group from Lewis Research Center in Cleveland.

Their job was to launch "Big Joe," an Atlas missile carrying the first Mercury capsule. The capsule was a test article, not the real thing—what they called "boilerplate," with the shape and weight of a real Mercury capsule but none of the electronics or environmental systems that would be installed for a manned flight. Lewis had assembled the bottom half, Langley the top half. The purpose of the test was to see whether the heat shield would keep cabin temperatures at an acceptable level during entry into the earth's atmosphere. The capsule would be shot to an altitude of 160 miles, then the Atlas would pitch over and accelerate the capsule back into the atmosphere at a top speed of 17,000 m.p.h. Thermocouples attached to the capsule's titanium shingles would measure the heat.

The Space Task Group had no facilities of its own down at the Cape;

the Air Force, which played host to the many agencies operating out of Canaveral, was supposed to provide them. Simpkinson reported to the Air Force people in charge of the Cape Canaveral Air Force Station, who took him over to Hangar S in the industrial area of the station.

Hangar S had been built by the naval research outfit that had belatedly put up the tiny satellite on Vanguard, and they were still there working on their missiles. Never mind, Simpkinson recalled the Air Force people telling him, "you're just a nose cone, you can go in the back." They put up some stanchions and roped off a portion of the floor of Hangar S for the Space Task Group to use. The Space Task Group's office space consisted of a narrow storage area barely wide enough to fit their desks end to end—whenever the draftsman at the far end had to get out (he sat by the only window, so that he would have enough light to draw), the other engineers all had to get up, push their chairs into their desks, and walk out into the corridor to let him by.

Down on the floor of Hangar S, conditions were even worse—"Those days were out of the Dark Ages," remembered Joe Bobik, a mechanic from Lewis who had jumped at the chance to come down to the Cape with Simpkinson. They had no White Room, the immaculately clean and protected area in which later spacecraft would be checked out and prepared for flight. The technicians were not yet dressed in the white smocks, hats, gloves, and booties that technicians always wore in the presence of a spacecraft in later years. The Mercury capsule just sat on the concrete floor of the hangar and technicians in mechanics' overalls worked on it.

None of them had ever experienced these conditions—the day-in, day-out hours, plus the heat, humidity, and mosquitoes of the Cape in midsummer. "They had an ammonia cooling system on the launch vehicle," Bobik recalled. "We were working sixteen, eighteen hours a day, as many hours as someone could work, and the mosquitoes were so bad at night that we'd get that ammonia and spray it around to kill those mosquitoes so we could work." In retrospect, Bobik would be appalled ("Ammonia is dangerous! It's toxic!"). But that's how bad the mosquitoes were. There were other improvisations—lacking a refrigerator in Hangar S, they used the fire extinguishers to cool their Cokes. As if to show them just how far behind they had left civilization, a Florida panther came up to the chain-link fence near Hangar S every evening and sat there, peering in—"like a zoo in reverse," one veteran commented.

The Air Force continued to treat the Space Task Group as a bunch of amateurs (which they were, in space flight) with a low official priority (which was all they had). This attitude had its positive side in averting tensions between the Lewis and Langley members of the team ("The Air Force made us get together, because they treated us like we were all incompetent," a Langley man said). But Simpkinson couldn't get any equipment from them. "The Air Force treated us like dogs down there," Simpkinson remembered. "About all the Air Force would give us was pencils." That got Simpkinson's back up, and he began taking matters into his own hands.

It was a natural role for him. A congenital loner in a business where teamwork was almost a religion, Simpkinson had been chosen to prepare the Big Joe shot because of a strange combination of qualifications. He had a long history of testing experimental engines. He had worked with rocket-driven supersonic shapes. He was versed in pyrotechnics. Also in his inventory of talents, though less obviously pertinent, he was the proprietor of an off-hours television repair business and played the trumpet in swing bands that toured the clubs around Cleveland.* Transcending these specific skills, Simpkinson possessed an unteachable ability to "smell a problem a mile away," as one of his colleagues put it. The technical term for his expertise was "failure-mode effects analysis," but at bottom what it meant was that Simpkinson was an engineer who was intimately at home with the hardware. He understood machines.

Simpkinson was also tenacious as a tick and didn't mind antagonizing people if he felt the occasion called for it. Rather than waste time trying to get along with the Air Force, he convinced Merritt Preston, his Space Task Group supervisor up at Langley, to arrange for Simpkinson to have his own $50,000 government checking account. When one of his technicians needed a stepladder or a hammer or a lathe, Simpkinson just went to a Cocoa Beach hardware store and wrote a check for it.

None of the members of the team that Simpkinson led to the Cape had ever checked out a spacecraft; none of them had ever launched one. With neither equipment nor experience, they made it up as they went along. There was the matter of getting the capsule from Hangar S out to the pad, for example. They didn't have a vehicle for that purpose. No one had thought of it. But Jack Kinzler, the shop foreman from Langley, had just driven down to the Cape in a flatbed truck to join Simpkinson's group. Simpkinson bought some mattresses and two sheets of plywood with his magic checkbook. They put the mattresses between the panels of

_____
* Simpkinson's work at Lewis sometimes seemed to be more his hobby than a job—his colleagues would run across his N.A.C.A. paychecks sitting in his desk drawer, uncashed, sometimes half a dozen of them at a time.

plywood and then lashed the capsule onto the top. Pictures still in the files attest to the story, showing a battered pickup with a spacecraft on mattresses pulling up to the pad where an Atlas rocket stands waiting. "Of course, we got a lot of ridicule," said Kinzler. "But it worked fine."

After they had hauled the capsule to the top of the Atlas with a crane, they found that the capsule wouldn't seat. Somehow, the heat shield had been made about half an inch bigger than the diameter of the Atlas. When they tried to lower it onto the rocket, it perched on the top like a bathtub plug slightly too large for the drain. By now it was the end of June and Big Joe was supposed to launch on the Fourth of July. If they sent the heat shield back to its maker, the General Electric Corporation in Philadelphia, it would be two or three weeks before they got it back again.

The solution that Simpkinson and Kinzler contrived in 1959 would have been unthinkable even a few years later. As they stood on top of the gantry and contemplated the oversized capsule, it occurred to them that only the heat shield was too big, not the capsule's metal frame. The heat shield was made of a kind of plastic material. It didn't contain any electronics. It wasn't carrying any structural loads. So Simpkinson and Kinzler called for the crane, lowered the capsule to the ground, and transported it back to Hangar S in the pickup. Simpkinson jumped in his station wagon, drove to the Sears store in Orlando, used his checkbook to buy a router, and drove back to the Cape. Meanwhile, Kinzler had made up a device with a pivot mount in the center of the heat shield. Kinzler attached the router to the pivot mount and hand-walked the router around the perimeter of the heat shield, measuring and routing, routing and measuring, until he had shaved off half an inch. They took the capsule back to the pad and hauled it up again. It fit, and they went on with the preparations for launch.

Despite these heroic improvisations, Big Joe was launched two months behind schedule after all. The power supply had been improperly designed: The first time they hooked up power to the capsule, transistors popped, readings were crazy—nothing worked the way it was supposed to. They had to send the capsule back to Lewis, and it wasn't until September 9, 1959, that the flight finally took place.

"And of course it failed," Simpkinson sighed. The first stage of the Atlas performed beautifully, lifting off into a clear night sky. But the Atlas didn't stage properly, the nitrogen thrusters on the capsule didn't get a chance to demonstrate that they could align the capsule properly for entry, the entry speed wasn't as high as it was supposed to be. Nothing went quite right. The preparation and launching of spacecraft at the dawn of the space age turned out to be a laborious business, and the participants were still amateurs sitting at the bottom of the learning curve. "We were green as thumbs," Merritt Preston used to say.

1

Two months after the flight of Big Joe, on November 2, 1959, President Eisenhower signed an executive order transferring Wernher von Braun's rocket engineers at the Army Ballistic Missile Agency in Huntsville to NASA. It was the indispensable step for making NASA legitimate, giving that young and uncertain agency an infusion of talent it could have gotten nowhere else.

The division of responsibility was clean: The Space Task Group had jurisdiction over the spacecraft while the people at Marshall had jurisdiction over the launch vehicle.* The development of the working relationship took time, and there would always be rivalry. But with the addition of the Germans, it became possible to go to the moon.

By 1959, von Braun's rocket team had been in the United States for thirteen years following their surrender to the U.S. Army at the end of World War II. At first they had been sent to White Sands in New Mexico, where they were put to work showing the Americans how the V-2 worked. Then in 1950 the Army decided that an inland site was too confining, a decision prompted in part by an unfortunate occasion when the German team put a V-2 into a cemetery south of Juarez. From then

---

* Here, in one place, are the genealogies and official nomenclature of the NASA centers principally involved in the space program: (1) For the spacecraft: The manned space program began at Langley Research Center, where the Space Task Group was an organizationally independent entity. In November 1961, the Space Task Group became the Manned Spacecraft Center (M.S.C.) located at Houston, Texas. In 1973, M.S.C. was renamed the Lyndon B. Johnson Space Center (Johnson, or J.S.C.). (2) For the launch vehicle: In the 1950s, von Braun's group was part of the Army Ballistic Missile Agency located at Redstone Arsenal in Huntsville, Alabama. In 1959, when von Braun joined NASA, it became the nucleus of the newly named George C. Marshall Space Flight Center (Marshall, or M.S.F.C.). (3) For launch operations: The area around Cape Canaveral served as a launch area for all of the armed forces during the 1950s. The initial launch group from Huntsville was called the Missile Firing Laboratory. In 1959, when Redstone Arsenal became Marshall, the Missile Firing Laboratory became the Launch Operations Directorate, still administratively under von Braun. In 1962, the Launch Operations Directorate became an independent center, the Launch Operations Center. It was renamed the John F. Kennedy Space Center (Kennedy, or K.S.C.) at the end of 1963, immediately after Kennedy's assassination.

on, the larger rockets were launched out over the ocean from Cape Canaveral. The von Braun team itself was moved to Huntsville, Alabama, where the Germans used the facilities of the Redstone Arsenal.

In contrast to the uneasy relationship between the engineers at Langley and the local Virginians, the Germans and the Alabamians got along fine from the beginning. The Germans were delighted to be out of the arid Southwest and in a wooded, hilly area that reminded many of them of the German countryside. The Alabamians were intrigued by these foreigners and entertained by their eccentricities. The assimilation was extraordinarily fast. By 1952, only two years after the Germans had arrived in Huntsville, former Luftwaffe sergeant Walter Wiesman was president of the Huntsville Junior Chamber of Commerce, elected by a membership that was 70 percent World War II veterans ("So am I!" Wiesman pointed out). As time went on, Huntsville became intensely proud of its Germans, and especially of its most famous German of all, Wernher von Braun.

Von Braun was the only non-astronaut in the space program who became a household name. In Congress, his prestige was enormous. Movie-star handsome, with an expansive smile and European charm to which he added a touch of Alabama folksiness, he could dominate a congressional hearing as easily as he dominated the media. Other senior people in NASA envied him and in some cases resented him ("That damned Nazi," one was known to mutter when he had had several drinks), intimating that von Braun spent too much time worrying about his public image and that the real work at Marshall was done by others. What was hard for some of his NASA peers to swallow was that von Braun was a natural. He was exceptionally good at being a public person, and none of the other engineers of Apollo could compete.

With the fame came a price. In some circles, von Braun was assumed to be a Nazi who had escaped judgment only because of his value to the United States. No such charges were substantiated. On the contrary, his history reveals a man who from his teens had a passion for rockets and space travel, as oblivious to politics in the 1930s as America's Apollo engineers were oblivious to politics during the 1960s. During the height of the war, von Braun was briefly jailed by the Gestapo for insufficient ardor in making weapons. But even among those who bore him no ill will, jokes were inevitable, given the contrast between von Braun's activities during the war and his transformation into an American hero.

When a movie about von Braun was entitled *I Aim at the Stars,* the underground version of the title quickly became *I Aim at the Stars but Sometimes I Hit London.**

Von Braun was not a creative genius—to that extent, the public image misrepresented him. It is not possible to think of "von Braun concepts" that changed the development of rocketry in the same way that "Faget concepts" changed the development of spacecraft. Instead, he was a natural leader and technical manager. Karl Heimburg, chief of the Testing Lab, explained how von Braun went around looking for new ideas, which he would then take to his associates. Heimburg would listen unimpressed and explain to von Braun why something wouldn't work. And then Heimburg would find himself saying, "But we could do it in this other way," and an innovation that he had not considered would have opened up before him.

In general, von Braun seems by the testimony of people who worked for him, ranging from senior colleagues to technicians on the pad, to have been just about as good as his publicity made him out to be. He won from them loyalty and affection as well as professional respect. "He was a noble type of man," said one American at Huntsville. "That's the only word I can think of to describe him."

Ultimately, von Braun won admiration, though sometimes grudging, even from his peers in the other centers. Robert Gilruth, who got far less attention from the press and who in the early days had to struggle to keep the Space Task Group from being eclipsed when von Braun was brought into NASA, had ambivalent feelings about his counterpart at Marshall. But once, after von Braun died in 1977, Gilruth was listening to another senior NASA official talk about a technical problem on the shuttle. "I wish Wernher were still around to ask about this," Gilruth broke in suddenly. "You know," the NASA official reflected, "that was probably the greatest tribute von Braun ever got." And then the NASA official proceeded to give von Braun the only compliment that really counted in the Apollo fraternity. When you ignored all the P.R. stuff, he said, "von Braun was actually a pretty good engineer."

* It was an awkward situation to which the Germans remained sensitive long after they had become naturalized American citizens. During the early years of racial integration, an American engineer once overheard someone asking Hans Gruene, head of Launch Vehicle Operations at the Cape, how he felt about having a black live on his block. "I have no problem," Gruene replied. "I wonder what he thinks of an ex-Nazi living on his?" By that time, many of the Germans had become more American than the Americans. Ray Clark, Kurt Debus's deputy for administration at the Cape, recalled gatherings after work at which Gruene and Debus would get into conversations about America's heritage and its future that more than once went on until dawn—they knew more about the United States than he did, Clark often thought.

*   *   *

The working end of the rocket team at Redstone Arsenal was divided into eight laboratories, and a German headed each of them. Another German, Kurt Debus, headed the Missile Firing Laboratory at the Cape. They ran these labs in a way that often seemed to caricature the precise, methodical rocket scientist of the popular imagination. Their colleagues were "Mister ———" or "Doctor ———," rarely "Chuck" or "Bill," even when they had been working together for years. Everything had to be done with a meticulous exactness and order. Debus, for example, was known for making the rounds of his subordinates' offices after hours, sweeping papers and books from desks that didn't meet his standard of neatness.

The Germans were extraordinarily conservative in their designs, with margins that were lavish even by aerospace standards—every component had to be able to bear far more weight, tolerate far higher temperatures, withstand far higher dynamic pressures than the rated performance of the vehicle required.* When preparing the hardware, the Germans for many years had no separate inspection function. Each engineer inspected the work of his own technicians, and he was expected to be a good enough machinist or electrician or hydraulics mechanic in his own right to be able to know whether the work was done correctly. The Germans' testing programs were excruciatingly thorough, "to the point of being ridiculous," said one American observer with both exasperation and envy.

What made this Germanic conservatism and precision remarkable was that by the late 1950s most of the people from Huntsville who were behaving this way weren't Germans at all, but the Americans who had been hired to work with them. Most of them were men from the small towns of the deep South, graduates of nearby engineering schools like Auburn and the University of Mississippi and Georgia Tech. The result was a combination of Germans like Eberhard Rees or Karl Heimburg or Walter Haeussermann—distinguished and courtly, talking about "ze vay ve do sings," very models of the rocket scientist—and Americans like Alexander A. McCool of Vicksburg, Mississippi, an esteemed propulsion

---

* When during Apollo the spacecraft kept exceeding its weight limits, NASA officials from headquarters went down to Huntsville to find out whether there was any way that von Braun's people could cut some weight from the launch vehicle and thereby raise its performance beyond its original rating. Sure, one American remembered them saying—"We can cut three thousand pounds, four thousand pounds—it won't make any difference." This was at a time when the spacecraft people were trying to shave ounces. The German margins also let them uprate the performance of the engines of the first stage of the Saturn V from 7,500,000 pounds of thrust to 7,650,000. This reservoir of extra performance that the Germans built into the launch vehicle came to the program's rescue more than once during Apollo.

expert, talking about "that thang" (meaning the rocket) which "hauled tail." Somehow, everyone understood what everyone else was talking about.

2

Coincidentally with the addition of von Braun's team to NASA, momentum behind a lunar program began to build. On the same day that Eisenhower signed the executive order creating Marshall, Bob Gilruth held a management meeting at Langley. It was time to begin the preliminary design of a multi-man spacecraft, he told them. They should think in terms of a crew of three. The mission would probably be a circumlunar flight, but they should keep other possibilities open. Bob Piland would direct the group.

A few months later, on the first Thursday of the new year, 1960, Keith Glennan gave his approval to the Goett Committee's final conclusion. Henceforth, the agency would take the position that, following Mercury, NASA should aim in the direction of a manned lunar landing. Just one week later, Glennan got a letter from Eisenhower: "You are hereby directed," Eisenhower wrote, "to accelerate the super booster program for which your agency recently was given technical and management responsibility." That meant the Saturn.* It was the first signal that Eisenhower might support a manned space program beyond Mercury.

That same January of 1960, the Apollo spacecraft was baptized by Abe Silverstein, head of the Office of Space Flight Programs. Silverstein had named Mercury a year earlier (Silverstein liked the image of a messenger in the sky), and since von Braun had named his new launch vehicle "Saturn," another Greek god seemed to Silverstein like a natural choice.† He remembered from his grade-school days the story of the god who rode the chariot of the sun drawn by four winged horses—Apollo, the child of Zeus. Silverstein, the meticulous research engineer, went back to his old book of myths and determined that Apollo hadn't done anything that "wouldn't be appropriate."

---

* "Saturn" was the generic name for the super booster. The Saturn came in many combinations from the Saturn I to the ultimate version, the Saturn V used for the lunar missions. "I" and "V" are Roman numerals, so "Saturn V" is pronounced "Saturn five." Similarly, the Saturn V's S-IC stage was "S-one-C," the S-II stage was "S-two," and the S-IVB stage was "S-four-B."

† Von Braun named his rockets after planets, which in turn were named for Greek gods. The German team had already built a Jupiter, and Saturn was the next outer planet in the solar system.

Soon after, Silverstein tried out his idea on Gilruth, Faget, and Charles Donlan, Gilruth's deputy director in the Space Task Group. The four men were discussing the new post-Mercury spacecraft over lunch at a little restaurant near Dolley Madison House. In the middle of the meal Silverstein said suddenly, "There ought to be a name for this that stands out in people's minds. You know, something like 'Apollo,' for example. I'm not saying you ought to name it 'Apollo' necessarily, but something like that." And then throughout the rest of the lunch, Donlan recalled, he kept calling this new spacecraft "Apollo," seeing how it would wear. It wore pretty well, and the spacecraft became Apollo. Silverstein didn't have to bother with things like public relations departments. "I had the whole program," Silverstein said simply. "I was naming the spacecraft like I'd name my baby."*

3

But even as NASA began to lay plans to send men to the moon, launching one man into earth orbit continued to be a struggle. The original date for the first manned flight, a suborbital flight once scheduled for January 1960, had come and gone, and the revised goal kept slipping later and later into 1960. Both Atlas and capsule were plagued with problems, and no one was promising much in the way of improvement. The chilling message that the Air Force continued to pass on to the Space Task Group was this: By mid-1961, when the first manned orbital flight was planned, the reliability of the Atlas would still be only 75 percent. The Space Task Group could expect to lose one out of four Atlases during the launch phase.

This gloomy projection seemed vindicated on July 29, 1960, when they tried to fly M.A.-1, the first time that a production Mercury capsule (not just a boilerplate model) was mated to an Atlas.† They launched after a heavy rain, under overcast skies. M.A.-1 lifted through the overcast, engines roaring, everything looking good. Then a minute into flight, out of sight beyond the clouds, it blew up during the period of "max q," maximum dynamic pressure, at an altitude of about 32,000 feet. No one

* As events transpired, Gemini, the two-man capsule, was the next spacecraft after Mercury. Gemini had yet to be conceived at the time Silverstein was naming Apollo.

† In Project Mercury, the missions were labeled according to the initial of the capsule (M for Mercury), the initial of the launch vehicle (R for Redstone or A for Atlas), plus a number indicating order. Thus M.A.-1 was the first launch of a Mercury spacecraft on an Atlas booster.

knew why. All they knew was that two signals to abort the flight had been sent, by sensors monitoring electrical power and thrust. Their assumption had to be that both the Atlas and the Mercury were still seriously flawed vehicles.

4

NASA continued to plan for the future in the teeth of present adversity. The very day that M.A.-1 blew up, steps toward Apollo were being taken in Washington. NASA had called together representatives from the aerospace industry to introduce their plans for the sequel to Project Mercury. The day before the launch, they gathered in a State Department auditorium that John Disher had borrowed for the occasion—NASA wasn't big enough to have an auditorium of its own—and Deputy Administrator Hugh Dryden used the name ''Apollo'' for the first time in public. George Low made a speech, telling his audience that during the 1960s NASA hoped to build a space station in low earth orbit and to conduct a circumlunar flight. Perhaps during the 1970s, if all went well, NASA would land on the moon.

On the second day of the conference, July 29, even as recovery teams were preparing to retrieve the pieces of M.A.-1 from the ocean floor, Max Faget addressed the group on the topic of a lunar landing. In the middle of his speech, Faget signaled to a confederate. The auditorium dimmed to a half-light that was somewhat darker than a heavily overcast day but brighter than a moonlit night.

Faget waited while the audience murmured. When there was silence, he told them that this was what earthlight would look like to an astronaut standing on the moon in a lunar night with a full earth shining above him. It was at that moment, sitting in the twilight of the auditorium, that John Disher realized for the first time that they really were going to go to the moon—some day.

Space station and a circumlunar flight: That was the immediate agenda. To get plans moving, NASA announced in August that three $250,000 contracts would be let for design studies of the Apollo spacecraft. The Request for Proposals specified that the spacecraft had to be compatible with the new Saturn and it had to be capable of a fourteen-day mission—

more than enough time to get to the moon and back. The proposals were submitted on October 9, 1960.

It was then, in the second week of October 1960, that a lunar landing moved from being an ambition to being a project. For one of the two men involved, Abe Silverstein, it was the natural next step—"the time had come," he said later. And perhaps it was as simple as that. The other of the two men, George Low, was asked directly about it less than four years later by an interviewer. What motivated him to act then? "I knew you would ask that question," said George Low, "and I don't know. . . . This was the time, of course, that we were beginning to discuss with industry what the Apollo Program was. . . . And we felt it would be most important to have something in the files, to be prepared to move out with a bigger program, should there be a sudden change of heart within the administration."

And yet it wasn't quite as simple as that either. For George Low, the most composed and deliberate of men, had an audacious streak, a fondness for the bold gesture that would break out repeatedly throughout his career. So probably it was a little of both. The time had come, but George Low also took it upon himself to give time, and history, a little nudge. Abe Silverstein's recollection was that Low brought him his proposition sometime during the second week of October. It was the kind of thing to which Silverstein had said no in 1959, but with planning for the new spacecraft under way and with the Saturn under development, enough had changed that they could go ahead now. The two of them came to an understanding and Low wrote it up in the form of a memorandum ("Paperwork was created to act as scenery for what we had already decided to do," as Silverstein once put it). Sixteen years later, when Low retired from NASA, the original was framed and presented to him; it hangs on the wall of the little study where his widow keeps some of his memorabilia. Presumably one day it will hang in the halls of the Smithsonian, for it is the closest thing the nation has to a birth certificate for the lunar landing program.

MEMORANDUM for Director of Space Flight Programs

Subject: Manned Lunar Landing Program.

1. It has become increasingly apparent that a preliminary program for manned lunar landings should be formulated. This is necessary in order to

provide a proper justification for Apollo, and to place Apollo schedules and technical plans on a firmer foundation.

2. In order to prepare such a program, I have formed a small working group, consisting of Eldon Hall, Oran Nicks, John Disher and myself. This group will endeavor to establish ground rules for manned lunar landing missions; to determine reasonable spacecraft weights; to specify launch vehicle requirements; and to prepare an integrated development plan, including the spacecraft, lunar landing and take-off system, and launch vehicles. This plan should include a time-phasing and funding picture, and should identify areas requiring early studies by field organizations.

3. At the completion of this work, we plan to brief you and General Ostrander on the results. No action on your part is required at this time; Hall will inform General Ostrander that he is participating in this study.

George M. Low
Program Chief
Manned Space Flight

Scrawled at the bottom of the memo is Silverstein's reply:

Low

OK

Abe

Low would later marvel that something so monumental could have been started so simply: "OK. Abe."

4

"He would rather not have done it"

It all depended on one's perspective. Abe Silverstein and George Low might think it was time to begin planning for a lunar landing, but for Bob Seamans, NASA's associate administrator, the prospects for the agency that October were looking distinctly gloomy. He was beginning to wonder whether leaving a secure and more lucrative job at R.C.A. had been a good idea after all.

When Low wrote his memo, Seamans, then forty-two, had been in NASA for only two months, brought to the agency by Keith Glennan to be its highest-ranking nonpolitical appointee. He came out of a different background from the one that had shaped the N.A.C.A. people. As a young man he had been a protégé of Charles Stark ("Doc") Draper, the genius of inertial guidance and the founder of the Instrumentation Lab at M.I.T. Seamans had been something of a prodigy in his own right, an M.I.T. lecturer at the age of twenty-two and the designer of aeronautical control systems that later became the basis for the guidance systems used in ballistic missiles. His first task upon arriving at NASA had been to learn something about the organization, and so Glennan had sent him off on a tour of the centers.

Seamans had been dismayed when he got to the Space Task Group at Langley. In Seamans's opinion, the manned space program was central to NASA's future, and yet it was pathetically underfunded and under-manned. Bob Gilruth's people "were just working their hearts out to pull the thing off," he recalled, but it was hard to imagine them succeeding unless headquarters could muster more support.

Getting that support was going to be no small feat. The Eisenhower administration continued to be unsympathetic that fall of 1960, and in fact slashed NASA's overall budget request for the next fiscal year. One of the casualties in the budget was the second stage of the Saturn, an especially disheartening turn of events. Just ten months earlier, Ike had directed Glennan to accelerate work on the super booster; now, he was reversing himself. Without the Saturn and its heavy lift capability, manned space flight was going to limp along in low-earth-orbit flights indefinitely.

Worse, there might be nothing at all after Mercury. A note attached to Eisenhower's last budget request for NASA said that Mercury was an experimental effort and added ominously that "further tests and experimentation will be necessary to establish if there are any valid scientific reasons for extending manned space flight beyond the Mercury program." And this was a softened version. Originally, Eisenhower had wanted to say flatly that there should be no commitment of any sort to manned space flight beyond Mercury. "Well, it's been very nice working for the government," Seamans remarked after one session with the Bureau of the Budget, "but it may not last much longer."

Then in November John Kennedy was elected President. Seamans, along with everyone else at NASA, looked anxiously to the new administration.

During the election campaign, John Kennedy had used America's failures in space as a campaign issue—a "space gap" to go along with the "missile gap"—but he had remained silent about what he had in mind for his own space program. Many in NASA had hoped for more. Space flight, and especially manned space flight, had the dash and drama that would have seemed to fit perfectly with the spirit of the Kennedy campaign. But Kennedy was only being honest. At that time, he really wasn't convinced that manned space flight had a place in his vision of the New Frontier.

Jack Kennedy would become an enduring hero to the men of Apollo. Twenty-five years after his death, Robert Gilruth would still speak emotionally of how important Kennedy was to the space program and to him personally. Max Faget would still have an embassy-sized official portrait of John Kennedy behind his desk. For Rocco Petrone, the decision to go to the moon would still be quintessentially Kennedy, emblematic of the spirit and style of the man. And there is no doubt that, once the decision to go to the moon had been made, Kennedy developed a lively interest in the space program. But to Bob Seamans, trying to read the tea leaves in the weeks after the election, it looked as if manned space flight was not only not at the top of the new President's agenda, it might not be on the agenda at all. And he was right.

Certainly Jack Kennedy the senator hadn't been interested in space. Doc Draper, Bob Seamans's mentor, remembered the time a few years before the election when a mutual friend had brought him together with John and Robert Kennedy for a social evening at Locke-Ober's restaurant in Boston, hoping that he could get the Kennedys excited about space flight. The meeting had been a disappointment. The Kennedy brothers had treated Draper and his ideas with good-natured scorn. According to Draper, Jack and Bobby "could not be convinced that all rockets were not a waste of money, and space navigation even worse." Hugh Sidey, watching the new administration in its first months from his vantage point as *Life*'s White House correspondent, would come to think that space was Kennedy's weak spot. In Sidey's opinion, Kennedy understood less about space than about any other issue when he entered the White House. The explanation was simple, said Jerome Wiesner, Kennedy's science adviser: "He hadn't thought much about it."

1

Against this backdrop of changing political tides, NASA approached the first launch of a Mercury capsule on a Redstone booster. M.R.-1, as the mission was labeled, would be the critical unmanned test preceding the first suborbital manned flight, now scheduled for early 1961. Like everything else that they had been doing at the Cape, it seemed to be jinxed. They had gotten within twenty-two minutes of launch in early October, when a malfunction in the reaction and control system forced a postponement. Then they aimed for November 7, the day before the presidential election, and missed that one.

By Monday, November 21, the men at the Cape were running on coffee and adrenaline. Marty Cioffoletti, then a young engineer with

McDonnell, remembered that his own record during that period, documented in his engineer's logs, was 135 hours in one week—an average of nineteen hours and seventeen minutes per day. Exhaustion was getting to be such a problem that Walt Williams, director of Flight Operations for the Space Task Group, had to issue a directive that no one was to work for more than twelve hours in a single twenty-four-hour period.

The capsule on this launch was no boilerplate, but a fully configured and equipped spacecraft off the McDonnell production line, and it was going to be checked out just as for a manned launch. The booster was to be the Redstone, a man-rated vehicle. The flight plan was identical to the one that would be used for the first manned flight. It was critically important that M.R.-1 should succeed—for the sake of public opinion, the opinion of the President-Elect, and the morale of an exhausted group of engineers and technicians who yearned for something to go right.

Everyone who was at Cape Canaveral for the brief flight of M.R.-1 seems to have remembered what happened in excruciating detail. No other event—not Al Shepard's first manned flight, not John Glenn's first orbital flight, not even the first lunar landing—engraved itself more deeply in the memories of those who watched.

On November 21, 1960, most of the leading figures in the American space program were gathered at the Cape. Over at the blockhouse beside the launch pad were Wernher von Braun, Kurt Debus, and the rest of the senior German officials from Huntsville. Over at the brand-new Mercury Control Center were Bob Gilruth, Chuck Mathews, Walt Williams, Chris Kraft, and the rest of the fledgling flight operations team. Outside, a few hundred yards away, were all the technicians and engineers who weren't needed for the launch. For many, it would be the first rocket launch they had ever seen.

The countdown was smooth. When the clock reached T − 0 ("T − 0" is spoken "T minus zero") and the launch sequence began, everything seemed at first to be going well. The thin umbilical tower fell backward, away from the Redstone, as planned; the engine ignited and smoke billowed up from the base of the rocket. And then something took off—fast. "I saw that thing go," remembered Joe Bobik, "with all that smoke and fire and a big *whoosh!* I shook my head and I said, 'That thing couldn't go that fast, could it?'" Marty Cioffoletti had the same thought—he had no idea those things climbed so quickly. Sam Beddingfield was standing beside Gene McCoy, their mouths gaping in astonishment.

Over in the new Mercury Control Center, they were watching the vehicle on a television camera. They saw the smoke and flames and an object shooting upward. The camera panned up, trying to track it. Then the camera stopped, searching for the missing Redstone; finally it panned back down to the pad. And there, wreathed in smoke, the Redstone still stood, the Mercury capsule still atop it.

In the blockhouse, Ed Fannin was on duty at the firing panel from which the firing command is given. His dials showed that the engine had ignited and was climbing toward full power. "But then I heard all this unusual noise," Fannin said. "I looked up and the rocket was still there. I could hear the other thing still going." The "other thing" turned out to be the escape tower attached to the top of the Mercury. It was streaking several thousand feet straight up into the sky. "Ed, get cutoff," ordered Ike Rigell, then the networks chief. Fannin pushed the cutoff button, which was supposed to safe all the electrical circuits. "All I got were crazy indications on the panel, totally abnormal," said Fannin.

What had happened, they later determined, was this: The main engine had fired normally. The Redstone had lifted off its supports (there were no hold-down arms in those days). As it rose the first inch, the tail plug—an electrical plug—on the umbilical connected to the base of the rocket pulled out, as it was supposed to, putting the Redstone onto internal power.

Terry Greenfield was the man who had to go out the next day and look up the bottom of the Redstone so they could find out what happened. "That was our first big lesson in configuration management," he reminisced years later: "how you make sure that the 'as-designed' drawings equal the 'as-built' drawings." The tail plug, which was reused for many launches, had two prongs. During the preparation of one of the preceding Redstone launches, a technician had found that one of the prongs didn't fit quite right, so he had filed about a quarter of an inch off the prong and didn't tell anybody. Thus one prong of the plug disconnected a few milliseconds earlier than the other prong. A quirk in the circuitry of the Redstone was such that the engine would shut down if only one of the prongs disconnected and if the rocket was not electrically grounded to the earth. With prongs of two different lengths, the conditions for sending a cutoff signal existed during that snippet of time as the rocket lifted off when the first prong had disconnected but the second prong, a quarter of an inch longer, hadn't. Because the tail plug had been used for many Redstone launches without a problem, no one recognized the potential danger.

On this launch, however, the extra weight of the Mercury capsule now riding on the Redstone slowed its acceleration slightly, just enough to add one additional, tiny fraction of a second to the time elapsed before the second prong disconnected. Now there was just enough time—about twenty-one milliseconds before the disconnect completed—for the ground circuitry to get the message through: Shut down your engine. So the Redstone obediently shut down its engine and after a journey of two inches settled back down onto its four support points, now completely unconnected with the ground.

Meanwhile, up in the capsule, the spacecraft's circuits had determined that the Redstone engine had fired normally (as indeed it had, for about three seconds). The conclusion, as far as the spacecraft was concerned, was: I'm flying. Then it got a message that the engine had shut down. Conclusion: Powered flight is over, and it's time to do the programmed sequence of actions for the spacecraft, beginning with the ejection of the escape tower. So, *zoom,* off went the escape tower. Thus the confusion among the onlookers on the nearby roads.

The more experienced hands grasped the implications of all this faster than the newcomers. Sam Beddingfield and Gene McCoy, recently arrived members of the Space Task Group, were marveling at the speed of the launch when one of the experienced Redstone mechanics standing nearby said, "Yeah, but that thing's gotta come back down," and crawled under a nearby truck. Cioffoletti was mesmerized by the disappearing escape tower. "Then I turned around and there was nobody standing with me. Everybody was running over the dunes, hiding behind cars." A few seconds later the escape tower smashed into the beach about 300 feet from where Beddingfield was standing.

Back at the launch pad, the capsule still thought it was flying. Presently, a package containing aluminum chaff (used to help in the radar search for the descending capsule) popped out of the top, as the capsule detected that it was in a gravity field (not noticing that it had never left one). Then another pop, and the drogue chute emerged. Then, at precisely the planned interval, there was still another pop, and the main chute deployed, hanging down to the ground from its long shrouds and then beginning to billow slightly as it filled with the morning breeze.

In the blockhouse, the launch team was coming to terms with the fact that somehow they had lost control of a fully fueled vehicle still sitting on the launch pad a hundred yards away—"Your worst contingency plan could never have accounted for that situation," Fannin said. Their embarrassment aside, they were thinking of the half-dozen different ways in which the vehicle might blow up—most immediately, if wind filled the parachute and pulled the stack over. But even if that didn't happen, the launch team was acutely aware that the fuel tank vent valve was going to open automatically when the onboard batteries decayed. If the bulkhead between the LOX (liquid oxygen) tank and the alcohol tank on the Redstone had been cracked when the Redstone settled back on its points, then when the vent valve opened the likely result would have been an explosion. And they didn't know how long they had before the batteries would decay.

They thought of all sorts of plans. Al Zeiler wanted to get a high-powered rifle and shoot holes in the LOX tank to depressurize it—he'd had to do something similar during World War II, at the Germans' rocket center at Peenemünde. Others considered bringing in a cherry-picker to cut the shrouds on the parachute. Finally, not sure what to do, they prudently decided to do nothing.

The launch team crept out of the blockhouse by twos to minimize the casualties if the Redstone blew. The day passed, and the night, without an explosion. The next morning, having determined (they hoped) that the bulkhead had not been damaged and that all the LOX had vented, Zeiler took Fannin and a group of pad technicians out to the Redstone and safed the vehicle. Greenfield went out and found the disconnected plug. Beddingfield went out and disarmed the pyrotechnics on the capsule. Then they all went home to read the newspapers and find out what was being said about the American space program now. At NASA headquarters, everyone had the same question: How would Kennedy react?

2

The President-Elect's staff had caused palpitations at NASA even before the failure of M.R.-1. By law, the President was supposed to be the chairman of a body called the National Aeronautics and Space Council. President Eisenhower had never liked the idea (he dozed off during the second meeting of the council when a debate about NASA's logo dragged on) and had planned to drop the council altogether. Theodore Sorensen, who began meeting with Eisenhower officials shortly after the election to plan the transition, agreed with the Eisenhower plan. The Kennedy administration would disband the National Aeronautics and Space Council. At one stroke, NASA was about to lose its one clear channel of communication into the Oval Office and its one clear symbol of political prestige as well.

Fortunately for NASA, Vice-President-Elect Lyndon Johnson didn't like this decision at all. It had been only a few months since Johnson, as majority leader in the Senate, had blocked exactly the same plan, and he wasn't going to put up with it now. Johnson, who was fascinated by space flight and had been the space program's best and most powerful friend in Washington since it began, met with Kennedy on December 20 and got the decision partially reversed. The Space Council would continue to exist, but its chairman would no longer be the president. Instead, the council would be headed by Vice-President Lyndon Johnson—good news for NASA insofar as Johnson was such an ardent friend. But pessimists pointed out that Kennedy was not going to give away the chairmanship of the Space Council if he expected space to be a center of attention.

The pessimists soon got more evidence that they were right. Kennedy appointed a number of ad hoc task forces to review policy areas. To head his Ad Hoc Committee for Space, he chose Jerome Wiesner of the Massachusetts Institute of Technology. Wiesner had served on the President's Science Advisory Committee (PSAC, pronounced "p-sak") since Eisenhower had created it in the aftermath of Sputnik, and PSAC had not been bashful about its opinion of manned space flight. "The Science Advisory Committee was very much against manned exploration of space," Wiesner recalled, "because it was our view that you wouldn't gain anything, and you'd pay a hell of a price. . . . You put a man in space, and you suddenly have to provide a life-support system. If you have an instrument, every time that the man moves, the satellite wants to react in another direction, so you have to put in an enormously complicated stabilizing system that you wouldn't have to even think about if you didn't have a man." Manned vehicles were more expensive than unmanned ones by orders of magnitude. If science was what you wanted, Wiesner told Kennedy, unmanned exploration was unquestionably the way to go. Privately, Wiesner imagined a NASA run by scientists, not by engineers; he was unimpressed by the N.A.C.A. people who still ran NASA.

Keith Glennan told his staff at NASA headquarters to put together briefing materials for Wiesner, which they did. But neither Wiesner nor any of his people asked to be briefed. From Wiesner's point of view, the reasons were innocuous enough—they had just a few days, the people working on it were knowledgeable about the subject, and it was a fairly low priority anyway. There were half a dozen major defense issues that Kennedy wanted Wiesner to get sorted out before the inauguration, and Wiesner just didn't have much time to think about NASA.

From NASA's perspective, Wiesner's disengagement looked like a calculated slight. As time went on and no one from the committee contacted the senior NASA people, Seamans began to fear the worst. In January, two weeks before the inauguration, the blow came. Wiesner and his committee submitted what came to be known as the "Wiesner Report," a devastating attack on NASA and especially on the manned space program. The only parts of the U.S. space program that Wiesner found reasonably satisfactory were the scientific, unmanned space probes.* Manned space flight, according to the report, was an expensive and risky public-relations gimmick. A confidential version of the report was especially biting on this point. It recommended that Kennedy be especially careful not to "endorse this program [Mercury] and take the blame for its possible failures." The task force put the conclusion in italics: "*We should stop advertising Mercury as our major objective in space activities. Indeed, we should make an effort to diminish the significance of this program.*" Upon accepting the report, Kennedy announced that he was appointing Wiesner to be his Special Assistant for Science and Technology, the President's gatekeeper for all matters having to do with space. Furthermore, Wiesner persuaded Kennedy to keep the same members for his Science Advisory Committee. PSAC would be as opposed to manned space flight under Kennedy as it had been under Eisenhower.

Morale in the upper echelons of NASA hit bottom. "I remember having lunch with Johnny Johnson [NASA's general counsel]," Seamans said, "where we just plain discussed, 'What is really going to happen?' And we discussed it not from the standpoint of who the next administrator might be, but was NASA itself going to continue."

NASA was going to continue, as the White House saw it, but manned space flight had only the most precarious of footholds. "We talked about [the manned space program] a lot," said Wiesner of the time immediately after the inauguration. "[Kennedy] couldn't put it aside, because he was

---

* In retrospect Wiesner would think that "we were too cavalier" toward NASA and the space program in the Wiesner Report. But nonetheless it would be incorrect to conclude from this account that the Wiesner Report was wrong. If Wiesner had been successful in getting his way, the manned program would have been much smaller, if it had survived at all. But Wiesner's job was specifically to give advice about science, and his arguments about the relative scientific productivity of manned and unmanned flights were never seriously contested.

being pressured. But I can tell you honestly that he would rather not have done it.''

The palace intrigue that was worrying Robert Seamans down on the ground floor of Dolley Madison House could have been taking place on Mars as far as George Low's little task force was concerned. Oblivious to the new administration, Low's team plunged ahead with its plans to go to the moon.

John Disher recorded events in the stenographic notebooks he carried everywhere. His book for the fall of 1960 reveals that the first meeting of George Low's working group convened on Friday, October 21, 1960, at 2 P.M. From then on, the pace steadily accelerated. By November 8, Disher's notes reveal, they were talking about spacecraft weights. On the twenty-second of that month, they met with Milt Trageser of M.I.T. about the requirements for the navigation and guidance system.

At the end of November, Faget called with encouraging news from his latest visit to the Marshall Space Flight Center: "M.S.F.C. are eager," Disher recorded, meaning that von Braun would throw his weight behind a lunar landing. By December, the group began discussing the question of artificial gravity. Would it be needed on a lunar mission? If so, should it be continuous or periodic? And then what about radiation from solar flares? Could a solar flare be predicted? What were the odds that a big one would occur during a mission? What would happen to the crew if it did?

Low's people also worried during these first weeks about the nature of the lunar surface. Was it firm and rocky? Or was the surface covered with several feet of fine dust, so that a spacecraft coming in for a landing would sink without a trace? There were advocates for both positions, and nobody really knew. At least finding a flat landing place didn't seem to be a problem, Disher noted: "Good pick of landing sites this side of moon."

Everything was new. Low's group held few formal meetings, but they would call in people who looked as if they knew something about the problem at hand, quiz them, and then add the information to their growing store of data. The committee itself kept expanding to include others in NASA who were becoming interested.

By the 13th of December, they had a piece of paper with a Proposed Flight Schedule on it. It was presumptuous, trying to define a time schedule for a task they still only barely understood. This first, almost completely uninformed but prescient estimation showed the circumlunar missions taking place in 1967–1968 and the lunar landings taking place in 1969–1970.

On Thursday, January 5, 1961, Low's group faced its first major review by top NASA officials. On the basis of the review, they would decide whether to pursue the lunar mission aggressively under the new administration or to table it instead and take a less ambitious, more self-protective stance. Low and his confederates gathered at lunchtime on Tuesday to rehearse their presentation on Silverstein.

First, they decided, no more of the "man-on-the-moon" language they had been using—it was too slangy, too likely to be ridiculed. "Manned lunar landing" would be the phrase for what they wanted to do. It was also obvious to Silverstein that they had better not pretend to be farther along than they really were. In fact, Silverstein said, they ought to emphasize that this was not a coordinated set of presentations. This was known around NASA as "the country-boy treatment," Disher explained later. "You could have been spending your life on it, but you go in and say, 'This is just something we threw together.' It helped disarm people sometimes."

Low followed his instructions (the minutes of the review specify that the nine presentations were only a "first cut"), and Glennan gave Low what he wanted, a go-ahead for lunar landing work to continue. Glennan even made the group legitimate, converting them into the "Low Committee" with a mandate to answer the question "What is NASA's Manned Lunar Landing Program?" For Low's people, it was exhilarating. At the same time that Bob Seamans was fretting about NASA's future as an agency and John Kennedy was wishing that he could get rid of Mercury, John Disher thought that they were halfway to putting a lunar landing program on NASA's schedule.

Bob Seamans knew things that other people didn't, and it didn't improve his spirits as the inauguration approached. Administrator of NASA is a political appointment. In new administrations, people are supposed to be eager to accept such appointments, but this was not the case for the NASA post. "The administration was getting a little desperate," Seamans recalled, "because quite a few people had turned the job down." By Lyndon Johnson's recollection, a total of seventeen men had said they weren't interested in running NASA. "Of course," Seamans continued, "we knew that too"—that people were refusing the NASA job—"and that was another thing that was demoralizing. You said to yourself, 'Why would anybody turn it down? It must mean that the plans for NASA are being pulled in.' ''*

The inauguration came and went, and still there was no NASA administrator. Finally, Kennedy called in Wiesner and told him to do something—this was getting to be an embarrassment. Wiesner, seeing that the pool of qualified scientist-managers had been pretty much depleted by the refusals, turned to a man who wasn't a scientist but had been in and out of Washington for more than twenty years, most prominently as Truman's director for the Bureau of the Budget and later as Undersecretary of State. His name was James E. Webb.

From that moment on, NASA seems to have been watched over by a solicitous Providence. Time and again, seeming misfortune turned out to be the best thing that could have happened, or the right person turned up at the right time, or a malfunction that would have been disastrous on one mission happened on another where it was not disastrous. The first instance of NASA's serendipity was that only because Kennedy was indifferent to space did Jim Webb end up in the administrator's position. If the earlier candidates had known that four months later NASA would become a custodian of the nation's honor, most of them would probably have snapped up the job. If the men in the White House had known, they would not have chosen anyone like Jim Webb.

For this was the era when Robert McNamara came to the Pentagon, Robert Kennedy came to the Justice Department, and McGeorge Bundy came to the White House—a new generation of leaders. Jim Webb could hardly have been more unlike these other New Frontiersmen. Stocky and voluble, Webb at fifty-five was from a different generation than most of the others in the new administration, and from a different world. Instead of Harvard and wire-rimmed glasses, clipped accents and dry wit, Jim Webb was University of North Carolina and rumpled collars, corn-pone accent and down-home homilies, a good old boy with a law degree.

Webb wasn't about to accept the NASA job just because it had been offered, any more than the first seventeen. After Wiesner called him, Webb flew to Washington and spent a weekend talking to trusted associates about the prospects for space under the Kennedy administration. "By the time Monday morning came I had a pretty good picture of what was going on"—namely, that Wiesner and the men around him

* Wiesner suggested that the real problem was Lyndon Johnson, the new chairman of the National Aeronautics and Space Council, who insisted that he was going to hire the new administrator. Johnson would call candidates, and they would turn him down. When Wiesner checked with some of them later to find out why, they seldom mentioned fears about the future of NASA. "Mostly, the people would report to me that they didn't want to work for Lyndon Johnson."

were dead set against the manned space program. Webb had decided that he "would not take the job if I could honorably and properly not take it."

But he had an appointment to talk to Lyndon Johnson that Monday morning at the office Johnson still kept in the Senate Office Building. "When I arrived at the Vice-President's office, Hugh Dryden was already there," remembered Webb. "He'd been invited by the Vice-President to meet me there. So I talked this over with Hugh. We got away from all the secretaries, and I said, 'Hugh, I don't really think this is the thing for me to do.' He said, 'I don't either. I don't really believe you're the right one, or that you'd want to do this.' " Then another acquaintance, Frank Pace, came into the anteroom for a meeting with Johnson. "Frank," Webb said, "Hugh and I don't think I'm the right one to do this. You've been in the missile business. Do you agree with that?" Pace vigorously agreed. "All right," Webb said to Pace, "you're the messenger. When the Vice-President comes in, instead of my keeping my appointment, we'll be outside. You tell him I'm not the right one."

Webb and Dryden retreated into the echoing corridor outside Johnson's office and waited. After a few minutes, Pace abruptly emerged. "He came out like he was being ejected from the office," Webb remembered. "The Vice-President just threw him out."

Webb was marched into Johnson's office and given the full Johnson treatment. As soon as he got out of there, he called his friend Clark Clifford, an influential Kennedy adviser, and said, "Clark, you've got to try to get me out of this." Clifford laughed and replied, "I'm the one who recommended you. I'm not going to try to get you out of it." Finally, when Webb said that he would not accept the job "except by direct invitation of the President," an appointment with Kennedy was arranged for that afternoon. Webb got no promises from the President that the United States was going to make a major effort in space. But in the end, Webb found he could not "honorably and properly not take" the job.

Though the manner in which Webb came to NASA was not auspicious, a case can be made that James E. Webb was the Kennedy administration's most effective appointment. He would have his share of problems and make his share of mistakes, but in Jim Webb NASA got that strange Washington hybrid, the politician-manager—a man who could run a large organization and know where all the bodies were buried, a man who could play congressional appropriations committees with the finesse, the hard-eyed calculation, and, when circumstances required, the deviousness of a Lyndon Johnson himself. The role he played from then until his resignation in the fall of 1968 was indispensable. Many of the engineers of Apollo, often men with little use for politicians, will tell anyone who asks that of all the people who got the United States to the moon by the end of the decade, Jim Webb was among the most important.

People could get badly burned if they paid too much attention to the North Carolina accent and the country-boy pose. "You must remember," a close colleague once said, "that Jim is a very complex fellow, and he has many hats. He jams on his lawyer hat, then he pulls his Marine flyer hat down over his ears. Then his businessman hat. What he said with one hat on doesn't always agree with what he said with another." Jim Webb was one of those men who could keep two contradictory ideas in his mind at the same time with no trouble at all.

Webb never tried to make technical decisions himself, but he would sometimes override a technical decision for nontechnical reasons. Despite this, and even because of it, he eventually won the respect of most of his engineers. To Ken Kleinknecht, an engineer's engineer, this politician ("a talker" is Kleinknecht's label for such persons) always remained a little exotic and incomprehensible. "I could listen to him talk for forty-five minutes, and when he got finished, you know, I really didn't know what he'd said. But you listened every minute!"* And Kleinknecht thought Webb made good decisions. Kleinknecht remembered when Webb canceled a seventh Mercury flight, a three-day orbital mission by Shepard. "We weren't very happy about it," Kleinknecht recalled. "But Webb argued, 'If you do it and it's successful, it doesn't mean a hell of a lot. If we were to have a failure, we couldn't recover. It might stop the manned space program.' And you know, he was a hundred percent right." For Kleinknecht, thinking back over a career that took him from manager of Project Mercury through troubleshooting the first launch of the shuttle, "Webb was the greatest thing that ever happened to NASA."

Jim Elms called Webb a "politician in the best sense of the word." He told a story in later years that encapsulated this for him. Elms was down in Houston as Gilruth's deputy after the lunar program had been approved and the Space Task Group had moved to Texas. They were trying to get the Control Center built. Fresh out of industry and not understanding how government funding worked, Elms and Earl Hilburn had laid out the Control Center on the assumption that I.B.M. computers were going to be installed—at that time, only I.B.M. was making the kind of equipment

---

* Webb enjoyed talking. A colleague from another agency who regularly had lunch with Webb once bet his secretary that he could get through an entire lunch without saying a word. He succeeded, substituting grunts for hello and goodbye. It was one of the best lunches they'd had together, Webb remarked cheerily as his guest left: "I learned so much."

they needed. But when the Control Center was half built, one of the other computer companies got wind of this and started to raise a fuss. This was all the more inconvenient because the General Accounting Office had just issued a directive saying that no one was to buy I.B.M. equipment except via a competitive bidding process. But if NASA went through the competitive procurement process, the entire Apollo program schedule would slip disastrously.

A nonpolitician would have bought the I.B.M. computers and embroiled NASA in a scandal, said Elms. A careful, cover-your-ass politician would have followed the rules, let the program slip a year, and blamed his subordinates. What Jim Webb did was to call in the chief executive of every major computer company in the United States, including the one who was complaining. As each came to his office, Webb told him (as Elms reconstructed it, in a convincing drawl), "I want to thank you very much for comin' here. You know, I got a problem I'm facin', and I want to share it with you. President Kennedy has said we are to get to the moon in the decade and get back, and Mr. Elms and Mr. Hilburn here have persuaded me that if we don't buy some computers from Tom Watson Junior, we ain't gonna be able to do it! And the G.A.O. said if I buy those computers from Tom Watson Junior they're gonna put me in jail!" Webb then put it to his guest: Was he prepared to come back in two weeks and sign a fixed-price contract for the computer? With that, Webb sent him off to be briefed on the Control Center's computer requirements by Elms and Hilburn, who were waiting next door. "And they came back in a couple of weeks," Elms chuckled, "including the chap who was so upset. They shuffled around a little bit and they said, 'What you'd better do is buy computers from Tom Watson here.' " The computers were installed on schedule and nobody's nose was out of joint. *That*, said Jim Elms, is a politician.

3

Back on the third floor of Dolley Madison House, Low was sailing ahead with his mandate to answer Glennan's question—"What is NASA's Manned Lunar Landing Program?" Although Webb was still barely aware of it, a growing cadre within NASA was assembling a detailed plan for the lunar landing mission. On St. Patrick's Day, March 17, 1961, they convened at Wallops Island for a strategy session.

They were now working toward the goal of convincing Webb and Seamans to approve a lunar landing and then seek permission from the White House to let a contract for spacecraft Apollo "A," the first phase of the program.* Disher's notes for the Wallops meeting take eighteen pages in his stenographic notebooks, a record. On the fifteenth page, he jotted down Silverstein's marching orders. The first key date was April 10, when the Apollo Technical Liaison Groups would convene to be briefed by the lunar landing people on the status of their work.

These "Liaison Groups" had been formed back in November. At that time, they had been a way of tapping into the expertise of people at Ames and Langley and Marshall who weren't part of the Space Task Group. Now, they were going to be used for the big push to get approval for a lunar program. So, Disher noted, between now and the meeting with Technical Liaison Groups, they were to do "missionary work." Then, if the meeting went well, they would use the resources of Liaison Groups to help them write the specifications for the Apollo spacecraft.

Only five days after the Wallops Island meeting, Webb and Seamans met with the President, requesting appropriations for several items on the NASA wish-list. A few, including funds for the second stage of the Saturn, were approved. But funds for the detailed design of the Apollo spacecraft were refused and put on indefinite hold.

The lunar enthusiasts, refusing to be deterred, gathered as planned a few weeks later for a three-day meeting of the Apollo Technical Liaison Groups. As they assembled in the old Administration Building, they could have had no idea how justified their optimism was. The day was Monday, April 10, 1961, and all the equations were about to change.

* By this time they were thinking in terms of a two-phase program: Apollo "A," a configuration of spacecraft and launch vehicle that would have the capability for lunar orbit, and Apollo "B," which would be able to land and return to earth. This division of the program would have an echo in the actual "Block I" and "Block II" versions of the Apollo spacecraft.

5

"We're going to the moon"

In the second week of April 1961, John F. Kennedy was enjoying a halcyon spring, a time that Arthur Schlesinger, Jr., would later call the hour of euphoria. In fact, things were going so well that some of the new president's admirers were beginning to worry. "The Kennedy buildup goes on," James MacGregor Burns wrote in *The New Republic* that week. "He is not only the handsomest, the best-dressed, the most articulate, and graceful as a gazelle. He is omniscient; he swallows and digests whole books in minutes; his eye seizes instantly on the crucial point of a long memorandum; he confounds experts with his superior knowledge of their field. He is omnipresent. . . . He is omnipotent. . . . He's Superman!" Burns disapproved. The buildup was too much, too fast. The drop, when it came, would be all the more precipitous. But Burns's was a lonely voice. Ted Sorensen, Kennedy's White House counsel, remembered the heady feeling that the new administration could do no wrong. To Sorensen, just twenty-six years old, it still seemed as if they had "the magic touch."

As that second week of April began, the glow was as bright and warm as ever. On Monday, the 10th, the President had invited the Boys' Club "Boy of the Year" who was visiting the Oval Office to come with him to the season's baseball opener at Griffith Stadium. The day was cold and overcast, and the hometown Senators lost to the Chicago White Sox 4–3, but the young man reported to the press that sitting next to Mr. Kennedy in the presidential box made him feel "fifty feet high." Despite the chill, the young President remained hatless and coatless, as was his custom. Meanwhile the First Lady was acting as hostess to a lunch for 200 women newspaper reporters, which—*Time* reported—swept those tough career ladies off their feet.

On Tuesday, April 11, the Kennedys dominated the nation's prime-time television ratings in an hour-long NBC special sponsored by Crest toothpaste. Reporter Sander Vanocur interviewed the First Lady; Ray Scherer interviewed the President. The First Lady was at once elegant and shy, talking earnestly about the difficulties of giving three-year-old Caroline and four-month-old John-John a normal childhood in the White House. The President talked about his hands-on management style and reflected on how much more subtle and complex judgments looked from inside the Oval Office than they had from Capitol Hill.

For the nationwide audience, it was one more moment in that magic Kennedy spring. In reality, this was its moment of apogee. Even as the program was being broadcast, Kennedy knew that his first major defeat as President was only hours away.

It was no secret that the Soviet Union was trying to put the first man into space. A month earlier, a *Newsweek* story had predicted that a Soviet manned shot would beat Alan Shepard's suborbital Mercury flight, which now had been postponed until early May. On this very day, the intelligence warnings had become quite clear. A Soviet launch vehicle was on the pad; it was being prepared for immediate launch; it would be manned.

That morning, press secretary Pierre Salinger had been so sure a launch was coming that he had prepared a special presidential statement. Later in the day, Wiesner asked Kennedy whether he wanted to be awakened during the night when the launch was detected. "No," the President answered. "Give me the news in the morning."

At 1:07 Eastern Standard Time on Wednesday morning, April 12, 1961, U.S. radar recorded the launch of an R-7 rocket from the Baikonur Space Center on the steppes of Kazakhstan in the south-central part of the U.S.S.R. Wiesner called Salinger at 1:30 to report the launch and again at 5:30 to tell him that Moscow had announced a successful recovery of the spacecraft Vostok I. Its passenger, cosmonaut Yuri Alexeyevich

Gagarin, had completed one full orbit of the earth and was said to be feeling fine.

As in October 1957 when the Soviets orbited Sputnik, the nation looked at the skies and saw a calamity. Perhaps Shepard would have beaten the Russians into space, were it not for a few minor technical delays. No matter. Appearances could have a reality of their own, *The Washington Post* editorialized in its Thursday edition: "In these matters, what people believe is as important as the actual facts, and many persons will of course take this event as new evidence of Soviet superiority." An influential congressman from New York announced that he was ready to call for a full-scale congressional investigation—the American people must be "properly alerted" to the need for wartime mobilization. Abroad, an independent newspaper in Manila reported that the people in its part of the world "see in all this the supposed superiority of the Communist way of life, economic system, and materialistic philosophy." Egyptian president Gamel Nasser had "no doubt that the launching of man into space will turn upside down not only many scientific views, but also many political and military trends."

Overnight, a gap in Soviet and American rocket technology that had been years in the making became Kennedy's personal failure. As a writer for *The New York Times* pointed out, such events would inevitably be compared with the President's efforts to present himself as a "young, active, and vigorous leader of a strong and advancing nation." As if to underscore how the youthful image could backfire, a political cartoon the day after Gagarin's flight showed a gleeful Nikita Khrushchev bouncing a rock-sized spacecraft off the head of a confused and boyish-looking John Kennedy.

A presidential press conference on Wednesday afternoon revealed to all the world the end of the honeymoon. People were tired of seeing the United States second, one reporter complained. What was Kennedy going to do about it? "However tired anybody may be, and no one is more tired than I am," the President responded, "it is a fact that it's going to take some time and I think we have to recognize it." The news was going to get worse before it got better, he went on. He hoped that there would be some other area besides space where the United States could be first. But not space. In fact, Kennedy said emphatically, the nation would *not* try to match Soviet achievements in space, choosing instead "other areas where we can be first and which will bring more long-range benefits to mankind."

But as the world's reaction to Gagarin's flight built, it became clear that it didn't make much difference whether there were other arenas in which the United States might be first. It didn't make much difference that unmanned flights with scientific instruments were a more efficient way to explore space than manned flights. In a way that transcended any detached scientific assessment, a historic event had occurred, a soul-stirring step—and it was the Russians who had done it. "This is the end of an uncomfortable day," Edwin Newman commented on Wednesday night as he closed a news special on Gagarin's flight. "No matter how you add things up, today belonged to the Russians."

Kennedy began to feel the same inexorable pull toward manned space flight that Eisenhower had felt in 1957. On Thursday afternoon, the President asked Sorensen to review the options for the space program. Sorensen scheduled a meeting for Friday, April 14, a day that Sorensen later came to regard as the single most crucial day in the decision to go to the moon. It was a small meeting in Sorensen's office overlooking the South Lawn, with just five men talking and arguing for several hours: Sorensen, Wiesner, budget director David Bell, NASA administrator James Webb, and Webb's deputy, Hugh Dryden.

Wiesner was opposed to trying to compete with the Soviets in what he regarded as technological stunts, just as he had opposed the attention that was being given to the United States' own Project Mercury. But this afternoon was no longer the time to make that case. A consensus emerged that if the United States was going to try to compete, the options were extremely limited.

For the next few years, the Soviet Union had a lead in launch vehicles that the United States could not possibly close. The Soviets could put more weight into orbit, and weight was critical. The American launch vehicles that were going to be available for the next five years were going to limit the United States to small spacecraft flying in low earth orbit. To the five men gathered in Sorensen's office, it seemed likely that the Russians were going to be the first to put crews of two and three into orbit, the first to establish a space station, the first to circumnavigate the moon. If the United States wanted to compete, it had to jump to the next step, to pick a goal that would use the next generation of space technology. As the meeting wore on it became clear to Sorensen that the United States had only one chance to be first, and that was a manned lunar landing.

Hugh Sidey, *Life*'s White House correspondent, was under the gun. His managing editor in New York, Ed Thompson, had demanded to know how the White House planned to respond to Gagarin's flight. Sidey, who enjoyed unrivaled access to the Kennedy White House, had asked to see Sorensen to try to find out. He arrived in the West Wing late Friday afternoon, shortly after the weary task group had assembled in the Cabinet Room to brief the President.

Sorensen came out to see him. No, he couldn't answer Sidey's question, he said, but he had some people who could; and he took Sidey with him back into the Cabinet Room. Kennedy rehashed the discussion for Sidey's benefit. "Now let's look at this," Kennedy said. "Is there any place we can catch them? What can we do? Can we go around the moon before them? Can we put a man on the moon before them?" The one hope, Dryden explained, was a crash program, like the Manhattan Project that had developed the atomic bomb. And even that would have only a fifty-fifty chance of beating the Russians. Sidey described the moment:

> Kennedy turned back to the men around him. He thought for a second.
> Then he spoke. "When we know more, I can decide if it's worth it or not.
> If somebody can just tell me how to catch up. . . ." Kennedy stopped
> again a moment and glanced from face to face. Then he said quietly,
> "There's nothing more important."

As the meeting ended and the others were leaving, Kennedy turned to Sidey.

"Have you got all your answers?" he asked.

"Well, yeah," Sidey said, "except, what are you going to do?"

Kennedy told him to wait, and disappeared into the Oval Office with Sorensen while Sidey looked over his notes in the secretary's nook that connects the Cabinet Room with the Oval Office. After a few minutes Sorensen emerged. "We're going to the moon," Sorensen said. It would have been an immense scoop, if Sidey could have used it. But he understood the rules well enough to know that this confidence was not expected to show up in the next issue of *Life*.

Vostok I was not the only crisis on Kennedy's mind that day. Even as he met with his advisers to talk about space, a brigade of soldiers composed of refugees from Castro's Cuba was assembling at jumping-off points in Guatemala, awaiting Kennedy's final approval for a landing at the Bay of Pigs. Two days later, at the last go/no-go decision point at noon on Sunday, he authorized the expedition to proceed to the beaches.

The Cuban brigade landed in the pre-dawn darkness of Monday morning, beginning what Pierre Salinger would remember as the three grimmest days of the Kennedy presidency. From the first hours, Castro's army responded to the invasion with unexpected efficiency. By the evening of the first day, the invasion forces were far behind schedule. By Tuesday morning, they were stalled and trying to hang on. By Tuesday afternoon, they were encircled by 20,000 Cuban army troops and the White House Situation Room began a death watch. At a midnight meeting that didn't break up until two o'clock Wednesday morning, the President rejected last-minute proposals for U.S. intervention and accepted the inevitability of defeat. It was just a week to the hour since the flight of Vostok I.

On that disastrous Wednesday, as the world learned of the full extent of the Cuban debacle, President Kennedy called Vice-President Lyndon Johnson to the Oval Office. The topic of the meeting was the space program. They conferred alone for about half an hour. The next day, Johnson received a memorandum from the President. "In accordance with our conversation," Kennedy wrote, Johnson was to bring him answers to five questions. Number one on the list was, "Do we have a chance of beating the Soviets by putting a laboratory in space, or by a trip around the moon, or by a rocket to land on the moon, or by a rocket to go to the moon and back with a man? Is there any other space program which promises dramatic results in which we could win?" Kennedy wanted answers "at the earliest possible moment."*

The next day, Kennedy held another press conference. This time when he was asked about space, his tone from the week before had shifted dramatically. Now he said, "If we can get to the moon before the Russians, then we should."

Jim Webb found himself on a roller coaster. Just days earlier he had been fighting for a bigger space program against the resistance of a skeptical White House. Now he was faced with a Lyndon Johnson who was racing ahead faster and farther than Webb was prepared to go, and Johnson was impossible to control. "He just picked up the phone and called everybody that he thought was tops, independently," said Webb. "He called in Wernher von Braun without even asking me yes or no about it." Johnson also sent a copy of Kennedy's April 20 memorandum to General Bernard

* The other four questions were: "2. How much additional would it cost? 3. Are we working 24 hours a day on existing programs? If not, why not? If not . . . make recommendations to me as to how work can be speeded up. 4. In building large boosters should we put our emphasis on nuclear, chemical, or liquid fuel, or a combination of these three? 5. Are we making maximum effort? Are we achieving necessary results?"

Schreiver and Vice-Admiral John Hayward, asking them for replies. Then he called in three close personal friends: the president of CBS, a vice-president of an electric utility, and the co-owner of a large Houston construction firm.*

Johnson brought them all together for a meeting on Monday, April 24. Von Braun, Schreiver, and Hayward made presentations. After they finished, the group talked for a while among themselves as Wiesner and Webb looked on in disbelief. Wiesner would laugh about it later: "Johnson went around the room saying, 'We've got a terribly important decision to make: Shall we put a man on the moon?' And everybody said yes. And he said 'thank you' and reported to the President that the panel said we should put a man on the moon." To Webb, this kind of pressure was disturbing. "Johnson had them in a meeting room . . . around the table talking about what should be done, and [Johnson] pressing for action," Webb remembered. "I'm a relatively cautious person. I think when you decide you're going to do something and put the prestige of the United States government behind it, you'd better doggone well be able to do it." To Webb, this wasn't the way to decide.

But as Webb talked to Seamans and Low about the pieces they had been putting together that spring, it became increasingly clear that they could in fact do it. If the Saturn booster worked, and von Braun said that progress was going well, then landing a man on the moon wouldn't require any new technological breakthroughs. All it required was a lot of money, and all that required was a lot of congressional support.

On May 2, the prospects for a lunar-landing program had reached the point where Webb and Seamans decided to appoint William Fleming to chair the Ad Hoc Task Group for a Manned Lunar Landing Study, a thirty-day crash effort that would take all that had gone before, synthesize it, and tell the President whether NASA could in fact get to the moon and how much money it would take.† If Webb had had his way, he would have waited for the results of that study before making a recommendation to the President. But he was not to get his way. By the meeting on May 3, according to Webb, "the Vice-President was very close to demanding that we come forward with definitive programs." The Vice-President would be leaving for a two-week tour of Southeast Asia on Monday, May

* The three were Frank Stanton (C.B.S.), Donald Cook of the American Electric Power Corporation, and George Brown of Brown and Root.

† Low was not chosen for this job because he had become too clearly identified as an advocate, whereas Fleming was presumably neutral.

8. He demanded that he have a set of formal and detailed recommendations to give to the President by then.

On May 4, 1961, John Disher was at Dolley Madison House, sitting in on a meeting the likes of which he could have only fantasized about six months earlier. He and the others were being briefed by Seamans, and he scribbled furiously.

> Next 30 days. Fleming will chair this effort. Rothrock 2nd. Administrator wants to be better informed as each day goes by. UN Conference Room, H Bldg. ABSOLUTELY DISCREET!

Seamans repeatedly emphasized the need for self-discipline. They were to come up with a bare-bones program based on what *must* be done to get to the moon, not on what they would like the program to do. The schedule was breathtaking. Twice a week, they would hold an intermediate program review. Once a week, they would have a full-blown program review with Abe Silverstein. They could expect Abe to make decisions and sign off on their work as they went along.

Next Seamans described the guidelines that were being passed down: They should assume direct ascent to the moon—one rocket, carrying a unitary spacecraft with the built-in capability of landing on the lunar surface and returning to earth. The target date for the first landing was 1967. The plan had to be honest (Disher wrote that down in capital letters)—no overexuberant promises, no wishful thinking. They weren't to throw up an abbreviated flight schedule. This was not going to be some one-shot deal, but a systematic series of flights with all prudent intermediate steps in between. They should not spend any time thinking about nuclear-powered engines; the technology was too immature. But they hadn't made up their minds yet on liquid versus solid fuel; they should look at both.

Disher's penciled scrawl got bigger and bigger, until finally these last two notes took up the whole page:

WHAT IS MINIMUM AMOUNT NEEDED IN FY 62 TO DO JOB?
9:00 AM TOMORROW DEADLINE ON EVERYTHING.

The remaining hurdle was the first U.S. space shot, Alan Shepard's suborbital flight. If that failed, coming after the one-two punch of the Gagarin flight and the Bay of Pigs, no one could anticipate what direction events might take.

It did not fail. Shepard was launched on Friday morning, May 5, 1961, and landed safely after a flight of fifteen minutes and twenty-two seconds. With the Shepard flight safely behind him, Kennedy's decision to go to the moon was for practical purposes a foregone conclusion. "Kennedy found himself confronted with three choices," as Wiesner summed it up later. "Quit, stay second, or do something dramatic. He didn't think we could afford to quit, politically, and it was even worse to stay second. And so he decided to do something where we had a chance of really beating the Russians."

But the merits of the decision continued to bother him. At a state dinner for Tunisia's president Habib Bourguiba the day after Shepard's flight, Wiesner was standing in a corner chatting with Bourguiba when Kennedy joined them. "You know, we're having a terrible argument in the White House about whether we should put a man on the moon," Kennedy said to Bourguiba. "Jerry here is against it. If I told you you'd get an extra billion dollars a year in foreign aid if I didn't do it, what would be your advice?" Wiesner watched as Bourguiba stood silent for several moments. Finally Bourguiba said, "I wish I could tell you to put it in foreign aid, but I cannot." "Kennedy went around probing like that all the time, to get a feel for what he was doing," Wiesner said. And the probes kept coming back with the same answer. The United States did not have the option of withdrawing from the space race.

Wiesner was resigned to the inevitable. The decision to go to the moon was "a political, not a technical issue," as he would later put it, "a use of technological means for political ends." Wiesner made certain that he met his professional responsibility as the head of the President's Science Advisory Committee, telling Kennedy that PSAC "would never accept this kind of expenditure on scientific grounds." Subsequently, he would take some satisfaction in noting that the President hardly ever tried to justify Apollo for its science.

Three weeks later, Kennedy stood before the joint session of Congress and asked the nation to go to the moon, and the engineers at NASA began to understand that the frenzied activity of that spring wasn't going to stop. During the summer and fall of 1961, NASA moved decisions through the system with a speed that today seems unbelievable.

In June and July, detailed specifications for the spacecraft hardware were completed. By the end of July, the Requests for Proposals were on the street.

In August, the first hardware contract was awarded to M.I.T.'s

Instrumentation Laboratory for the Apollo guidance system; NASA selected Merritt Island, Florida, as the site for a new spaceport and acquired 125 square miles of land.

In September, NASA selected Michoud, Louisiana, as the production facility for the Saturn rockets, acquired a site for the Manned Spacecraft Center—the Space Task Group grown up—south of Houston, and awarded the contract for the second stage of the Saturn to North American Aviation.

In October, NASA acquired 34 square miles for a Saturn test facility in Mississippi.

In November, the Saturn C-1 was successfully launched with a cluster of eight engines, developing 1.3 million pounds of thrust. The contract for the command and service module was awarded to North American Aviation.*

In December, the contract for the first stage of the Saturn was awarded to Boeing and the contract for the third stage was awarded to Douglas Aircraft.

By January of 1962, construction had begun at all of the acquired sites and development work was under way at all of the contractors.

The speed was startling even to the people who were doing the pushing. John Disher finished the statement of work for the guidance contract and turned it in just one day before he left for a brief vacation. Not a week later, reading a newspaper on a beach at Lake Michigan, he saw an article reporting that M.I.T. had been awarded the guidance contract for the Apollo moon project. "I could hardly believe we could move that fast," Disher recalled. "In those days, you could do things with a half-page memo."

* The gumdrop-shaped spacecraft in which the crew rode was called the "command module" (C.M.). The cylinder behind it, carrying the rocket and propellants, oxygen tanks, and the rest of the environmental and support equipment, was called the "service module" (S.M.). The unit was called the "command and service module" (C.S.M.).

BOOK II

BUILDING

When they said, "Let's go to the moon"—hell, everybody didn't stand around saying, "What am I supposed to do?" or "Send me a directive," or "What's the procedure for going to the moon?"

—Thomas (Jack) Lee

6

"The flight article has got to dominate"

For Rocco Petrone down at Cape Canaveral, June and July of 1961 were so many lost weeks. It was a time so frantic, he recalled, that meetings were *scheduled* for 2:30 and 3:00 in the morning. "I only wish we'd had someone sitting in the corner taking notes," he said wistfully. Petrone, a history buff, wanted to reconstruct the events of that period, but couldn't "because so much happened and it happened so fast. It was almost like being in an accident." For Petrone himself, the summer of 1961 was the pivotal moment in his career when he emerged from obscurity to become Kurt Debus's point man for Apollo.

When the Huntsville organization had joined NASA, Debus, head of von Braun's launch team, had become head of Marshall's Launch Operations Directorate at the Cape. Now he put Petrone in charge of the newly created Heavy Space Vehicle Systems Office. Petrone, the Army major who gave up his military career to stay at Canaveral, was the son of Italian immigrants and a West Pointer. Ebullient and blunt-spoken, he looked the way his name sounded—big and strong and Italian. His new assignment was to oversee the planning and construction of the ground facilities that would be required to put a man on the moon. As of June,

Petrone suddenly had three gigantic tasks before him: Decide on a site, decide on a launch system, and then get everything built.

1

The launch vehicle—the "flight article," in Cape parlance—would come to seem anthropomorphic to Petrone, as if it were a giant to whom Petrone and the thousands of workers at the Cape were bound like servants to a demanding master.* "You can't be saying to him, 'I'm sorry, you can't have that much propellant,' or 'You can't have that much juice or that much wiring,' " said Petrone. "He's going to get what he wants. The flight article has got to dominate." And because Goliath's demands were going to be so outrageous, the machines for tending him would have to follow suit. Launching Saturn Vs involved the management of extremes—the biggest and the smallest, the hottest and the coldest, wispy fragility and colossal strength—and in the design of the Launch Operations Center, form followed function. But they were bizarre forms to fit an outlandishly extravagant function.

After Kennedy's speech, finding a site was the first item on the agenda. NASA already had use of the Air Force facility at Cape Canaveral; the obvious course would be to graft the Apollo facility onto it. But in many ways Cape Canaveral was not a good place to launch large rockets. In 1961, the local labor force was too small to supply the manpower that would be required for the construction. The Cape coast was prone to rain, lightning, and strong winds throughout the year and hurricanes during the fall. Worse yet, the weather was unpredictable, a nasty problem with a large vehicle sitting out in the open on a launch pad. Even when the weather was sunny, the humidity and the salt air corroded metal and insinuated gremlins into delicate electronics. Cape Canaveral was not an ideal place.

But alternative sites with better weather and larger labor forces had to be considered with this in mind: The destructive equivalent of a fair-sized nuclear weapon might one day explode on the premises.† This meant, first, that the site had to be large enough for its facilities to be built at safe

* A "test article" is a mockup of the real thing, suitable for testing certain systems but never intended to fly. A "flight article" is the real thing.
† The Saturn V had the explosive potential of a million pounds of T.N.T.

distances from the launch pads. It had, in fact, to be huge. Second, the site had to be distant from major population centers, so that if an errant Saturn did "get outside the gate," as the engineers euphemistically put it, it would not immolate a large number of people. Third, because things could go wrong at any time in the first few minutes of flight, the site had to border on several hundred miles of uninhabited space down-range—an area so vast that only an ocean would do. Moreover, the ocean had to be to the east of the launching site, because, for reasons of orbital mechanics, it was much more efficient to launch eastward against the rotation of the earth than to launch westward. Also for reasons of orbital mechanics, the closer to the equator the site could be, the better.

There were more restrictions. For security reasons, the site had to be in an area under the control of United States. For cost reasons, it had to be within reasonable proximity to the manufacturing facilities that would be producing the launch vehicles and support equipment. For logistical reasons, it had to be adjacent to a deep harbor or large river, so that it could receive barges carrying the huge components of the Saturn.

On April 25, as the administration shoved NASA into high gear, Bob Seamans told Debus to begin looking for some place better than Canaveral. During May, the criteria for a site were developed and Cumberland Island in Georgia came under serious consideration. In June and July, Petrone and the other staff members of what was known as the "Debus-Davis Committee" analyzed the options.* On August 1, after another of their all-night marathons, they flew up to Washington with their report.

By this time, they had considered eight locations: Merritt Island (adjacent to the existing Canaveral facilities); an offshore site near Cape Canaveral (to be built over the water, like an oil-drilling rig); Mayaguana Island in the Bahamas; Cumberland Island off the coast of Georgia; a mainland site near Brownsville, Texas; White Sands Missile Range in New Mexico; Christmas Island in the mid-Pacific south of Hawaii; and South Point on the island of Hawaii. The Debus-Davis report said that Merritt Island was by far the best of the lot—or, as Jim Webb stated at the time, Merritt Island was the worst place in the world except for everywhere else. The facilities on Merritt Island would be known as the Merritt Island Launch Annex, or MILA (pronounced "mile-a").

* * *

* "Davis" was General Leighton Davis, commander of the Air Force Missile Test Center.

The more difficult task facing Petrone's Heavy Space Vehicle Systems Office was figuring out how to prepare and launch a Saturn. It was a complex task made still more complicated by Debus's and Petrone's determination to use a radically new launch system.

A year before Kennedy's speech, von Braun's Future Projects Office in Huntsville, envisioning a large space program and costs-per-launch that might drop as low as $10 million, had told Debus that he should be prepared to launch as many as a hundred C-2 Saturns a year. In retrospect, the prediction seems absurdly high. At the time, Debus thought it was at least possible; and if not a hundred vehicles, then the total might easily be thirty or forty. The lowest plausible figure in the Future Projects Office's projection was twelve per year—and even that number could barely be handled by the Cape's current and planned facilities.

This prospect set Debus to thinking about how the Launch Operations Directorate might cope, and that in turn led him to reminisce. Just over fifteen years earlier, in Germany, Debus had been launching large numbers of V-2 rockets quite efficiently despite continual Allied bombing. He had done this by preparing the rockets in a hangar. After a rocket had been checked out, it was loaded horizontally onto a *Miellerwagen* and trucked out to the launching pad where it was erected, fueled, and launched in a few hours.

Debus had continued to do some of the preparatory work in hangars with Redstones and Jupiters in the United States, but the process had gotten slower as more and more work had to be done on the pad. His team could set up a Redstone on the pad and get it off in about a month, maybe three weeks if everything went well. But for the Saturn I they were planning to do all the assembly and checkout on the pad itself, and they were looking at a launch cycle of twelve weeks. For the larger launch vehicles to follow, they had to expect that the cycle would slow still further. Debus wondered whether something like the mobile launch system at Peenemünde might not be used at the Cape, for the Saturns.

Then in December of 1960, around Christmastime, Debus's musings were jolted by intelligence reports that somehow the Soviets were recycling their large launch vehicles on a much more rapid schedule than the United States. This got under the skin of Debus's young Army associate, Major Petrone, who had not been at Peenemünde but had seen a lot of airplanes take off, and who knew that you didn't try to check out the airplane on the runway. You checked it out in the hangar, and only when it was ready did you take it out to the runway, rev it up, and send it off. Petrone wanted to do the same thing with rockets—including the Saturn.

The idea of a mobile launch system for the Cape began to move from informal shop talk to a rudimentary plan. In February 1961, Debus called his team together and assigned one of his associates, Georg von Tiesenhausen, to prepare a detailed discussion of mobile launch alternatives. It was a lucky thing, Petrone thought later. If the mobile launch system had still been a brand-new idea on May 25, they would never have considered it—mobile launch would have been "introducing another unknown into what already was a big unknown" at a time when it seemed imperative to simplify. But when the lunar frenzy began in late April and Debus was summoned to Huntsville to confer with Bob Seamans and von Braun, he was able to go to Seamans with von Tiesenhausen's plans in his briefcase.

Debus briefed Seamans on April 25. The associate administrator was receptive to the mobile concept. The atmosphere was dramatically different from earlier meetings, when every new plan had to be scaled back because of the difficulties of getting funding from the administration. Now, when Debus went through the cost comparisons of mobile and fixed systems, Seamans waved them aside. The way things were shaping up, he told Debus, the most important thing was to come up with a system that was going to be able to launch large boosters on a rapid schedule. Cost was secondary.

By summer, Petrone was pushing ahead with plans for a mobile launch system, but by doing so he had locked himself into a set of requirements that were increasingly absurd. The flight article was getting bigger and bigger, more and more demanding, and harder and harder to satisfy.

One requirement led to another. The point of a mobile launch system was to prepare the rocket away from the pad. For V-2s and Redstones, this meant a low building where these vehicles could lie horizontally while technicians bustled around them. That was now out of the question. "With the size of the vehicle we were looking at—over three hundred and sixty feet long—and the umbilical tower it had to have," Petrone said, "we just could not see a way to prepare it horizontally and then in its entirety put it up vertically." There would be bending effects, all kinds of stresses on the vehicle. And they had no choice but to put it up in its entirety, because if they disconnected the umbilicals, then they would lose much of the work they had done in preparing it.

So they needed a place that would accommodate not just a 363-foot vehicle standing vertically, but its assembly as well. Huge pieces of hardware would have to be lofted to great heights by a crane that itself would have to be considerably above the height of the entire assembled stack. ''The height was really dictated by what we call the 'hook height,' the need to have that hook [of the crane] at a distance of four hundred and sixty-five feet above the floor,'' Petrone said. Add in the structure and the roof, and you've got no choice: ''Once you decided to do it vertically, that set the height. So we've got a five-hundred-and-twenty-five-foot building.''

But what kind of building? ''We asked ourselves the question—must it be enclosed?'' Petrone continued. ''Could we just build large derricks to which you could bring the launcher?'' No, that wouldn't work. They would be working where winds of fifteen, twenty, twenty-five knots were common, and such winds made it difficult to perform delicate assembly operations on a 363-foot vehicle—never mind rain and thunderstorms. Furthermore, it had to be a building with elaborate shops and support facilities, for this would not be a simple matter of hooking together some stages and setting the spacecraft on top.

''You must remember that the stages never see each other until they arrive at the Cape,'' Petrone said. ''You've got stages and hardware coming from all over the country.'' The stages had to do more than just fit together (as the old Mercury heat shield and the Atlas in M.A.-1 had not); they had to talk to each other. The Instrumentation Unit of the Saturn, its brain, was located at the top of the third stage. Thousands of signals and sensors and microswitches from the top of the stack down to the bottom had to make their connections without electrical interference or change of signal strength. All this was bound to take a matter of a few months, including the checkout and test, even after they got good at it.

So some kind of enclosure was essential. How many Saturns should they be able to handle at one time indoors? Here, Kennedy's goal forced the issue. Reaching the moon by the end of the decade had clear implications for test schedules and launch rates: The building had better be big enough to prepare at least four Saturns at a time.

Hence Rocco Petrone found himself recommending to Kurt Debus and headquarters that NASA construct the largest enclosed space in the world on top of sandy soil in a place periodically swept by hurricanes. The flight article had dominated.

The result—the Vehicle Assembly Building, the V.A.B.—is one of the man-made wonders of the world, but at the time of Apollo it was celebrated mostly for things that weren't true. The most widely accepted legend was that the V.A.B. was so huge that it had to be air-conditioned to prevent clouds from forming inside. Wrong on both counts: The V.A.B. was not air-conditioned, and clouds didn't form anyway. It was also generally believed that the V.A.B. was designed to accept the Nova, which is why it was so much taller than any rocket that ever occupied it.* Wrong again. It was assumed in 1961, when the V.A.B. was being planned, that the Nova would be assembled on a fixed pad out on the beach, feasible because only one or two Novas would be launched per year.

It is ironic but true that the most accurate legend, the building's sheer, unbelievable size, is at once both the most factually correct and appears to be the least so. The building is immense. Its floor covers eight acres. Its walls extend upward 525 feet (the Statue of Liberty is 305 feet high). Each of the four doors of the V.A.B. is 456 feet high, tall and wide enough to admit the United Nations' headquarters building. The V.A.B. is secured to Florida bedrock by 4,225 piles driven down 160 feet to prevent it from taking off like a box kite in a hurricane.

Yet conveying this size in ways other than numbers turns out to be impossible. Ray Clark, who ran the Cape's support facilities in the 1960s, recalled that as the V.A.B. was nearing completion, NASA wanted to show the public how big it really was. But nothing their photographers produced captured the immensity. NASA got in touch with *Life* magazine, whose editors replied loftily that NASA need not worry, *Life*'s photographers knew how to do that sort of thing. The *Life* photographers flew down and set up shooting positions from every conceivable angle. They shot hundreds of rolls of film and went back to their labs in New York. "A while later they came back to see us," Clark remembered. "They said, 'We don't know how to do it either. There's just no way to portray photographically, or any way we know of, what the building's really like.' "

For there is no scale. The V.A.B. sits out on the flatlands of Merritt Island, surrounded by a few small buildings. Approaching it, the human mind converts what it sees to a manageable set of dimensions. On a small part of one side of the V.A.B., an American flag was painted; and while visitors are told that each stripe in the flag is wide enough for a Greyhound bus to drive on it, the comparison doesn't register. "Nope," said Clark, "there's no scale. If you could see a man in the photo, that might do it—but to get the whole building, you've got to get back so far you can't see the man."

* The Nova was the vehicle to be built for a direct ascent to the moon, with eight F-1 engines in the first stage (the Saturn V had five), developing 12 million pounds of thrust. Throughout the first year of planning the MILA facilities, Petrone had to make allowances for the possibility that he might have to launch this monster.

2

The logic that Petrone and the other planners had to bring to their job had an Alice in Wonderland quality to it. They were required to believe any number of seemingly impossible things every day. If they wanted a vehicle big enough to send a spacecraft to the moon, then it had to be the size of the Saturn. If they wanted a vehicle the size of the Saturn, then they had to have a building with the size and capabilities of the V.A.B. And if they wanted to use a mobile launch capability for this vehicle, then there was the one remaining problem that had been on everyone's mind from the beginning. It was elementary, as Petrone pointed out: "You gotta move the bird." More specifically, the system required that they take a vehicle the size of a navy destroyer, stand it on end, transport it three and a half miles, then set the whole stack on the launch pad.

Rocco Petrone went to an engineer named Don Buchanan and told him to design the appropriate equipment for doing all of this. "And let me tell you," Petrone said later, "he'd rather not have drunk from that cup." Buchanan's first reaction was that Petrone must be out of his mind to want to move that thing. "Well, Don," Petrone said to him, "we've got to do *something* different." He urged Buchanan to consider the glory of it all: Nowhere else in the world would there be anything like it. Buchanan suggested perhaps there was a reason for that. "Don was a bit negative," acknowledged Petrone, but Petrone kept at him, for he considered Buchanan the best designer of ground support equipment anywhere. Finally Buchanan accepted what Petrone thought of as the heart of the burden in preparing the Cape for the Saturn: the design of the crawler and of the launcher.

Don Buchanan was a great rarity, perhaps unique, in being both a Langley man and a Marshall man. He worked at Langley from 1949 to 1956, and then transferred to Huntsville where he remained until he became part of K.S.C. in the 1960s. A Virginian—the name is pronounced "Buck-anan"—he was a gentle man who looked as if he might be the president of a small-town bank and whose only eccentricity was an astonishingly large repertoire of limericks, mostly unprintable. In his professional life, Don Buchanan was the kind of engineer of whom other engineers say, "He works with a sharp pencil." No matter how large his office became, it always had a drafting table in the corner where Buchanan spent most of his time. He was a man with a vocation, utterly absorbed in the practice of designing workable machines for unlikely tasks.

He was still living in Huntsville, forty-one years old, when he got his assignment from Debus and Petrone. Altogether, Buchanan had the responsibility for designing the launch platform itself (Buchanan called it simply "the launcher"), the umbilical tower, the flame deflector, the support arms, the hold-down arms, and whatever it was that would transport the Saturn V from the V.A.B. to the pad.*

As with a Chinese puzzle, one piece was the key to assembling the rest. In the case of the launcher, the key was the flame deflector. When a rocket was launched, something had to deflect the engine's flames so that their back-pressure (a bouncing effect) did not destroy the base of the ascending rocket. For the Redstone, the flame deflector was a little iron pyramid with four sides which a strong man could pick up and carry. For the Saturn V, the flame deflector was a metal wedge almost forty-eight feet wide, seventy-eight feet long, and more than forty feet high. The size of the deflector then dictated the dimensions of the elements in the launch complex which had to fit over it: umbilical tower, mobile service structure, and launcher.

On top, stretching into the sky alongside the Saturn V, were the umbilical tower on one side and the mobile service structure on the other. The mobile service structure, as tall as the launcher, had five work platforms which gave access to the launch vehicle and also contained several enclosed rooms in its base. It was moved up to the vehicle after the launcher was in place, then moved back again to its parking area more than a mile away from the pad eleven hours before launch.

The launcher itself was the platform on which the Saturn sat, a steel box 25 feet deep, 160 feet long, and 135 feet wide. The engines of the Saturn were positioned over a 45-foot-square hole in the middle of it. Beneath the whole structure was the wedge-shaped flame deflector, which sat in a trench between embankments on either side. So from ground level on up, the addition went like this: 48 feet for the flame trench, 22 feet between that and the bottom of the launcher (to make room for the transporter), and 25 feet for the launcher, which finally put Buchanan, staring at his drawings, at the Zero Deck—95 feet above the ground, roughly the height of a seven-story building.

Then above the launcher was the 380-foot umbilical tower, structurally integral with the launcher. The tower had two high-speed elevators to carry

* Later, he would also take over what proved to be one of the most difficult tasks in the entire inventory of ground support equipment (G.S.E.), the swing arms that connected the umbilical tower to the launch vehicle.

men and equipment to the nine swing arms that carried electric, propellant, and pneumatic lines to the Saturn. The swing arms themselves were massive, wide enough to drive a jeep across and averaging about 24 tons in weight—but also movable, designed to disconnect from the Saturn and swing away from it in the first few seconds after the engines ignited. Atop the umbilical tower was a 19-foot-high, 25-ton hammerhead crane, and atop that was a 48-foot retractable lighting mast. Thus, the total height from the base of the flame deflector to the top of the lighting mast was 542 feet.

As they began to work on the designs for the launch pad in more detail, Buchanan began marking up the side of their office building, a converted cotton mill in Huntsville, putting yellow tape where the different levels of the structure would be. It got his people ''thinking in the right proportions,'' Buchanan said. Then, to make sure that they really understood what they were up against, Buchanan got a large photograph of the tallest building in Huntsville, the Times Building, 107 feet high. Buchanan superimposed on the photograph a sketch of Launch Complex 39. The top of the Times Building was only 12 feet higher than Zero Deck of the launcher. Buchanan would show the picture to his Huntsville friends, and they reacted more or less as he had when Petrone first gave him the job. ''Well, you're *probably* crazy to even build that thing,'' they would say to Buchanan, ''but then we *know* you're crazy when you say you're going to move it.''

In the beginning, Debus and Petrone hadn't been that worried about how to move the Saturn to the pad. They knew they wouldn't be able to move it horizontally, but they had always assumed that some sort of barge system would be the solution. Barges were a good way to carry great weight. And so from the earliest planning through the summer and fall of 1961, the plans called for a canal to be built from the V.A.B. out to the launch pad. A huge motor-powered barge would carry the assembled Saturn and its umbilical tower.

But after Buchanan was put to work on the problem in October 1961, he found that the more they worked on the barge design the more problems they encountered. The onboard propulsion system they had planned consisted of a set of large outboard motors. But tank tests revealed that they couldn't use such a system in a canal of the planned width using a barge of the required dimensions. There would be so much backwash that more motors would be needed than they had places to put them, and they couldn't widen the canal enough to get rid of the backwash effects. The next version of the plan scrapped the outboard motors for a rail system running alongside the canal—a twentieth-century version of the mules that used to pull barges along canals a century earlier. It would work, but it was also making the system much more complicated and expensive.

Then they discovered that it was going to be impossible to steer the barge: The Saturn and the umbilical tower would act like a huge sail, forcing the barge to the banks in a cross breeze. The rail system would have to be modified so that it not only pulled the barge, but kept it clear of the canal banks. And there were other little things. If they had a three-and-a-half-mile canal cutting through the middle of the Merritt Island facility, they would have to build a system of drawbridges, which would be complicated and expensive. A canal going into the V.A.B. would create high humidity in a place where they wanted as little humidity as possible. Contending with that would also be complicated and expensive. And after figuring out solutions to all these problems, Buchanan had to decide what to do when the barge and its stack reached the launch pad. No matter how Buchanan fiddled with different ideas, it came down to the same choice. Either he had to jack the whole launcher up a hundred feet so he would have room to put a flame deflector under it, or he had to evacuate the water out to a depth of a hundred feet, like a big bathtub. "Neither option was all that attractive," Buchanan said dryly. As he looked long and hard at the barge option, he began to think that too much had been assumed too quickly.

By the end of January 1962, Debus and Petrone were still committed to the mobile launch concept for reasons that by now had nothing to do with the speed of launch. They wanted the flexibility that a mobile system gave them, the protection from the elements, the room for growth. But the underlying technical rationale for the system had originally been that it would be the most efficient way to launch large numbers of vehicles, and the break-even point had generally been calculated as twelve. By this time, realism had prevailed. There was no more talk of a hundred launches per year or of $10 million per-launch costs. On the contrary, headquarters was now confident that the lunar program would require considerably fewer than twelve large Saturn launches per year.

Two studies by NASA contractors had already come out recommending a fixed system. Debus knew further that a systems analyst at headquarters had put a three-man team to work reassessing the launch system and that their thinking was moving in the direction of a fixed system. If the mobile system was going to survive, they must pin its pieces in place, which meant reaching a firm decision on the mode of transport. At a meeting at Huntsville on Tuesday, January 30, 1962, Debus made it clear that he wanted a decision, he wanted it soon, and unless they had something better, then they would have to go with the barge.

On Friday of the same week, one of the men who had been in the meeting, O. K. Duren, got a call from a man named Barry Schlenk. Schlenk was at Huntsville that day talking to some people about an overhead crane for the Titan system. Schlenk had overheard someone talking about Marshall's problem with moving the Saturn down at the Cape. Did Duren know that Schlenk's company, the Bucyrus-Erie Company, had a rig that was being used for mining in the Kentucky coalfields? It crawled along the surface of the ground on tank treads, stripping the overburden from the mining area. It was huge and it was self-leveling. Would Duren be interested in hearing about it? Duren would, and the two spent the afternoon looking at pictures of the machine. Duren was enthusiastic and called his boss, Albert Zeiler.

Zeiler told Debus about the shovel, adding that it was the dumbest thing he'd ever heard of. But Debus told him to take a couple of people with him and go up and look at those shovels and come back and tell him why they shouldn't consider them. Reluctantly, Zeiler took Theodor Poppel, head of the Launch Facilities and Support Equipment Office, Duren, and Buchanan up to Paradise, Kentucky—a most inaptly named place, in Buchanan's opinion—to see this machine. The story that was passed down in Cape lore in later years was that they went out to the coalfield and climbed onto the vehicle and examined it, and then Zeiler and the others sat down, grabbed onto the nearest handholds, and said, "We're ready! You can start!" And their puzzled hosts said, "What do you mean? We've been moving for a couple of minutes now."

The shovel was too slow, only twenty feet per minute, and the leveling technology wasn't up to aerospace standards. But the fact remained: Here was this great big machine transporting an extremely heavy weight with no vibration. Buchanan began to look into it more carefully. Studies were conducted, technical comparisons were made, and only four months after the NASA engineers had so skeptically gone to Paradise, Debus approved Buchanan's recommendation to discard the plans for a barge and to proceed instead with a gigantic transporter-crawler.

Because of its much larger scale, the final result bore only a passing resemblance to the big stripping shovel in the Kentucky coalfields. But Buchanan used its basic design—a platform with a "truck" at each of its four corners. Each truck looked rather like a tank, with a double set of belted links, or "shoes" (known to the engineers familiar with the budget as "them golden slippers"). Each link, or shoe, was nearly eight feet long and weighed a ton. Each belt—eight in all—consisted of fifty-seven shoes.

Describing the scale of his creation, Buchanan used to suggest that his listeners imagine the infield of a major-league baseball diamond being cut out and made into a platform. Then they should imagine the platform, elevated on its monster treads, moving along a track for three miles. Now all Buchanan's listeners had to do was imagine the elevated baseball infield moving along a track for three miles carrying two-thirds of the Washington Monument on top, coming to a five-degree slope, and then climbing it. Last of all, they must imagine that as the assembly climbs the slope, the baseball infield is hydraulically jacked so that it remains level throughout the climb. That was the crawler.

It was a marvel for which Buchanan was honored with some of the most prestigious awards in the engineering fraternity, but the ways in which it was a marvel are for the most part too technical to be appreciated by anyone except other engineers. For the lay reader, perhaps this is enough: During its voyage from the V.A.B. to the launch pad, including its climb up that five-degree slope, the tip of the Saturn, 363 feet above the platform, never moved outside the vertical by more than the dimensions of a basketball.

All this Buchanan promised Debus that spring of 1962, and he got his chance to make good on it. Debus and Petrone were not to be stopped: They overcame the objections of the Office of Manned Space Flight and Congress alike and received permission to proceed with the mobile launch system.

"We often looked at the launch site and those things on the ground that don't fly as Stage Zero [of the booster]," Rocco Petrone said. "You had to have all the intricacies of a stage, things like swing arms, hold-down arms, feeding the gases in, all the propellants. When you've released, your Stage One is flying, but if you haven't done all these things on Stage Zero, Stage One would never get a chance to fly." The Cape's task was to design all this even as the Apollo-Saturn stack itself was being designed. "That was a hell of a challenge," Petrone continued, "and, I think, a challenge not very well understood. Here you have the [launch vehicle] stages going down the road with the spacecraft, and they develop needs for more juice, more wire, more propellants. You've got to be in the right rhythm, because when they come together you must all be at the crossroads at the same time."

CHAPTER

7

"We had more harebrained schemes than you could shake a stick at"

In the months following Kennedy's speech, Bob Gilruth would occasionally remark to the others in the Space Task Group that everything Kennedy had said in that speech was fine except for that one little word "safely," as in "and returning him safely to earth." "That's not simple," Gilruth would say to the others. "We don't know how to do that." That was why he had been aghast when he heard Kennedy's speech. Gilruth was confident that NASA could get a man to the moon safely—eventually—or that they could get a man to the moon within the decade if they were willing to take a pretty high chance of killing him. But as of 1961 they still didn't know how to do it both safely and within the decade.

It was hard enough just deciding what "safely" meant. Caldwell Johnson and Bob Piland, who wrote the work statement for the Apollo spacecraft, decided they had to have a number stating how reliable the spacecraft must be. That one demand would make a huge difference in the cost of the program because, as Johnson pointed out, "if you can afford to lose half of the spacecraft and half the men, you can build them a damn sight cheaper." On the other hand, Johnson continued, "in this country you just don't go around killing people. So we had to pick a number. Piland, he didn't dare pick it. And God knows I wasn't going to pick it. Nobody wanted to pick the number. So one day we walked down to see Gilruth and we said the time had come to bite the bullet."

Max Faget was there. Faget didn't have much respect for this reliability analysis—it was okay, he said, except that it depended upon data that didn't exist. In this straightforward objection lay a controversy that would fester for the next two years.

The process for computing reliability was simplicity itself. If a machine has a part that fails twenty out of a hundred times, its reliability—the probability that it will work—is .80. If the machine has a second part that will fail ten times out of a hundred, its reliability is .90. The probability that both parts will work is the multiple of the two individual reliabilities, $.80 \times .90$, or .72. The complete machine (if it consists of just the two parts) will work seventy-two times out of a hundred.

In a spacecraft, there would be tens of thousands of parts that all had to work in order to complete the mission successfully. To compute the overall probability of mission success, it was necessary only to know the reliability of each of those individual parts, and then calculate their combined product, which could be done in a comparatively short time on an ordinary calculator. Some smaller number of parts all had to work just to get the crew back safely (a less demanding task than completing the mission successfully). For the probability of getting the crew back, just multiply the reliability of all those parts.

The problem facing the Space Task Group people who were supposed to be estimating the reliability of their spacecraft was that, to know whether a part truly functions eighty times out of a hundred, they had to test it several hundred times. In fact, the individual parts for the spacecraft were supposed to have reliabilities of .99999, or .999999, or sometimes .9999999. On a probability basis alone, there was no way to make such claims on the basis of statistical evidence unless the engineers tested the parts millions of times.

Testing a part millions of times wasn't economically possible. It wasn't necessary either, in the view of most of the engineers in the program: Design it right and fabricate it per the print, and the component will work, every time. Engineers of this persuasion, Faget among them, argued that their time was much better spent searching for design flaws than mindlessly running tests. On the other hand, when you have tens of thousands of parts and are working with high-energy systems and are trying things that have never been tried before, the chances that something is going to break down somewhere along the road are high. "Max was sort of all for saying, 'Gee, if we are successful half of the time, that would be well worth it,' " Caldwell Johnson remembered. Gilruth said that 50 percent was a little too low. When it came to completing the mission successfully, nine out of ten seemed about right to him. As far as the safety of the crew was concerned, he hated to put a number on it, but to lose one crew out of a hundred sounded reasonable, given such an intrinsically hazardous mission.

About that time, Walt Williams, the Space Task Group's deputy director for Flight Operations, joined the discussion. He thought that they were being ridiculous. Losing one crew out of a hundred was a reasonable goal but they couldn't afford to say so—it didn't put enough pressure on the contractors and the operators. The Range Safety people use one in a million, Williams pointed out, in deciding what chance they would accept of a rocket landing in a populated area. Why not use that?

Gilruth thought that was unrealistic. "You ought to say what you honestly want," he reasoned. "In the long run, that's a better way than to try to kid yourself. It's like setting your clock up when you know you're going to be late." They compromised. The probability of getting the crew back safely was set at three nines (.999), or 999 times in a thousand. The probability of completing the assigned mission was set at two nines, 99 times in a hundred.

And that, Johnson recalled, is how it happened, in a ten-minute talk. "We wrote those numbers down, and they had a most profound effect on the cost of the program. If you took one decimal point off of that thing, in theory you could probably cut the program cost in half. If we'd added one more, there's no way in the world we could ever have done it —there's not enough money in the world to do it."*

But having set the requirements for the spacecraft's reliability in getting to the moon and back did not help to answer the deeper question: How the hell were they actually going to do it?

* The joke that made the rounds of NASA was that the Saturn V had a reliability rating of .9999. In the story, a group from headquarters goes down to Marshall and asks Wernher von Braun how reliable the Saturn is going to be. Von Braun turns to four of his lieutenants and asks, "Is there any reason why it won't work?" to which they answer: *"Nein." "Nein." "Nein." "Nein."* Von Braun then says to the men from headquarters, "Gentlemen, I have a reliability of four nines."

1

The truism about NASA during the Apollo years is that thousands of people were involved in everything, and this was as true of the design of the Apollo spacecraft as anything else. Still, at the core of the spacecraft design was the unique presence of Max Faget, and always at Max Faget's elbow was his closest collaborator, Caldwell C. Johnson, Jr. "Collaborators" isn't really an adequate word. They were sometimes Wilbur and Orville Wright, sometimes Tom Sawyer and Huckleberry Finn, and sometimes Don Quixote and Sancho Panza.

Caldwell Johnson—the first name is pronounced "Cadwell" in the Virginia manner—was small and slight like Faget and just as blunt and opinionated. They were both men whom other engineers consistently called "intuitive." But Johnson was a man of the Tidewater peninsula from birth, the sort of fellow who looked at first glance as if he ought to be whittling a stick on the porch of a country store. Twenty-five years after moving to Houston, he would still talk in a rich Virginia accent and an untamed down-home vernacular.

Johnson, who didn't have an engineering degree (he dropped out after his first year at the University of Virginia, partly for financial reasons and partly because he was bored with school), had grown up within shouting distance of Langley. He had watched the Brain Busters and imitated them, and in the process he became an unexcelled builder of model airplanes. With a minor reputation even in aeronautical engineering circles as the kid who had won a variety of awards in model-building competitions, he was hired as a model-builder for P.A.R.D. at the age of eighteen. In later years, it was a recurring headache for Bob Gilruth to promote Johnson to the next Civil Service level. The Civil Service people kept insisting that because he was not an engineer he couldn't go any further. After all, it was right there on his résumé: just a high school degree.

That sort of obsession with credentials made Johnson cranky ("The kind of stuff you learn in school ain't worth a pinch of shit anyway"), but in any case his résumé never seemed to bother the engineers who worked for him. As Owen Maynard, who had become one of Johnson's section chiefs, said, "You didn't go around stumping Caldwell on some little piece of physics or aerodynamics that you'd learned in college." But mostly, people remembered Johnson's ability to take an idea and translate it into an elegant design. "Caldwell's the sort of guy who's the artistic designer as opposed to the engineering designer," remembered one colleague. "He says, 'Well, dammit, if it doesn't look right, it's not right.' And he'll come up with something that looks right and it'll work."

Working for Johnson was not like working for a run-of-the-mill engineer, especially during the early days when the Space Task Group was still back on the Chesapeake Bay and Johnson's home fronted on the James River. Owen Maynard used to get to work and find a note on his desk. "There's three ducks and five flounders in the refrigerator for you," it might say, and alongside that would be another note listing half a dozen technical tasks for Maynard. Johnson had been out on the water before daybreak, doing his thinking on the boat. Often, the notes were in the form of drawings. Johnson's ability with a pen was legendary in NASA—Johnson drew freehand sketches that looked like engineering drawings and engineering drawings that were works of art.

Johnson detested higher authority every bit as much as Faget did and he instinctively disagreed with anyone whom he considered a smartass. When they were youngsters together in P.A.R.D., Faget discovered how to get Johnson to do his best work for you. If you just gave him a sketch and asked him to work it up into an engineering drawing, your work would languish on the bottom of his stack—Johnson was swamped with engineers wanting him to do their drawings for them. What you had to do was to design something yourself in detail—down to the last nut and bolt—and then tell Johnson, "I've got a pretty good design here, just draw it up the way it is." Johnson couldn't stand that. "It would inevitably bring out the best in him," Faget recalled, as Johnson set out to show Faget how he should have designed the goddamned thing. He was, Faget concluded, "undoubtedly one of the orneriest guys you could work with."

Of his long and spectacularly successful partnership with Faget, Johnson once said, "We need each other. He has some pretty good big ideas, and he really isn't too red-hot in turning them into good engineering things. And I'm pretty good at turning big ideas into good engineering things. I don't like to go around dreaming up new big ideas." Many observers have been surprised that two people so opinionated and so ready to argue were able to work together for more than forty years. But Owen Maynard, who watched them during Apollo, shrugged off such considerations. What you have to understand, he said, is that Faget and Johnson couldn't have a real disagreement. "Their minds are almost interconnected."

*   *   *

Caldwell Johnson was once reflecting on the accounts he had read of how the Apollo spacecraft came to be. "The way the history books say things came about," he said, "they didn't come about that way. The official records and all, that's a long way of explaining a lot of things. It turns out that the thing was done by people, not by machines, and people have a way of coming to a very rational conclusion in a very irrational manner."

By the time they began to design the Apollo spacecraft, Faget and the others who had been doing the preliminary work were told that this time, unlike the early Mercury days, they couldn't just lay out the design they liked and let a contract for somebody to build it. This time, they would be more systematic. This time, they would take advantage of industry's accumulated expertise.

First, they would complete their own preliminary studies, developing some broad guidelines. Then NASA would put out study contracts with three private aerospace firms, and these firms, each with its own team of engineers, would independently develop their own more detailed configurations. And then, with all these options in hand, Faget's division would assess the situation, choose the best design, and prepare the Request for Proposals (R.F.P.) that would lead to the final design and production contract.

What actually happened adhered to the letter of the process, but just barely. Faget's division produced a preliminary design for a spacecraft. Then NASA awarded the three study contracts to Convair, General Electric, and Martin. But while these firms were earnestly going about their work, Faget's division was continuing to develop its own design.* The others never had a chance. "We had no intention of ever using the other three, as far as I know," said Johnson. "At least, I didn't." Of course, Johnson added, "we couldn't lose. We could watch all three of them but they couldn't watch each other or us. So everything good they did, we would steal. But everything good that we had, they didn't know about."

When it came to the choice of the Apollo shape, they didn't even wait to see what the other contractors came up with. Before the study contracts had been awarded, Faget called them together and said, "Let's quit messing around and arguing about lenticular shapes, or this shape, or that shape. We're going to pick one." And the one they picked was the gumdrop-shaped spacecraft known to history, with its three-man crew,

---

* Faget and Johnson agree that it is impossible to specify precisely who designed the Apollo spacecraft, but, still omitting many who played an important role, these other men were also central to the effort: Bob Piland, Kurt Strass, Owen Maynard, Bob Chilton, Jack Heberlig, Alan Kehlet, and Bryan Erb.

gently rounded corners, and gleaming metallic skin. By the end of October 1960, Johnson had drawn a picture of it that to a layman is indistinguishable from the spacecraft that flew to the moon.

When after six months the three firms presented their painstakingly developed proposals, Max Faget accepted them, thanked the contractors for their hard work, and announced that the Space Task Group would use its own design. "It was kind of a gutsy thing to do," Faget acknowledged, "to say, 'Well, we paid them all this money to tell us how to build this thing, and they each came out with a different design, and we still want a design just like the one we've been using all the time.' But that's what we did."

Not everyone was happy with that bit of Faget chutzpah. "It was a real shock to the people who had worked on the studies," said one man close to the process. "Their Apollo shape had no real maneuverability. As it turns out, we didn't need it. They didn't realize that then, though. . . . A lot of these things were just gut decisions on the part of people like Caldwell Johnson and Max Faget, saying, 'We don't have time for that' or 'It's going to cost too much money' or 'We don't like it' or 'Congress won't buy it' or something. But they had no technical basis."

Max Faget told anyone who wanted to listen why this kind of talk was nonsense. On maneuverability, he said, they had good evidence from preliminary guidance studies that the spacecraft would be entering the earth's atmosphere within a narrow ten-mile corridor. "Matter of fact, it was a damned good thing we didn't put more [maneuverability] in it than we did," Faget said. "We really spent most of the time getting rid of the damn lift-to-drag ratio by banking back and forth." And through the backward-looking lens of history, Faget had the strongest of all arguments on his side: The spacecraft he wanted got built and flew successfully within the time schedule. Still, Johnson's own point about history and the way decisions got made ("People have a way of coming to a very rational conclusion in a very irrational manner") also had a lot to do with the process.

The choice of a three-man crew, for example, turned out to be perfect: Two men were needed for the lunar exploration activities and a third man was needed to operate the command module while the other two were on the lunar surface. But at the time they were making decisions about the size of the crew, the designers hadn't thought about such things as lunar modules. They just figured that they would run the duty shifts as the Navy did, four hours on, eight off, which meant they needed three astronauts to ensure that an astronaut would be on duty all the time. By the time Apollo flew, the flight controllers on the ground could monitor the cabin systems continuously and the astronauts could follow a more natural pattern of sleeping and working at the same time, so they no longer needed the third man to keep watch—but they did need him now to carry out the lunar-orbit rendezvous.

An even better example of Johnson's principle is the way that Apollo came to have those gracefully rounded corners. "You'll talk to some aerodynamicists, or some heat transfer people, and they'll explain to you the marvelous characteristics of this rounded corner, and why it was this way, and all," said Johnson. "That's a bunch of nonsense. When I first laid that thing out, it was a cone like Gemini and Mercury. And there's a good reason for it, too. That's a nice clean separation of flow on those sharp corners." And, imperatively, the diameter of the bottom of the spacecraft was no bigger than the diameter of the third stage of the launch vehicle—160 inches—that Marshall was developing.

According to Johnson, things were going fine when they got a call from von Braun saying that Marshall had reduced the diameter of the third stage to 154 inches. "We said, 'Jesus Christ, now we can't leave this sonofabitch hanging out over the edge like that, and we can't change the mold lines, because that'll cut in on the interior space.' So we said to each other, 'Let's just round the corners, nobody'll ever know the difference.' And that's how that command module got rounded. I don't doubt that the rounded corners are very beneficial and all that, but that was not the driving reason when it was done."*

To add to the irony, all that fiddling would have been unnecessary had they procrastinated a few more months. After struggling to make the capsule fit a Saturn that had been reduced from 160 inches to 154 inches, Marshall changed its mind again and replaced the planned third stage with another, larger one. The Space Task Group could have made the volume of the spacecraft just about as big as it wanted, after all; but by that time the design had been locked into the 154-inch configuration. This is why the adapter atop the third stage of the final Saturn/Apollo stack slanted inward—to make the rocket diameter small enough to fit the spacecraft diameter that Johnson and Faget had reached.

* * *

* Owen Maynard, who was also involved in deciding on the configuration of the spacecraft, thought the rational aerodynamic reasons were more persuasive than Johnson acknowledged in this remark. But Maynard, too, assumed some irrationality in the evolution of the design: One of his nonaerodynamic reasons for rounding the corners was that he knew the design would inevitably accumulate weight as time went on. He assumed that sooner or later the spacecraft would "have the density of water," and that the only way to limit the weight was to limit the amount of space that the engineers could stuff new things into. Rounding the corners helped.

Because Apollo's basic design was laid out before the lunar landing mission had been given to it, the spacecraft did not have the capability to get down to the moon's surface and up again. As talk about a landing mission became more explicit, Johnson went to Owen Maynard one day and told him to put his Systems Integration Section to work devising a spacecraft capable of a lunar landing and liftoff. But as Maynard and his team went to their drafting tables and began to try out ideas, it soon became apparent that designing such a vehicle would be a bafflingly difficult thing to do. The usual assumptions about how man was going to land on the moon didn't seem to work very well.

2

For many years, small boys pretending to be Buck Rogers knew exactly how to get to the moon. First you got a big rocket with fins on it which took off from earth, and then when the rocket got close to the moon you turned it around and used its rocket engine as a brake. You got out and climbed down a ladder, explored for a while, climbed back up the ladder, and took off for earth. When you got close to the earth, you did the same thing all over again. In the parlance of Apollo, small boys pretending they were the first man on the moon opted for the "direct ascent mode."

In 1952, a new idea was shown in spectacular full-color pictures in the *Collier's* issue of March 22. A panel of scientists headed by Dr. Wernher von Braun told how a space station, a great spinning shining ring, filled with workshops and living quarters and observation posts, with artificial gravity, serviced by a fleet of space tugs, could be built within the next ten to fifteen years.* Among other things, such as promoting world peace ("It would be the end of Iron Curtains wherever they might be"), the space station would be an intermediate step for getting to the moon. A spacecraft would be assembled in space, using components launched separately from earth. This new way of getting to the moon (minus the

---

* It was part of a series of articles about space travel which had a significant effect on promoting interest in the space program. Many NASA people dated their interest in the space program from the time they read these articles as teenagers or young engineers. Andy Pickett, who would be a senior engineer at the Cape during Apollo, remembered reading the *Collier's* series and dropping everything to drive to Huntsville to apply for a job.

great spinning space station) would become known later as the "earth-orbit rendezvous mode," or E.O.R.

The advantage of E.O.R., when NASA came to study it, was that it required a smaller launch vehicle than did direct ascent. The vehicle that landed on the moon (it was assumed) had to be a large, self-contained rocket system, complete with crew quarters, on-board life-support systems, equipment for exploring the lunar surface, enough fuel to escape the moon's gravity for the return to earth, and a heavy heat shield for surviving the 25,000-m.p.h. entry into the earth's atmosphere. Lifting that much weight in a single launch called for a mammoth vehicle. If instead the spacecraft could be launched from earth in two or three components and assembled in earth orbit, the lunar enterprise immediately became much more realistic.

The disadvantages of E.O.R. were considerable, however. First of all, how easy was it to rendezvous in space? In the early 1960s, no one knew yet. But presumably those difficulties could be overcome "Rendezvousing in earth orbit was not the problem," Faget explained, "but [rather] the business of, after you rendezvous, how do you put together the wherewithal to go to the moon?" At the time of Kennedy's speech, two main variations were being considered. In one of them, the propellants for the lunar journey would be put into orbit by one Saturn V, then a complete but unfueled spacecraft would be launched by another Saturn V. Another variation called for the spacecraft itself to be launched in segments and assembled in space.

These sounded straightforward enough, but actually doing them would be a tricky business. If the E.O.R. scheme called for rendezvousing with the propellants, then exactly how could they transfer these large quantities of volatile liquids from the storage tanks to the spacecraft in the weightlessness and vacuum of space? And if instead the E.O.R. scheme called for putting the spacecraft up in two or three segments, what were the actual engineering devices whereby these segments were to be connected in outer space, ready for a lunar voyage? "Every time you'd tell them what was wrong with one way of doing it, they'd tell you, well, they were going to do it the other way," Max Faget recalled. "As far as I know, those problems never got solved." For a long time, direct ascent still seemed simpler than E.O.R.

Furthermore, by the early 1960s direct ascent was no longer an impossible dream. The F-1 engine for powering the Nova, the same engine actually used for the Saturn V, was already in development by the time of Kennedy's speech. Von Braun's rocket scientists had the booster itself, called Nova, on their drawing boards. It was mammoth. In one typical configuration, Nova was planned to have eight engines clustered in its first stage (the Saturn would have five) developing 12,000,000 pounds of thrust; a second stage with four engines developing 4,800,000 pounds of thrust; and a third stage with 200,000 pounds of thrust.* Many people in NASA thought the Nova should be built (Faget was an especially enthusiastic proponent), and as late as June 1961, when the Fleming Committee submitted the first post-speech plan for Apollo, the recommendation was for a Nova configured in three stages, capable of boosting 160,000 pounds into orbit.

But there were daunting obstacles, too. "It would have damn near sunk Merritt Island," one engineer observed of Nova, not entirely kidding. The designers of Nova weren't even sure that they could launch Nova from a land-based launch pad because of the noise and vibration it would generate. They were thinking about launching it from barges built several miles offshore from the Cape. It was this kind of complication that kept getting in the way of making the decision for Nova. It seemed to be an unearthly vehicle in many ways.

Oddly, it was Marshall that took the lead in opposing Nova and supporting E.O.R. in its place. Nova had an obvious attraction for the people at Marshall—it would have been the ultimate new toy for people who loved to design rockets. Nonetheless, they preferred the Saturn. Earth-orbit rendezvous would still require a huge booster, but Saturn-sized. Von Braun argued, with growing support, that E.O.R. would be a faster, surer way of getting to the moon.

Some of the people in the Space Task Group, which was favoring direct ascent, suspected ulterior motives for Marshall's position. Earth-orbit rendezvous would multiply the number of launch vehicles that would be needed, thereby increasing Marshall's role. But the argument openly used by von Braun and his people was plausible: Going directly from Redstone and Jupiter vehicles to the Nova was too big a step. Building Saturn seemed to be a much more orderly and prudent advance in rocket technology.†

* By way of comparison, the shuttle has a total liftoff thrust of about 6,400,000 pounds, only a little more than half the thrust of the Nova's first stage alone.

† If they hadn't been talking about a Nova-class vehicle, Owen Maynard once pointed out, the huge Saturn would itself have seemed impossible: "The fact that we could postulate Nova with somewhat of a straight face automatically made the five-eighths-size Saturn credible." It's one of the tricks engineers sometimes use on themselves, he said, to maneuver themselves into taking on tasks that would otherwise be terrifying.

3

Whichever way the choice went, direct ascent or E.O.R., the problem facing Maynard and his team was the same. Whether it was lifted in one piece or assembled in earth orbit, the spacecraft had to be able to escape Earth's gravitational field, cross 240,000 miles of empty space, execute a landing on a surface of uncertain terrain and composition, execute a liftoff from the lunar surface without ground support, survive entry into the earth's atmosphere, and then serve as a boat after it landed in the ocean. It was one thing to talk about such a vehicle in the abstract; it was quite another to create one that an astronaut could actually fly.

"As an example," Maynard said, "for launch, you wanted the guy lying on his back so that the escape system could haul him away from the launch vehicle at as high an acceleration as he could stand." They also had to be watching their displays, so that they could monitor all the command module's systems. They had to be able to look out a window, so that they could see a horizon and have secondary sources of information about the attitude of the spacecraft.

The difficulty the designers faced was to make the resulting layout (about which they had virtually no choice) work when the spacecraft came in for a landing on the moon. At this point in the flight, as Maynard described it later, "I want the guy to be able to see where he's landing, to be sitting upright with his spine aligned with the engine axis. That means I've now got to turn him around ninety degrees, and re-orient his displays and controls. This made you have either very complex arrangements of displays or two sets of displays." Maynard and his people tried everything, including swiveling couches and displays. Nothing worked. "There's no way you could see the surface all the way down without putting the astronaut out on a porch," said Faget. "So it was a hard spacecraft to land if you wanted to eyeball it, and there was certainly every reason to believe that the astronauts would insist upon having eye contact with the surface at the time of landing."

As the options diminished, they finally were driven to thinking about giving up eye contact with the surface, but it never got far enough to take to the astronauts—none of the designers really wanted to do it. "But that was the only other alternative," Faget said. "So we were really at an impasse." Maynard felt he was trying to overload the spacecraft with functions.

They did manage to come up with some designs that were supposed to be able to do the job, but all of them were getting too big and awkward and heavy. "It was like landing a Mack truck instead of a small sports car," said Maynard. And landing was only the half of it. Assuming you got down in one piece, there was still the problem of lifting off. "It just got dicey as hell," said Caldwell Johnson. "That stage that landed was a big clumsy thing. . . . If you land with the cylinder upright so you can take off again, the sonofabitch will fall over. And if you land it flat, it won't take off again. So it wasn't only the crew's position but all the mechanics of the whole thing. We had more harebrained schemes than you could shake a stick at."

For Walt Williams, in charge of trying to launch the Mercury astronauts on the Atlas even as the Apollo concept was being developed, the whole thing began to take on an ominous parallel. The spacecraft that Johnson and Maynard were coming up with was about the size of an Atlas, ninety feet long, and Williams shuddered at the notion of "backing an Atlas back down on the pad." They were having enough trouble getting the Atlas to go the other way, he thought, without trying to land one.

Finally they moved toward a scheme that solved some of the technical problems. Instead of trying to land a ninety-foot-long cigar on the surface of the moon, they would build a "lunar crasher." The assembly would consist of a command module with a comparatively small engine and propellant tank, along with an additional stage especially designed to take out most of the velocity on the descent. The additional stage—the "crasher"—would slow the rest of the spacecraft until it had reached an altitude of about 10,000 feet above the surface and a velocity of 1,000 feet per second (about 680 m.p.h.). Then—and here things got a little tricky—the crasher would separate and crash onto the lunar surface, leaving the rest of the spacecraft to use a smaller, lighter descent engine to get the rest of the way down on its own.

Hairy as it sounded, the lunar crasher was the most plausible arrangement they came up with. "We went to headquarters and argued long and hard for the lunar crasher," said Faget, "and as a matter of fact we did sell the concept." But it was so cumbersome, so inelegant, so downright ugly. None of their schemes, including the lunar crasher, gave Maynard that "warm feeling" that he liked to have about a design.

8

"Somewhat as a voice in the wilderness . . ."

Tom Dolan's team at Vought Astronautics was the first to come up with the notion of using a second spacecraft to descend to the lunar surface.* Dolan, a farsighted fellow, read the accounts of the new Project Mercury back in December of 1958 and decided that it would surely be followed by something more ambitious. He assembled a team of engineers at Vought to try to get a head start on the competition. By the middle of 1959, they were already concentrating on how to get to the moon.

The chief barrier to getting to the moon and back again was the "energy budget." Putting a pound of payload into earth orbit takes a lot of energy coming off the launch pad, which means a lot of propellant; taking a pound all the way to the moon requires that much more; taking a pound all the way to the moon and back again requires the most of all. Dolan and his men developed a solution: Design the spacecraft so that you can throw away parts of it as you go along. The Dolan team called

---

* Historians have found earlier references to the lunar-orbit rendezvous concept. Yuri Konratyuk, a Russian rocket theoretician, had written a paper suggesting an analogous scheme back in 1916, and H. E. Ross, a British scientist, had described one in 1948. The people who worked on the lunar-orbit idea in the 1950s do not seem to have been aware of this work. Dolan's team gets credit for being the first to propose lunar-orbit rendezvous after a lunar landing became a live technological possibility.

it a "modular" spacecraft, in which different segments were designed exclusively for certain tasks. When the task of one module had been completed, it would be detached from the remainder of the spacecraft and discarded, ending its drain on the energy budget.

In exploring how to make the most efficient modular system, someone in the team came up with an idea one day that was not at all obvious: After you had gotten the spacecraft out to the moon, suppose that you didn't land the whole thing? Suppose that instead you designed a second spacecraft exclusively for the purpose of going down to the lunar surface and returning to the mother craft?

They ran the idea through their slide rules and came out with revolutionary results. A second spacecraft built specifically for the lunar landing would not have to carry the heavy heat shield necessary for entering the earth's atmosphere—a big savings. It wouldn't have to carry the propellants for returning home—another big savings. A lunar lander could be constructed of light materials, because it wouldn't have to contend with much gravity (only one-sixth of earth's) and would encounter no atmosphere at all. The weight savings of this method were enormous. There was one difficulty, however. To take advantage of this method, the astronauts on the moon would have to rendezvous successfully with the mother craft orbiting overhead, and no one knew how to do that yet. Furthermore, the rendezvous would have to be conducted 240,000 miles out in space, far from any hope of rescue.

At about the same time Dolan was putting his team together at Vought Astronautics, a Langley engineer named Clint Brown believed, mistakenly, that a new little rocket called the Scout might be used to put a two-pound payload into orbit around the moon. He and his colleagues began to play with lunar trajectories. They soon realized that they could never use the Scout for that purpose, but they had become intrigued with the problems of lunar trajectories and continued to work on them. "We had a very nice competition going between the two divisions at Langley," Brown recalled, with John Bird over in the Flight Research Division and Bill Michaels in Brown's own Theoretical Mechanics Division working through the theory of getting into and out of lunar orbit. After a while, they began to postulate manned spacecraft in their theoretical analyses.

Early one morning in February 1960, Bill Michaels came into Clint Brown's office with some new calculations on what they were calling a "lunar parking orbit" for use in a manned landing. Michaels's figures indicated weight savings far beyond anything they had realized before.

This looked really interesting, they agreed. They ought to get out and tell someone right away. And at ten o'clock that very morning, Brown would recall ruefully, they were scheduled for a presentation by a team of engineers from Vought led by Tom Dolan. "And, darn, they got up there and they had the whole thing laid out. . . . They had scooped us. Bill was going around with his face hanging down to the floor."

Michaels went ahead and wrote up their analysis in a Langley working paper entitled "Weight Advantages of Use of a Parking Orbit for Lunar Soft Landing Mission," which Langley circulated in May 1960. At about this time, another name for the technique began appearing—"lunar-orbit rendezvous," or L.O.R. But at that time only a handful of people in NASA were thinking seriously about a lunar landing anyway, and no one paid much attention.

Even after NASA did begin to think seriously about a lunar landing in late 1960, no one would take L.O.R. seriously. An engineer from the Instrumentation Laboratory at M.I.T. remembered being down at Langley during this period. Someone there described to him this idea of detaching a little bug from the main spacecraft. The engineer went back to M.I.T. and told his colleagues "about this nut who had suggested such an outlandish thing, and we all chuckled." That's what everybody thought the first time they heard it: Those guys couldn't possibly be serious. "It's like putting a guy in an airplane without a parachute and having him make a midair transfer," John Disher scoffed to John Bird. No, Bird replied, it's like having a big ship moored in the harbor while a little rowboat leaves it, goes ashore, and comes back again. But no one was persuaded.

1

If it hadn't been for John Houbolt, that might have been the end of it.* The Langley engineers on Brown's team had done what Langley did best.

* Herein lies contention. In 1969, at the time of the first moon landing, *Life* magazine ran a long article that portrayed John Houbolt as both the creative force and lonely crusader for L.O.R. This was an exaggeration, perhaps customary in glossy magazines but scandalous to Langley engineers, who were always finicky about questions of scholarly attribution. Subsequently, Robert Gilruth was to become the leading proponent of an alternative interpretation of history, that Houbolt's contribution really wasn't that important—L.O.R. would have won on its engineering merits anyway. It is of course impossible to be sure what would have happened without Houbolt, but the text reflects our own reading of this history: Houbolt was not the originator of the L.O.R. concept (nor did he claim to be), but his advocacy was crucial, probably decisive, in leading to the adoption of L.O.R. There is a fascinating doctoral dissertation yet to be written on this episode, however.

They had explored an interesting research problem, prepared a technically careful discussion of it, and published the results in a Langley paper. If the practitioners cared to use their work, they were welcome to it. If not, that was their business. But by that time John Houbolt had gotten involved.

Houbolt was a Langley engineer who headed up the Rendezvous Panel, a group established in 1959 at Langley to support the planning for a space station. Houbolt had been working on the general subject of space rendezvous since 1957. In the course of these investigations, Houbolt too had begun to think about rendezvous in lunar orbit, and at about the same time that Dolan and Brown were doing their work, Houbolt independently did what he called "back of the envelope" calculations that revealed the large weight savings to be gained by use of a second spacecraft.

For Houbolt, what happened next was like a conversion experience. "Almost spontaneously," he wrote later, "it became clear that lunar-orbit rendezvous offered a chain-reaction simplification on all back effects: development, testing, manufacturing, erection, countdown, flight operations, etc. All would be simplified. The thought struck my mind, 'This is fantastic. If there is any idea we have to push, it is this one!' I vowed to dedicate myself to the task."

The scholarly John Houbolt became a crusader. Beginning in the summer of 1960, he wrote memoranda and letters, talked informally to anyone who would listen, and tenaciously tried to get across to NASA that a wonderfully simple way of getting to the moon was open to them, if only they would take a fresh look. Nobody paid much attention.

In December 1960, when a lunar landing was beginning to be taken seriously, Houbolt finally got his first chance to brief an important conference. Virtually everyone was there, including Glennan, Seamans, and von Braun. Houbolt and Brown put up a large chart and went through their arguments. When Houbolt finished presenting his estimates of weight savings, Max Faget spoke up from the audience, even blunter than usual. Houbolt remembered his words as "His figures lie, he doesn't know what he's talking about." Others recalled "a prominent Space Task Group engineer" saying "Your figures lie." In either version, the words were shocking. Even in a private bull session they would have been provocative. In an open meeting, in front of Houbolt's peers and supervisors, they were a brutal thing for one Langley engineer to say to another.

When von Braun (who ultimately was converted to L.O.R.) was once asked about Faget's alleged remark, he shrugged and answered, "They [the figures] did lie. The critics in the early debate murdered Houbolt.'' The crux of the problem, von Braun said, was whether the savings that L.O.R. promised were worth the price. ''John Houbolt argued that if you leave part of your ship in orbit and don't soft-land all of it on the moon and fly it out of the gravitational field of the moon again, you can save takeoff weight on earth,'' von Braun said. ''That's pretty basic. But if the price you pay for that capability means that you have to have one extra crew compartment, pressurized, and two additional guidance systems, and the electrical power supply for all that gear, and you add up all this, will you still be on the plus side of your trade-off?''

In these early presentations, Houbolt and Brown were talking about three possible lunar modules, ''plush,'' ''economy,'' and ''budget.'' The ''budget'' model was a stripped-down, 2,500-pound version in which an astronaut descended on an open platform, and it attracted ridicule—the astronaut would descend ''with a silk scarf around his neck,'' it was said. Nobody in the Space Task Group seriously considered the stripped-down version. Faget argued that the weight of even the plush model was underestimated. Later, he enjoyed pointing out that the configuration that actually flew to the moon had a total weight (command module, lunar module, and service module combined) which approximated their initial calculations for direct ascent.*

Apart from the questions about the weight savings, there was that rendezvous 240,000 miles from home to worry about. ''We had never demonstrated a rendezvous,'' George Low pointed out later. ''L.O.R. involved doing a rendezvous a quarter of a million miles from home. It seemed like an extremely far-out thing to do.'' Far-out, and dangerous. Low's early task group thought it had to be more dangerous than either direct ascent or E.O.R. Direct ascent, if they had a big enough booster, would be the safest of all by virtue of being the simplest. And if direct ascent was not possible, then the rendezvous in earth orbit had to be safer. If the rendezvous failed in an E.O.R. mission, the astronauts in the manned spacecraft could fire their retro-rockets and return safely to earth while the unmanned segment continued to orbit. Or if for some reason the manned spacecraft was stranded in earth orbit, it was at least conceivable that a second spacecraft could be launched to rescue them. If a rendezvous failed in lunar orbit, the astronauts in the lunar lander were dead.

* Houbolt thought the weight of the actual Apollo lunar module was excessive. But he also made the point that, if they'd started with direct ascent, the weight of that configuration would have grown just as drastically, because of the ''back effect'' that keeps forcing up weights in all developmental efforts.

Furthermore, even in those days, NASA was already beginning to formulate the primal rule of manned space flight: Do not commit yourself any earlier than you absolutely have to. If there is any way to give yourself a way to back out, keep it. E.O.R. kept all sorts of options open longer than L.O.R. did. In earth orbit, the astronauts and ground controllers could check everything out while they still retained the option of canceling the mission and coming home. With L.O.R., some of the most critical maneuvers of the flight would have to be done after committing the spacecraft to a circumlunar trajectory. Still others would have to be performed after committing the spacecraft to a lunar orbit. There was no question about it: Maybe L.O.R. would save some weight, maybe it wouldn't, but it was still too crazy for serious consideration.

For Houbolt and the other Langley people who were continuing to work on the L.O.R. problem, the criticism proved to be a stimulant. From the end of 1960 through Kennedy's speech in May of 1961, Houbolt and his colleagues worked to find answers to the objections they were hearing. They prepared studies on lunar landers, landing gear, descent and ascent trajectories, and rendezvous maneuvers. The more they explored L.O.R., the more convinced they became that it was not just a faster way to get to the moon, it was also safer.

Houbolt persisted. Just a week before Kennedy's speech, Houbolt wrote a letter directly to Seamans, skipping over half a dozen bureaucratic layers in the process, pleading for a fair hearing for L.O.R. A few weeks later, when Seamans appointed Bruce Lundin from Lewis to investigate alternative modes for getting to the moon ("mode" is the NASA word for it, so that the whole lengthy process is now remembered as the "mode decision"), Seamans made sure Houbolt was on the committee.

The Lundin Committee, however, was another rejection for Houbolt. He explained L.O.R. to the other members and got what was by now a standard response. "They'd say, 'Oh, that's good,' " Houbolt recalled, "and then the next day they'd laugh." In its report, the Lundin Committee put L.O.R. at the bottom of its list along with "lunar-surface rendezvous," a scheme whereby the astronauts would land on the moon with empty tanks and then do their own refueling from a fuel cache landed earlier. In Houbolt's view, lunar-surface rendezvous was "the most harebrained idea I'd ever heard of."* That the committee put it in the same category as L.O.R. was degrading.

---

* The most harebrained idea of all, however, was to land astronauts on the moon with a few years' worth of food and oxygen and leave them there while the people back on earth figured out how to get them home again—actually proposed to NASA, though never seriously considered.

Next came the Heaton Committee, which had a charter to explore earth-orbit rendezvous. "I was in a strange position," Houbolt recalled later, "because I was one of the staunchest believers in rendezvous in the country. I'm not against earth-orbit rendezvous; I am in favor of it." It was just that Houbolt was so sure that L.O.R. was an even better way. "What upset me on the Heaton Committee," Houbolt said, was that E.O.R. "was becoming a beast. The configurations [that the other members of the committee] were coming up with involved putting together five pieces of hardware. It was getting to be a great, big, long cigar." Houbolt could hear engineers reading the Heaton Report and saying to themselves, "The guys on rendezvous are nuts like I always thought they were." So Houbolt pressed on.

"That summer of 1961 was probably the busiest summer I've had in my life," Houbolt said. "I was living half the time in Washington, half the time on the road, dashing back and forth." In the early fall, he was asked to make a presentation on L.O.R. before the Golovin Committee, the group that was supposed to recommend the boosters to be chosen for Apollo. For the first time, a crack appeared in the united wall of opposition, as Nick Golovin (who, ironically, would later become a major antagonist to L.O.R.) saw some merit in Houbolt's arguments. But it was only a crack, and Houbolt decided that they weren't going to get anywhere until they put together such a thorough documentation of the argument for L.O.R. that it couldn't be ignored. Houbolt, John Bird, and Arthur Vogeley put together a two-volume report describing the concept in full technical detail.

Finally, in November, Houbolt wrote another letter to Seamans, once again leapfrogging bureaucratic channels. "Somewhat as a voice in the wilderness . . ." he began this nine-page plea from the heart. It was written in haste and ignored the staid conventions of NASA internal memoranda. "It is conceivable that after reading this you may feel you are dealing with a crank," Houbolt conceded in the opening paragraphs. "Do not be afraid of this. The thoughts expressed here may not be stated in as diplomatic a fashion as they might be, or as I would normally try to do, but this is by choice and at the moment is not important. The important thing is that you hear the ideas directly, not after they have filtered through a score or more of other people, with the attendant risk that they may not even reach you." Houbolt proceeded to make an impassioned plea for NASA to rethink the mode issue, and attached to it a copy of the two-volume report they had prepared down at Langley.

*   *   *

Seamans had been keeping track of L.O.R. ever since he had first heard Houbolt's spiel on a visit to Langley a year earlier. Seamans had never forgotten Houbolt's conclusion that 50 percent of the launch weight of the spacecraft could be saved by using L.O.R. Before joining NASA, Seamans himself had done rendezvous work at both M.I.T. and R.C.A., and he was convinced that the rendezvous techniques for L.O.R. were within the state of the art.

He passed the letter on to Brainerd Holmes, who had just taken over from Abe Silverstein as head of Space Flight Programs, now renamed the Office of Manned Space Flight (O.M.S.F.). Holmes didn't like the letter at all, and he told George Low so when he passed it to Low for his opinion. He didn't like the style and he didn't like Houbolt jumping over channels like that. In a memorandum of his own, Low wrote back diplomatically that he of course agreed that Dr. Houbolt hadn't followed procedures. "Nevertheless, I feel that Houbolt's message is a relatively sound one and I am forced to agree with many of the points he makes," Low told Holmes. Among other things, "I agree that the 'bug' approach may yet be the best way of getting to the moon and back." Low concluded by recommending that Holmes invite Houbolt to come to Washington to present the L.O.R. scheme in detail—and that "you might also consider asking him to join your staff."

It was December 1961 when Holmes pondered this politely mutinous response. The hardware contract for the Apollo spacecraft had been let. The facilities at the Cape were being laid out on the drafting tables. And the Office of Manned Space Flight still didn't know how they were going to go to the moon. A decision had to be made soon.

2

About this time, Holmes hired a new deputy director of O.M.S.F. Since Holmes had come on board in October, he and Seamans had been looking for a specific kind of expertise in systems engineering, someone who could offer what Seamans called a "systems competence"—a person with the technical expertise to keep up with the Fagets, the pertinacity and ego to deal with the von Brauns, and the capacity to make the many pieces of Apollo interlock both at the grand managerial level and down in the trenches. To put it even more roughly, Holmes needed a technical foreman for the sprawling lunar program and, when required, an enforcer.

For the next two months, O.M.S.F. cast about for such a person. Finally one of Holmes's advisers remembered someone he had known a few years ago at Bell Labs who fit the requirement perfectly. He was young, he had the right kind of experience and the right temperament. And when people talked about him, they all said the same thing—"a brilliant engineer," they said, as if "Brilliant Engineer" were his title. He was hired shortly after Christmas 1961. For his first assignment, Holmes and Seamans decided, why not let him sort out the mode decision?

And so Joe Shea came to Apollo.

At the end of 1961, Joseph Francis Shea was thirty-five years old, within months of the same age as George Low, the man whose career in Apollo would be so intertwined with his. Shea had grown up in a working-class Irish neighborhood in the Bronx, the oldest son of a mechanic with the subway system. He had gone to a local Catholic high school and graduated at sixteen, exceedingly bright but with no aspirations beyond getting a secure job. He was a good runner, and mostly he dreamed of being a track star.

It was 1943 when he graduated, wartime, and Shea heard about a special Navy program that would send him to college. He applied, was accepted, and the Navy sent him off to Dartmouth College in the mountains of New Hampshire. The young man from the Bronx discovered hills and forests and deep, pristine snow that squeaked when you walked on it. He also discovered engineering, and that he liked it. Then the Navy sent him to M.I.T., and after that to the University of Michigan. Gradually he began to realize that the engineering courses that his friends found so difficult were, for him, simple.

For the next several years Shea moved back and forth between Michigan, where he eventually obtained his engineering doctorate, and Bell Labs. It was an educational odyssey that took him from engineering mechanics to electrical engineering to theoretical mathematics to physics to inertial guidance. "The nouns change but the verbs remain the same" became one of Shea's sayings as he went from one specialty to another.

Then in 1956 Shea found out how it all fit together. At the age of twenty-nine, Shea was named systems engineer for a radio guidance project connected with the Titan I. "I didn't know what 'systems engineer' meant,'' Shea said, but he learned quickly, traveling around to the subcontractors on the Titan I, becoming a member of the small fraternity of engineers who were coming of age in this new field. At night after work they would gather at a bar near the plant where they had been working that day. They didn't even drink that much, Shea recalled, they were so busy talking—about testing, grounding, vibrational spectrums, weights, stability, electrical interfaces, guidance equations, all the myriad elements of the system that some lucky guy, like a systems engineer, got to orchestrate.

By 1959 Shea had acquired enough of a reputation within the ballistic missile fraternity for General Motors to hire him to run the advanced development operation for its A.C. Sparkplug Division, which was trying to wedge its way into the missile business. Shea was in charge of preparing a proposal for the inertial guidance contract for the Titan II. After the proposal won, Shea went back to administering the advanced development office. But a year later, in September 1960, the contract he had won was six months behind and Shea was called away to rescue it.

Shea began to discover that he had a knack for leading. His was not a gentle style, but if he was tough on people who fell short, he was generous and loyal to those who didn't. And for engineers tired of working for bosses who had forgotten their engineering, working with Shea was refreshing. It didn't make any difference what your specialty was. Shea's maxim was that if you understood it, you could make him understand it—and once he did, you never had to explain it again. The only problem was keeping up.

It was about this time that Shea discovered the uses of what he would come to call his ''controlled eccentricity.'' When he was still at Bell, his wife had bought him a pair of red socks as a joke. One day in a meeting he absentmindedly put his feet up on the table, getting some laughs and loosening up the meeting. So Shea started wearing red socks, not all the time, but to important meetings. Eventually the socks were accepted as a good-luck charm to wear to presentations. Even senior management at General Motors, where putting one's feet on a desk was discouraged and wearing red socks was unthinkable, got used to the idea.

Shea had other eccentricities as well. Puns, for example. Shea loved plays on words, and puns would come spilling out every few minutes, or sometimes seconds. Good puns and bad puns, subtle and obvious, double entendres, triple entendres—Shea's punning subroutine (his phrase for it) was not discriminating, it just kept producing them, and it became another way to keep things loose.

Armed with his red socks and his puns and an emerging sense of how good he was getting to be at this sort of engineering, Shea set out to rescue the lagging Titan contract. He moved into the plant, and for five days a week, all three shifts, he was there, catching catnaps on a cot set up in his office. It was a pattern he would repeat later, during Apollo. The reasons were partly motivational—people work harder when they see the boss working all three shifts. "But it also lets you find out everything that's going on," Shea said. "Things I'd find out at night, I'd get corrected during the daytime." Shea began handing out red socks as an award for good performance. His enthusiasm and energy were infectious.

Shea pulled it off, making up the six months, bringing in the contract on budget and on time. In the world of military contracts, that kind of performance got attention, and soon afterward, in August, T.R.W. lured Shea away from G.M. and out to the West Coast.

With the move to California, Shea had promised his wife and five daughters that they'd stay put for a while. But when he got a call from NASA in early December, asking him to interview for the job as Holmes's deputy, he couldn't resist. "I could see they needed good people in the space program," he recalled, "and I was kind of cocky in those days." He flew to Washington ready to listen. There was some patriotism in it, but also a lot of Joe Shea. The experience with the Titan II had created in him an appetite for high-pressure jobs where great things had to be accomplished and the schedule was falling behind and Joe Shea was told to rescue the whole unwieldy, out-of-control mess. The Apollo Program sounded as if it had that kind of job available, and Shea wanted the action.

CHAPTER

9

"What sonofabitch thinks it isn't the right thing to do?"

"I came into the agency the first week in January of '62," Joe Shea reminisced. "I still didn't know how disjointed the program was. Hell, I was still learning the names of the Saturn stages. One day Brainerd came into my office. Somehow we wound up with Seamans, who'd gotten the famous letter from John Houbolt, and Seamans said to me, 'Anything to this?' I said, 'Well, I don't know.' He said, 'You know, I don't think we really yet know how we're going to go to the moon.' I said I was beginning to get that same suspicion." And so Shea decided to go talk to the people down at Langley to find out what was going on.

When Shea walked into the Langley conference room in Building 58, he had a casual preference for earth-orbit rendezvous. He hadn't worked it out in detail, but the Nova looked unnecessarily big. You didn't need that much booster to get to the moon, Shea thought. And if you didn't go direct, then earth-orbit rendezvous was the obvious alternative. But Shea didn't have a commitment to E.O.R. or an aversion to any of the other modes. Shea prided himself on going wherever the data took him and, as

124

he listened to the briefing, it seemed to him that the data for L.O.R. weren't so bad.

John Houbolt watched Shea closely as he went through his pitch yet one more time, and he began to hope—"Shea grasped the thought and seemed to be receptive." Houbolt recalled Shea saying to the Space Task Group, "Look, what's going on here? Why haven't you been thinking about this thing longer? Looks pretty good to me."

The reaction Shea got was not quite what he expected. Gilruth and Faget did not rise up to denounce L.O.R. Instead, as the day wore on, it became increasingly apparent to Shea that their message, still tentative, was that, actually, they had been doing some more thinking about lunar-orbit rendezvous and, as a matter of fact, they were beginning to think it was a good idea.

1

In later years, Gilruth minimized Houbolt's role in promoting L.O.R. "John Houbolt just assumed that he had to go to the very top," he once said. "He never talked to me, you know. . . . Actually, we at the Space Task Group had not decided on any mode to go, and I was very much interested in the lunar-orbit rendezvous, and we were the guys that really sold it. It wasn't Houbolt's letter to Seamans."

This attitude mystified and hurt Houbolt. "I talked to Gilruth many times," he recalled. "If just once he had said, 'Look, John, I'm on your side. You can stop fighting, we'll take it from here,' then I would have been satisfied. But he never said anything like that." Nor did Seamans have any indication throughout 1961 that Gilruth was interested in L.O.R. In his mind, the notion that the Space Task Group really sold L.O.R. was "baloney." In any case, it seems clear that throughout 1961, the Space Task Group's public position remained firmly hostile to any sort of rendezvous. In September, for example, Gilruth wrote a letter to headquarters arguing that all rendezvous schemes were suspect. Rendezvous of any kind would "degrade mission reliability and flight safety," he wrote, and he warned that "rendezvous schemes may be used as a crutch to achieve early planned dates for launch vehicle availability, and to avoid the difficulty of developing a reliable Nova-class launch vehicle.''*

And yet, hidden from public view, minds at Langley were changing. Insofar as it can be pieced together from memories consulted a quarter of a century later, the reality seems to have been that in 1961 the Space Task Group was increasingly divided on the mode controversy, but kept quiet about it. The Space Task Group's official position throughout the year was to favor direct ascent. In the meantime, the Lundin and Heaton Committees up in Washington were coming out in favor of earth-orbit rendezvous. Through all this, the people closest to the design of the spacecraft were doing pretty much as they pleased.

This kind of independence was a hallmark of early Space Task Group behavior. In the early 1960s, NASA headquarters was theoretically like headquarters in any large bureaucracy, which is to say that it established policy and enforced it. In practice, from the viewpoint of the centers, headquarters was an object of disdain. ''You'd do everything you could to keep headquarters from knowing what the hell was going on,'' one Space Task Group engineer said, ''or letting them participate in it, if you could help it.'' Later, Owen Maynard couldn't even remember what the official headquarters policy toward the mode was supposed to have been during that period. ''Well, headquarters did what they did,'' he said pugnaciously. ''And I didn't give a shit. I had no respect for these people in faraway places.'' He would design whatever seemed to work, as long as Max Faget and Caldwell Johnson didn't mind, and what Maynard knew after struggling with this beast of a spacecraft was that earth-orbit rendezvous was a lousy way to try to get to the moon and direct ascent wasn't much better. He wasn't looking at trajectories and propellant loads and liftoff weights in the abstract. He was trying to design a real, functioning flying machine, and that dose of reality made a big difference.

Maynard was in that frame of mind when, in January of 1961, Caldwell Johnson told him to go listen to a presentation by Houbolt. Johnson didn't say much more than that, positive or negative—''They

* There were also inter-center rivalries involved in these calculations about the mode. As part of a management struggle at headquarters, Abe Silverstein left his job as director of Space Flight Programs in the fall of 1961 and went back to be center director of Lewis, at just about the same time that the lunar crasher looked as if it would be the winner. The word around NASA was that Lewis would be given the ''crasher'' stage of the lunar crasher as part of an effort to placate Silverstein. But neither Gilruth nor Faget liked the idea of splitting management responsibility for the command and service module with a second center. Also, Silverstein had made no secret of his opinion that Gilruth was not the right man to become director of the new Manned Spacecraft Center to be built at Houston. Gilruth was not inclined to do Silverstein any favors, such as giving him part of the responsibility for the spacecraft, if he could help it.

didn't tell me to do it, or even give it special consideration,'' said Maynard. ''It was just one more of sixteen thousand ways of getting to the moon.'' So Maynard and Kurt Strass and Bob O'Neal went to one of Houbolt's many briefings. ''I guess that a minute after the meeting we probably thought it wasn't a very good idea,'' Maynard said, and Bob O'Neal wrote a memo afterward saying so. But Maynard, struggling with his sketches, kept coming back to the L.O.R. idea, and he found that ''when you take all of the different ways of doing things into consideration, when you've got a chance to digest them, and ask a few thousand more questions, it begins to make more sense.'' As the weeks passed, he and Houbolt chatted from time to time.

Maynard began to realize that lunar-orbit rendezvous was not so much harder than some of the things he was already considering. He had lately been playing with the idea of leaving the propellant tanks in lunar orbit—they were heavy, so why not leave them up there instead of carrying them all the way down to the surface and back up again? He had liked the idea, except that as he explored it he found that if he wanted to rendezvous with a propellant tank, he had to devise some sort of docking mechanism, plus a lot of other complexities. And if you were going to have to add that much complexity anyway, why not go Houbolt's route? You might as well build a completely independent second vehicle with a separate set of displays and controls that were just for landing on the moon and coming back up again. It wouldn't need a heat shield. It wouldn't have to survive aerodynamic pressures. It wouldn't have to have any aerodynamic controls. It could be small and light and maneuverable— that sports car he was looking for, instead of the Mack truck.

''I got convinced pretty quick,'' Maynard said, ''within about a month of that date in the Bob O'Neal memo.'' He was much too far down the ladder to worry about selecting the mode, so he kept making sketches that eventually led to the lunar crasher, but he was sketching some lunar modules on the side.

Maynard also remembered Gilruth listening sympathetically to his problems in designing a workable spacecraft, and that Gilruth was becoming intrigued by L.O.R. soon after Maynard himself did. That jibes with Gilruth's own memory of an exchange with Faget shortly after the Kennedy speech. Gilruth had discovered that Faget had told people at headquarters that the Space Task Group wasn't interested in lunar-orbit rendezvous, and he was uncharacteristically abrupt: It wasn't Faget's prerogative to tell them that, Gilruth recalled telling the designer—and besides, they ought to be looking hard at L.O.R. themselves. Later,

Gilruth would say that it wasn't until he became convinced that lunar-orbit rendezvous would work that he truly began to believe that they could get to the moon by the end of the decade.

Al Kehlet and Chuck Mathews began some new in-house analyses for the Space Task Group on the lunar-orbit rendezvous option and convinced themselves that there was a lot to it. As Faget watched Caldwell and Owen struggling with ways to make one spacecraft do all the different functions of a lunar landing, he too began to appreciate the merits of having a second, specialized spacecraft to make the landing. He still thought that Houbolt's figures on weights were wrong, but lunar-orbit rendezvous was turning out to be the more elegant engineering solution. If Faget was opinionated and stubborn, he was also irresistibly attracted to elegant solutions. Faget, formerly the apostle of direct ascent, the man who had so grievously insulted Houbolt, began to come around to the conclusion that, in his own words, "lunar-orbit rendezvous really looked like the thing to do." Houbolt could never get over Max's gall—not only did he become an advocate, but a few years later at cocktail parties, Faget would come up to him and say, "Oh, anyone who thinks about it for five minutes can see that lunar-orbit rendezvous is the way to go."

2

These conflicting views about the mode were still being untangled when Caldwell Johnson drove out to Langley on Tuesday, September 19, 1961. It was overcast, a gloomy fall day. Hurricane Esther was moving up the coast toward Hatteras and storm warnings were out for the Tidewater area. But Johnson was in good spirits. He had recently moved into his dream home, situated on a bluff over the mouth of the James River, looking out over eight miles of water, with good fishing off his dock and his duck blind five minutes away.

He was scheduled later that morning for a briefing by some engineers working on Surveyor, a project to soft-land an unmanned spacecraft on the moon and send back information about the lunar surface. The engineers had come to Langley to ask what the Space Task Group needed from Surveyor. As it happened, Johnson needed a great deal. Regardless of whatever mode was eventually chosen, the craft that touched down on the moon was going to need landing gear, and Johnson had been fretting about how to design it.

Strange roles for the Apollo cast. J.F.K., young and modern, was the skeptic about manned space flight. L.B.J., the deal-making politician, was the visionary. Jim Webb (seated far left), self-described country boy, was a superb administrator of high-tech NASA. They are with two of the most important figures at the Cape: Kurt Debus, the director of NASA's launch operations, and Rocco Petrone (standing far left), then in charge of building the Saturn V's support facilities. (*NASA*)

Two of the Founding Fathers. The man on the right is Robert Gilruth, head of the Space Task Group, in many ways the George Washington of America's manned space program. With him is George Low, who in 1960 pushed NASA into planning for a lunar landing and in 1967 took over ASPO after the fire—on everyone's short list of people who got the United States to the moon. (*NASA*)

For some reason, the Air Force down at Cape Canaveral thought that the people in the Space Task Group were a bunch of amateurs. . . .

"Those days were out of the Dark Ages." Space Task Group technicians on the floor of Hangar S, working on the boilerplate capsule to be launched on Big Joe. Jack Kinzler is at far right. (*U.S. Air Force, Pat Shea*)

Scott Simpkinson's team, along with visitors from Langley. Scott Simpkinson is to the right of Emily Ertle, the only woman in the team. Directly behind him is his boss, Merritt Preston. Standing above Preston is Chuck Mathews (striped shirt). Max Faget is in the T-shirt, far left, third row. Jack Kinzler is in the bottom row, far left. (*U.S. Air Force, Bob Special*)

America's first spaceship on its way to the launch pad, cushioned by a mattress, perched on the back of a Langley flatbed truck. (*U.S. Air Force, Fred Santomassino*)

The father of the Apollo spacecraft, and Mercury and Gemini as well, this is Max Faget—cheerfully cocksure, endlessly creative—holding the blunt-body spacecraft he pioneered. "Max," Joe Shea used to say, "with your personality, every one of your spacecraft designs is going to be blunt." (*NASA*)

"Their minds are almost interconnected." When Faget arrived at Langley in 1945, he found this young man, Caldwell Johnson, college dropout, master model-builder, unequalled draftsman, who would be his closest collaborator for more than four decades. (*Courtesy of Kitty Johnson*)

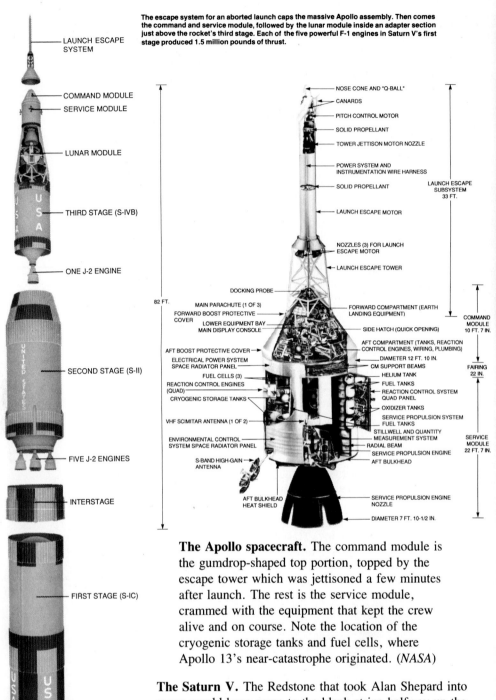

The escape system for an aborted launch caps the massive Apollo assembly. Then comes the command and service module, followed by the lunar module inside an adapter section just above the rocket's third stage. Each of the five powerful F-1 engines in Saturn V's first stage produced 1.5 million pounds of thrust.

LAUNCH ESCAPE SYSTEM

COMMAND MODULE

SERVICE MODULE

LUNAR MODULE

THIRD STAGE (S-IVB)

ONE J-2 ENGINE

SECOND STAGE (S-II)

FIVE J-2 ENGINES

INTERSTAGE

FIRST STAGE (S-IC)

FIVE F-1 ENGINES

82 FT.

NOSE CONE AND "Q-BALL"

CANARDS

PITCH CONTROL MOTOR

SOLID PROPELLANT

TOWER JETTISON MOTOR NOZZLE

POWER SYSTEM AND INSTRUMENTATION WIRE HARNESS

SOLID PROPELLANT

LAUNCH ESCAPE MOTOR

NOZZLES (3) FOR LAUNCH ESCAPE MOTOR

LAUNCH ESCAPE TOWER

LAUNCH ESCAPE SUBSYSTEM 33 FT.

DOCKING PROBE

MAIN PARACHUTE (1 OF 3)

FORWARD BOOST PROTECTIVE COVER

LOWER EQUIPMENT BAY

MAIN DISPLAY CONSOLE

FORWARD COMPARTMENT (EARTH LANDING EQUIPMENT)

SIDE HATCH (QUICK OPENING)

COMMAND MODULE 10 FT. 7 IN.

AFT BOOST PROTECTIVE COVER

ELECTRICAL POWER SYSTEM SPACE RADIATOR PANEL

FUEL CELLS (3)

REACTION CONTROL ENGINES (QUAD)

CRYOGENIC STORAGE TANKS

AFT COMPARTMENT (TANKS, REACTION CONTROL ENGINES, WIRING, PLUMBING)

DIAMETER 12 FT. 10 IN.

CM SUPPORT BEAMS

HELIUM TANK

FUEL TANKS

REACTION CONTROL SYSTEM QUAD PANEL

OXIDIZER TANKS

FAIRING 22 IN.

VHF SCIMITAR ANTENNA (1 OF 2)

SERVICE PROPULSION SYSTEM FUEL TANKS

STILLWELL AND QUANTITY MEASUREMENT SYSTEM

ENVIRONMENTAL CONTROL SYSTEM SPACE RADIATOR PANEL

RADIAL BEAM

SERVICE PROPULSION ENGINE

S-BAND HIGH-GAIN ANTENNA

AFT BULKHEAD

SERVICE MODULE 22 FT. 7 IN.

AFT BULKHEAD HEAT SHIELD

SERVICE PROPULSION ENGINE NOZZLE

DIAMETER 7 FT. 10-1/2 IN.

The Apollo spacecraft. The command module is the gumdrop-shaped top portion, topped by the escape tower which was jettisoned a few minutes after launch. The rest is the service module, crammed with the equipment that kept the crew alive and on course. Note the location of the cryogenic storage tanks and fuel cells, where Apollo 13's near-catastrophe originated. (*NASA*)

The Saturn V. The Redstone that took Alan Shepard into space would have come to the black stripe halfway up the first stage. Total length, 363 feet—like launching a navy destroyer. (LIFE IN SPACE–*Artwork by George Bell © 1983 Time-Life Books Inc.*)

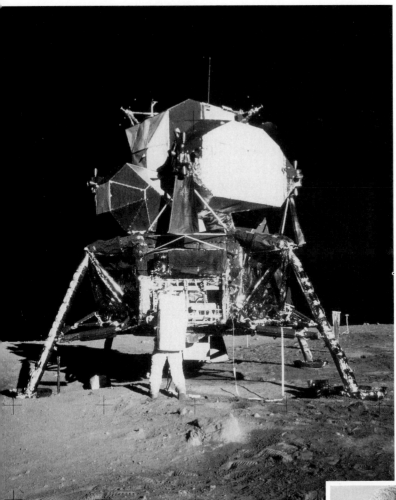

"A voice in the wilderness." Trying to land the command and service module on the lunar surface would have been like trying "to back an Atlas back down on the pad," and no one had figured out how to do it. But there was a strange and probably crazy idea called lunar-orbit rendezvous, promulgated relentlessly by this man, John Houbolt. (*Arthur Schatz*, Life. *Copyright © Time Inc.*) [BELOW]

The result was this unearthly machine, the lunar module. This one is Apollo 11's *Eagle,* on the lunar surface with Buzz Aldrin standing to the side. Note the landing gear, designed by Caldwell Johnson and Owen Maynard, who obstinately assumed that the moon had to be just like Arizona. (*NASA*) [ABOVE]

"He was a noble type of man," said one American who worked for him, and just about everyone who knew him seemed to agree. Wernher von Braun stands beside the business end of the first stage of the Saturn V with its five F-1s looming above him. This is one of the few pictures that conveys the size of the machine that von Braun's vision had wrought. (*Space and Rocket Center, Huntsville, Alabama*)

Paul Castenholz, co-director of Rocketdyne's Combustion Devices Team. His goal was to make the F-1 continue to function even when the team exploded a bomb in it. (*North American Aviation*)

The inferno within the F-1 defied subduing. Jerry Thomson "somehow got picked" by Marshall to see that the combustion instability problem was solved. Thomson, at the left, stands beside the J-2 engine that powered the second and third stages of the Saturn. With him is Elmer Ward, one of Karl Heimberg's propulsion engineers in the Testing Laboratory. (*NASA*)

Intense and brilliant, Joe Shea was the catalyst for the decision to go with lunar-orbit rendezvous and the ramrod for moving the spacecraft from design to hardware. (*NASA*)

Then came the fire. . . . Four top headquarters people in the Apollo Program, in an unhappy moment, testifying before a Senate investigating committee after the fire. From left to right: Bob Seamans, Jim Webb, George Mueller, and Sam Phillips. (*NASA*)

"Nope, there's no scale." This may help: On the road beside the crawler is a large fire truck—almost too small to be picked up in this enlargement. The launch vehicle is the first Saturn V, A.S.-501, on its way from the V.A.B. to Pad 39. (*NASA*)

"He seemed like a bull." The gentlemanly engineers of Langley needed some help in getting Flight Operations organized, and they got it from Walt Williams, a legend in the flight-test business even before Mercury. (*NASA*)

The embodiment of Flight Operations. When Mercury began, Chris Kraft was a frustrated young Langley engineer with an ulcer. By the time the Apollo flights began, he was synonymous with Mission Control and the model of flight-controller cool that everyone else tried to emulate. (*NASA*)

The first flight director after Kraft. Blue Flight, John Hodge, from England by way of AVRO. During Apollo, he moved out of the MOCR into planning for the J Missions. (*NASA*)

A covey of Flights, in a shot taken just before the launch of Apollo 11. The three in the center defined the job for succeeding generations. From left: Cliff Charlesworth, the riverboat gambler; Glynn Lunney, the quickest mind in the MOCR; and Gene Kranz, "General Savage." They are flanked by two of the next additions to the flight directors' corps, Milt Windler (left) and Gerry Griffin (right). (*NASA*)

Three of the inner sanctums.

One second before launch. Kurt Debus and Rocco Petrone awaiting the launch of S.A.-8, a Saturn I flight in 1965. They are in a blockhouse next to the launch pad, similar to the arrangement at Pad 34 where the fire occurred. Wernher von Braun is at the right periscope. (*NASA*)

SPAN. This unprepossessing room in Building 30 is where the most senior technical people in the Apollo Program orchestrated the nationwide support system for the Apollo flights. SPAN manager Scott Simpkinson talks to Owen Maynard while Dale Myers (eye patch) examines a schematic. (*NASA*)

"We didn't need any fancy damn consoles." This is the MER, the Mission Evaluation Room, unknown outside NASA but famous within it, a place for diagnosing and fixing hardware problems a quarter of a million miles away. Don Arabian, "Mad Don," is on the phone. (*NASA*)

"To always be aware that suddenly and unexpectedly we may find ourselves in a role where our performance has ultimate consequences."
—*Foundations of Missions Operations*

When lightning struck Apollo 12, John Aaron, EECOM, knew exactly what to do even though no one knew quite what had happened. (*NASA*)

When a program alarm sounded on the first lunar descent, the man standing in the center, Steve Bales, Guido, had to decide whether to tell Flight to continue or to abort. A good view of the Trench, this picture also shows Gran Paules (closest to the camera), who was Yaw during the first landing and, at the far end of the Trench, Retro John Llewellyn, "Butch Cassidy born a hundred years too late." (*NASA*)

When the oxygen tank exploded on Apollo 13, Sy Liebergot, EECOM, was at the focal point of the crisis in the MOCR, with no way to stop the hemorrhage of cryogenics. (*Courtesy of Seymour Liebergot*)

The Firing Room at the Cape, a few minutes after the launch of Apollo 12. Unlike the blockhouses used for earlier rockets, the Firing Room for the Saturn V launches was three and a half miles from the pad. (*NASA*)

Mission Control to the outside world, the MOCR to the controllers, as seen from Management Row. The windows at the right look into the sim room. This picture was taken during the Apollo 13 crew's telecast five minutes before the oxygen tank exploded. Astronaut Fred Haise is on the screen. Gene Kranz's back is to the picture. The top of EECOM Sy Liebergot's head and G.N.C. Buck Willoughby's profile are visible in the row ahead of Kranz. Their quiet evening was about to turn into a nightmare. (*NASA*)

During the crisis on Apollo 13. Those surrounding Flight Glynn Lunney include Chuck Deiterich (behind Lunney), Jerry Bostick (face partially hidden), Bill Tindall (seated), and Chris Kraft (with cigar). (*NASA*)

Odyssey **splashes down safely at the end of Apollo 13.** Gerry Griffin, Gene Kranz, and Glynn Lunney lead the cheering. Behind, in Management Row, Chris Kraft lights the traditional splashdown cigar for Bob Gilruth. (*NASA*)

He had been getting bizarre advice from high places. One eminent lunar scientist had come forth with the opinion that the mare—the flat "seas" on the lunar surface—were deep pools of dust: If anything lands in one of these, Johnson remembered being told, "it'll just sink down to the bottom. *Boom,* and it's gone." Another scientist, more eminent yet, a Nobel laureate no less, had speculated that the mountains were nothing but friable webs of rock that would crumble at even light pressures. "Well, that's a hell of a note," Johnson said. "How in the hell were we gonna design a landing gear if the mare are nothing but pools of dust and the mountains are nothing but blown-glass fairy castles?"

Furthermore, Johnson realized, it was hopeless to try to get the scientific community to straighten the thing out. They would argue for 500 years. "So Owen and I got together one morning and we said, 'It's got to be like Arizona! The moon has just *got* to be like Arizona! Can't be nothin' else. So let's design a landing gear like it was.' " That's what they had been doing when the Surveyor people arrived on that Tuesday. They remained worried, however, at what might happen to the astronauts if they were wrong and the moon wasn't just like Arizona.

Before the Surveyor briefing that morning, there was one other meeting to attend. Brainerd Holmes, head of the Office of Manned Space Flight, had flown in from Washington, and Johnson joined the rest of the senior staff of the Space Task Group in the cramped conference room on the second floor of Building 58. What Holmes had come down to tell them was, first, that the Space Task Group was going to become a full-fledged center, to be called the Manned Spacecraft Center. And then he told them that this center was going to be located twenty-five miles south of Houston, Texas.

Still digesting this awful news, Johnson was not at his most gracious when he got to the Surveyor briefing. The poor Surveyor fellows were asking what they could do to help the Apollo people, Owen Maynard recalled. "They were sitting there with their pens poised ready to write our answer down, and Caldwell said, 'Crash into the moon and smash all to hell and then at least we'll know we don't have several meters of dust on it.' " It was pretty deflating for the Surveyor guys, who were just trying to be helpful, "but Caldwell was really torqued off, let me tell you."

Why Houston? just about everybody in the Space Task Group asked plaintively. The word got around quickly that the new site in Texas was a flat, treeless plain. The temperature was in the nineties from February to November. "I was so upset about going to Texas, I wouldn't even let them send me the free subscription to their goddanged newspaper," said one of the Space Task Group engineers. "It was my intention *never* to go to Houston." Hampton might not be perfect, but most of them had homes there, and children in school. There was the Chesapeake Bay nearby, and Norfolk across the new bridge. And if they couldn't stay in Hampton, why not go to Florida, or Denver, or somewhere nice?

Why Houston? Well, Tom Markley once explained with a straight face, there were these criteria that NASA had established for the site of the new Manned Spacecraft Center, and, lo and behold, Houston won. For example, the site had to have water transportation. "Clear Lake had a foot and a half of water in it, so Houston met that one." And it had to be, as the site selection committee put it, in a climate "permitting year-round, ice-free water transportation." No problem—"I guess Houston, with ninety-one degrees average temperature and a hundred percent humidity, met that criterion," said Markley.

There was more. For example, another criterion was the availability of 1,000 acres of land—and, by what the official Apollo history calls a "politically arranged gift," an oilman just happened to offer precisely 1,000 acres of salt-grass pastureland south of Houston to Rice University if Rice would in turn agree to donate it as a site for the center. The oilman, as it happened, also owned several other parcels of land adjoining the site. But it wouldn't cost a penny out of the taxpayers' pockets.

In truth, there were eight criteria and a solemn selection process. But it was also true that Albert Thomas, the chairman of the House Independent Offices Appropriations Committee—NASA's appropriations committee—came from Texas, and his district included Houston. And the Vice-President of the United States, the space program's staunchest ally, was named Lyndon Johnson. Yes, acknowledged Charles Donlan, deputy director of the Space Task Group, they dutifully went through this site selection process, "but it's as though you went through a maze knowing all the time what door you were going to come out."*

So Houston it was, and off they went, 700 engineers and their families, from Virginia to the Gulf Coast of Texas, in the middle of the most frenzied period of Project Mercury (they were trying to launch John

---

* Rumors that Thomas and Johnson made money out of the choice of Houston through land deals circulated in NASA from the day the decision was announced but they remain only rumors. Thomas's influence over the decision is generally conceded to have been enormous, however, and was probably even more important than Johnson's. George Low later recalled a meeting in Abe Silverstein's office where Silverstein (who was the behind-the-scenes progenitor of so much in the space program) said that "a program like Apollo comes along once in a lifetime, and only once in a lifetime will we have an opportunity to build a new center like this." And then, when they had decided to try to pull it off, the first thing Abe Silverstein said was, "I wonder where Albert Thomas's district is?"

Glenn on the first manned Atlas), and just as the Apollo Program was beginning to grow exponentially. Caldwell Johnson went too, leaving his native waters and his beautiful new home behind. For one thing, he had overheard a senior Langley engineer speculating that Johnson would never agree to leave Hampton. Johnson didn't like being predictable. Also, he figured, "I'd eat my heart out if I stayed there and let all these other guys come to Houston and do this. I would've kicked myself fifty thousand times." So Johnson went to Houston. Almost everyone in the Space Task Group did.

Two days after Christmas, Tom Markley arrived in Houston to oversee the transition. "Okay, Tom," Gilruth had said to him, "you get down there and take over the site while we get our arms around Mercury." Gilruth and the others would stay at Langley until the Glenn flight had gotten off. Markley had said okay—since his house had already been sold anyway and his household goods were being packed up for shipping to Baltimore. It wouldn't be hard to reroute them to Houston.* "What are we doing?" his family asked. A change in plans: "We're going to move to Houston, Texas!" Markley told them. "What'll we do with your stuff?" the Mayflower moving people asked. "I don't know! Take the truck down there, and I'll call you when I get there!" said Markley. "So we took off and went to Houston, Texas," as Markley recounted it later. "Got down there, and the place was chaos. I'd never seen anything like it in my life."

In 1962 alone, M.S.C. would hire 2,000 new people, nearly quadrupling its size. When Markley arrived, hardly any of the supervisors were in place to interview the candidates; people were being selected by matching their academic records against the qualification lists. By comparison, it made the previous hectic hiring at Langley look like a careful, reasoned process—"Anybody who knew how to spell 'Apollo,' we took them," said one veteran of those days.

Meanwhile, the skeleton staffs that had been sent down to Houston were independently rummaging through the Houston real-estate market trying to find places to put all these new bodies. When Markley sat down for his first day at the job, he kept getting calls from people he'd never heard of saying something like, "Well, you just acquired two thousand more square feet of rental space out here in. . . ." Chaos. "My God, what's going on?" Markley asked himself. "So I go back to my *diktat*

* For the story of why Markley had thought he would be moving to Baltimore, see Chapter 12.

from Gilruth and it says, 'Take charge!' I called a staff meeting in one of the hangars down there and said, 'Will you guys all introduce yourselves? Tell me who you are and what you're doing here.' God's truth. So they did, and I said, 'Okay, well, first thing is, where do we have office space now?' They started showing me a map, and I couldn't believe it. We were all over that town.'' During February and March 1962, as the rest of the Langley staff migrated down to Texas, Markley sometimes felt like the wagon master for a particularly unruly train of settlers.

The twist to this story is that Houston was by all accounts a fine place to put the Manned Spacecraft Center. ''Moving it to Houston was magnificent,'' said the same person who refused the subscription to the Houston newspaper. ''We got away from the old fogies. We got down there and created a whole new thing with a bunch of twenty- and thirty-year-olds. Holy smokes, that move to Houston is what made the program.''

Some of the selection criteria had been absurd (especially the one about ''water access,'' which was unnecessary and eliminated otherwise attractive landlocked sites), but one feature of Houston that wasn't on the list turned out to be crucial: Houston was about halfway between the coasts, only three hours from North American in Los Angeles or M.I.T. in Cambridge, Massachusetts. And if the climate was scorchingly hot, Hobby Airport in Houston didn't get snowbound or clouded over in the way that, say, Stapleton Airport in Denver did—not a trivial point for people who were to live on airplanes for the next several years.

As to the most obvious question of all—why not just put the Manned Spacecraft Center at the Cape?—the answer is that NASA didn't want the facilities of the M.S.C. at the Cape. The launch facilities had little to do with the work of the Manned Spacecraft Center. The flight controllers spent their time preparing for and running flights. The people in the engineering and design division and the program offices spent their time working at drafting tables or in laboratories or traveling to visit contractors—many of whom were on the West Coast, a continent away from Canaveral. When the astronauts trained, they used simulators, and the people they needed to talk to before, during, and after the flight weren't the launch teams, but the flight-control people.

It must be added that Houston helped its own case considerably when the newcomers arrived. One longtime Virginian vividly remembered his first startled impression of Houston, that everything out there—the houses, the stores, the cars, the streets—glistened. And Houston piled on hospitality in Texas-sized portions. The old N.A.C.A. people kept comparing it with the indifference of the Hamptonians toward Langley. "They didn't think we were worth *schmatz* back in Hampton, Virginia," mused one of them. "In Houston, there were big billboards welcoming us. This little jerkwater town hadn't cared for us at all and the sixth-biggest city in the U.S.A. was welcoming us." Houston really was, after all, a pretty good place to put the Manned Spacecraft Center.

3

Shea's mandate from Brainerd Holmes was twofold. He was supposed to wring a mode decision out of the centers. And he was supposed to do it in a way that helped unify NASA. As things stood, the tensions among the centers—and the erstwhile Space Task Group was now a full-fledged center—were terrible. They were united only in their mutual distrust of Washington and the new breed of systems engineers who seemed to be taking over.

Joe Shea knew that he was an emblem of this new order, and so he set off in early February to visit the centers to try to build alliances and allay fears. If the folks down in Huntsville and Houston were worried that Shea was going to build a Washington empire, he would defuse this anxiety by peopling his own staff with personnel from the centers—a few from Houston, a few from the Jet Propulsion Laboratory (J.P.L.), a few from Marshall, a few from Lewis. They wouldn't have to come to Washington. They would report to Shea, they would be his technical resource, but they would remain at the centers, under the eyes of the centers. It sounded to him like a great idea.

He went first to Marshall to meet Wernher von Braun. It was Shea's first personal contact with him. Von Braun was extremely gracious and courteous to the newcomer from Washington, Shea recalled. Both von Braun and his chief of staff, Eberhard Rees, briefed Shea on what Marshall was doing, took him on a tour of the Center, listened to what he wanted to do in Washington. And when Shea asked for some technical support from Marshall, they said fine, and came through with Arthur Rudolph, one of von Braun's most senior people. "Overall it was a nice day and the opening up of what turned out to be a good relationship," Shea said. He left Marshall impressed by von Braun's grasp of the program and a little surprised to find that von Braun wasn't just a pretty face after all. Shea flew on to Houston to meet Gilruth.

Gilruth and von Braun played the same generic role at their respective centers but interpreted it in different ways. Each was the paterfamilias, presiding over a family. Each of them was indulgent to a degree. But whereas von Braun was the German father who expected everyone to be at the dinner table every night to give an account of his day, Gilruth was more like the American dad who was satisfied if they called home now and then. When Shea proposed his plan, it was not Gilruth's style to call in Max Faget and tell him that he thought it was a good idea and Max should take care of it. The technical resources that Shea wanted were under Faget, and therefore Shea should go talk to him. Shea did.

It was the first encounter between these two outsized talents and egos. They got together over in the old Farnsworth Building where Faget had set up his temporary headquarters. Caldwell Johnson and the director of the Apollo Spacecraft Program Office, Charlie Frick, were also present. "I said, 'Max, I need half a dozen, maybe eight or ten guys, who can pad out my operation,' " recalled Shea. "And Max, God bless him, said to me: 'Look, we don't need Washington, we don't need any technical strength in Washington, we're bigger than you are now, we'll always be bigger than you.' In effect, get lost." Thinking back to the same meeting, Faget acknowledged that, yes, he explained to Joe "in great detail" that "Washington didn't do anything very useful."

At this point, memories of the meeting diverged. Caldwell Johnson recalled Shea making some rejoinder to the effect that "Just because you guys have done Mercury, you think you know everything." Shea didn't remember it quite like that, but he did remember remarking that his shop didn't need to be bigger than Faget's. Just smarter. And he added that being smarter than Faget and his bunch shouldn't be too hard.

"And God, we got hotter than a hornet then," said Johnson. "Max said something like, 'Well, what the hell have *you* ever done?' About that time, Charlie Frick jumps in and says, 'Max, I'd like to talk to you a minute, it's something private,' and then he turns to me and he says, 'Why don't you go on back to the office?' And so he broke the thing up. You might say it never did start right." Or as Shea put it in an official interview with NASA historians, "We started off with an antagonistic interface."

It was in this context that Shea set out to satisfy Brainerd Holmes's mandate both to midwife a mode decision and to bring the centers together. "And I scratched my head," Shea recalled later, "and said, 'Well, how do you do it? These guys are so parochial'—and they really were!'' For while he had gotten along a lot better with Marshall than with Houston, they seemed equally stubborn on the issue of how NASA should proceed with Apollo. By this time, M.S.C. was openly behind lunar-orbit rendezvous while Marshall remained attached to earth-orbit rendezvous. After his visits Shea reported to Holmes:

> Most of the MSC people seem enthusiastic about LOR. However, I don't feel they have a good understanding of the rendezvous problem, and their weight estimates for the LOR operation seem quite optimistic. . . . MSFC [Marshall] has not paid any attention to LOR and was not in a good position to comment on the mode. Their instinctive reaction, however, was negative.

Even the two centers' plans for earth-orbit rendezvous were at odds: "In essence, each center has its equipment doing most of the work, and completely ignores the capability of the other's hardware."

Shea decided to hire a contractor to do an independent study of how heavy a lunar module would have to be—until then, the main source of contention between the mode adversaries. Meanwhile, he also set Houston and Marshall to work on additional studies directed by his office. To guard against puffery, the analyses of L.O.R. were assigned to Marshall and the analyses of E.O.R. to Houston.

Shea worked fast. He had the Request for Proposals out on the street, proposals submitted, and a winner chosen within about a month. Chance-Vought won, an offspring of the same organization that had done the earliest work on L.O.R. back in 1960. They came back within another month with their preliminary results, which showed a lunar module weighing about 29,000 pounds, significantly heavier than Houston's most recent estimate of 20,000. There remained a substantial weight difference between the two modes, however, and it came down to this: If NASA went via earth-orbit rendezvous, they would need two launches of the Saturn V. If they went via lunar-orbit rendezvous, they could do it in one.

By this time, Shea was also pushing the centers to come to grips with the mode issue through a series of meetings that he sponsored. The principals from Houston and Marshall would come together in the same room, Shea presiding, and thrash out a set of technical issues related to one of the modes. Then they would be sent back to their respective centers to conduct more detailed studies, which Shea continued to assign according to the rule of hostile testimony: Marshall did the studies

involving L.O.R.; Houston did the studies involving E.O.R. The
meetings did not produce consensus, but they at least permitted people to
confront their differences and, more important, raised everyone's aware-
ness that a decision about the mode had to be made soon.

Houston decided to take a bold step. The Manned Spacecraft Center
would talk directly to von Braun and Marshall, and try to convince them
to accept L.O.R.

"It's hard to characterize how formal the relationship was," said Tom
Markley, speaking of the early relationship between the Manned Spacecraft Center and Marshall. The tensions between the two centers
resonated with the defensiveness of little brothers to older brothers.
People like Wernher von Braun and Eberhard Rees had devoted their
lives to rockets. The exploration of space had been their dream long
before anyone at Langley even gave it a thought. Their record of success
was glittering. And yet the aeronautical engineers of Langley had been
given the prize of designing and operating the spacecraft.

To Tom Markley, the message from the Marshall people came
through loud and clear. When he was sent down from the Space Task
Group (still at Langley then) to Huntsville in the fall of 1960 to act as
a liaison with Marshall, the Germans were faultlessly courteous, but it
was obvious that they "really wanted to run the whole mission." "You
could pick that up," Markley said, "through all the courtesy and colle-
giality: 'We're more qualified, we've done all these flights. . . . We can
do this, we know how to make things work. Watch us.' "* The fact was,
they had missed their chance to run the whole mission when they had
stayed with the Army for the first year after NASA was founded. They
could have come in, after NASA's civilian supremacy was conceded, if
they had been willing to desert the Army unceremoniously, but von
Braun had not been willing to do that and he had reconciled himself to the
consequences—von Braun never tried to take over the spacecraft operation. But Marshall still thought they were the people who should have
been doing it.

For their part, the Space Task Group people were sensitive to being the

* Besides their two decades of experience with rockets, von Braun's team had a more recent card
to play: They were the ones who had put up the first American satellite. Bob Gilruth was quick to
pick up on the politics of such events. Back in 1958, Markley ran into Gilruth in the airport at Boston.
"We were flying back," Markley recalled, "looking out at the moon, and Gilruth said, 'You know,
someday we're going to be there.' I said, 'Oh.' He said, 'I'll tell you a little-known thing. We just
agreed today that von Braun and his team are going to put the Jupiter up. Don't be surprised if it
happens much sooner than anyone expects.' I said, 'What do you think that means?' He said, 'It
means they're going to be tough to live with.' "

novices in the space-flight business while the Marshall people were the old pros. They perceived von Braun's group as arrogant, even as they recognized, as Markley said, that "they had a right to that kind of arrogance—they'd earned it over a long period of time." Thus, Houston, a little shaky at first in its new role as a center, tended to be defensive, standoffish, quick to see slights and bureaucratic threats. In the spring of 1962, when Houston still had only three short manned flights to offer as proof of its own capacity, these feelings were at their height. For Houston to go to Marshall and say to von Braun "We're going to persuade you that we have a technically superior idea about space flight" was complicated in ways that were as much psychological as bureaucratic.

But that's what Charlie Frick, then manager of the Apollo Spacecraft Program Office—ASPO—decided to do. An earlier effort they had made to present the arguments for L.O.R., at one of Shea's meetings up at headquarters, had been disappointing, and Frick announced that the next time they would do it right—perhaps with "a bit of showmanship," too. They would aim to convince von Braun that lunar-orbit rendezvous was both less costly than E.O.R. and better from an operational standpoint—which, they planned to emphasize, was something they did have a lot of experience in. "Charlie Frick's Road Show," they called it. On April 16, it came time to play its Huntsville date.

They held nothing back. Along with Maynard and Johnson and the others who would make the technical presentations came Faget, and Walt Williams and Chuck Mathews and Bob Piland and Bob Gilruth himself—the entire senior staff of M.S.C. Frick even brought along John Glenn, just returned from his successful orbital flight. All the senior staff from North American, the spacecraft contractor, attended as well.

Von Braun treated the presentation as a major event, assembling all of Marshall's senior staff. They were on the tenth floor of Marshall's headquarters building, known locally as the Von Braun Hilton, in the conference room that adjoined von Braun's office. "There were maybe fifty people at this meeting," Caldwell Johnson recalled, "every person you could pack in that big conference room at Marshall."

Charlie Frick was a little surprised at the reception they got. He came expecting some hostility, but found himself admiring the way that the Marshall people confronted a technical problem once they had decided to do so. They had occasional questions, but for the most part the Marshall engineers, a mix of Germans and Americans, listened attentively as the briefing continued throughout the day.

In the late afternoon, the last of the Houston presenters put away his transparencies and the lights went up. The packed conference room was quiet. To Caldwell Johnson, it was the kind of uneasy silence that falls after a long argument is over and no one knows how to say the obvious. To him, the meeting had made it clear that "we had to go lunar-orbit rendezvous; it just wasn't gonna work any other way, no matter what the politics were." But no one knew how to say it, how to end the meeting.

John Paup, North American's program manager for the Apollo spacecraft, found himself in an awkward situation. North American, holding the main contract for the command and service module, had a corporate interest in earth-orbit rendezvous. If earth-orbit rendezvous was chosen, presumably North American would build the expanded space-craft. If, on the other hand, lunar-orbit rendezvous was chosen, there was a good chance that a separate contract for the lunar lander would be written and that some company other than North American would get it. Paup and Harrison Storms, general manager of North American's Space Division, had been stubbornly arguing for earth-orbit rendezvous and resisting every suggestion that the command and service module might have to be adapted for an L.O.R. mission. But now Paup had to wonder whether it made sense to keep on fighting it. Abruptly, he spoke into the silence: "I've heard all these good things about lunar-orbit rendezvous," he said. "I'd like to hear what sonofabitch thinks it isn't the right thing to do."*

There was another long pause. Owen Maynard remembered seeing one of the Marshall people at the back of the room stand up and von Braun quietly gesture for him to sit down. Charlie Frick remembered Storms saying, "No, everything looks great. That's the way to go," and von Braun making "a very gracious speech thanking us and saying he understood the advantages of the system." Faget came away with a vivid memory of von Braun "very generously throwing in the towel," without any rancor at all. Caldwell Johnson remembered just the long pause, and then Paup saying, "I guess that's the way we're going to go, lunar-orbit rendezvous." To Johnson, everybody seemed happy and relieved that it was over with.

*   *   *

* When Charlie Frick recalled Paup's words for the NASA historians, they came out as "It looks to me like we're all convinced. Is there anybody that wants to speak against this, anything that's negative?" The version given in the text is Owen Maynard's recollection. Caldwell Johnson independently recalled Paup's words almost identically as "Who's the sonofabitch in here that's not for lunar-orbit rendezvous?" Johnson added that we would never find any official minutes to confirm those words, but that's what they were.

The Houston people were a little optimistic in their reading, for Wernher von Braun had not yet decided finally in favor of L.O.R. The data presented at Marshall that day represented many of the facts von Braun had been waiting for, however, and he meditated on them. He himself dated his commitment to lunar-orbit rendezvous from a trip to Houston a few weeks later when his Marshall team presented its most recent thinking on earth-orbit rendezvous. "I think that was the occasion when some people felt that I had walked away from our proposal," he recalled later. "On that flight back from Houston there were a couple of disappointed guys on the plane, that I remember." But if that was actually when von Braun changed his mind, it was a closely held secret. A month later, in a long afternoon's meeting with Joe Shea, von Braun gave no hint that he had decided in favor of lunar-orbit rendezvous.

Like Gilruth, von Braun would always strongly deny that he had changed his mind about anything. "I wasn't committed to earth-orbit rendezvous very strongly," he later said. "I'd always taken the position that we in Marshall would investigate E.O.R. and Houston would investigate L.O.R. I just wasn't ready to vote at all until I had the facts." Of a published report that "von Braun changed his mind" about the mode, von Braun said shortly, "That's a lot of crap."

The final act in the drama of NASA's decision on the mode occurred on June 7, 1962. The place was once again the conference room at the Von Braun Hilton, this time at a review arranged by Shea's office. For Shea, who had had a long discussion with von Braun about the mode decision just a week earlier, it was a singular experience, one that taught him a lot about how von Braun ran Marshall. For six hours, members of the Marshall staff stood up and gave presentations that were markedly pro-E.O.R. "And finally, at the end of the day," Shea recalled, "von Braun stood up and said, 'Gentlemen, it's been a very interesting day and I think the work we've done has been extremely good, but now I would like to tell you the position of the Center.'" And then he announced, to Shea's surprise and to the apparent stupefaction of most of the men from Marshall, that while all of the leading mode alternatives were feasible, lunar-orbit rendezvous "offers the highest confidence factor of successful accomplishment within this decade." Engineering elegance had won out: "A drastic separation of these two functions [lunar landing and entry into the earth's atmosphere] is bound to greatly simplify the development of the spacecraft system [and] result in a very substantial saving of time."

*   *   *

A few weeks later, John Houbolt was at NASA headquarters on other business and saw a number of M.S.C. people gathered in the hallway. What's going on? he asked. Didn't you know? they replied. They were having a rehearsal briefing for Seamans before presenting the L.O.R. case to Webb. Houbolt, stunned that he had been so completely ignored, asked if he could watch, and Charlie Frick said of course he could.

Shea was the presenter. It seemed to Houbolt that Shea and the Houston group were managing to discuss all the advantages of L.O.R. without any reference to the people who had been voices in the wilderness. But at least at the end, there was Bob Seamans turning to him and saying, "Well, John, how does that answer your letter?"

Shea and his staff prepared the technical, budgetary, and scheduling analyses that once and for all laid out the alternatives side by side in complete detail. Their conclusion, like von Braun's, was that all of the alternatives were feasible but L.O.R. was the best choice. Holmes was convinced, Seamans was convinced, and Jim Webb accepted the judgment of his technical people. By early July, NASA had made up its mind, and called a press conference for July 11. Four men would be on the platform representing NASA: Webb, the administrator; Seamans, the associate administrator; Holmes, the head of the Office of Manned Space Flight; and Joe Shea.

4

NASA had made up its mind, but NASA had not reckoned with Jerome Wiesner and the President's Science Advisory Committee. They had been watching from the sidelines, and were perplexed and disturbed by NASA's decision.

At the outset, PSAC's response was similar to the initial reaction that everyone had toward lunar-orbit rendezvous: It had to be more complicated and more dangerous than earth-orbit rendezvous. Wiesner assigned one of his own people to conduct a study of the mode problem, and he chose for that job Nicholas Golovin, a mathematician and former NASA employee who had preceded Shea as Holmes's deputy a year earlier.

Two problems complicated the situation. One was personal: Golovin had been openly encouraged to resign, and had left NASA with what one of his colleagues called "bitter gall" in his throat. It worried Seamans when he heard that Wiesner was going to hire Golovin. When Wiesner asked him for a recommendation, Seamans cautioned that Nick "wasn't always wrong, but his thinking could be colored by some of the relationships over here in NASA." The other problem was professional: Golovin and Shea were diametrically opposed in their stances toward that recurring issue in the Apollo Program—what is the meaning of "safety" or "reliability" when going to the moon? Golovin believed in a statistical method (test the components until you have a statistical basis for assessing reliability), while Shea assigned reliability estimates to components on the basis of design considerations (given the mechanism, the materials, and the environment in which they are operating, what's a reasonable estimate of the component's vulnerability to failure?) and then compared the results across different modes.

Joe Shea first heard that something had gone awry in the approval of the L.O.R. decision on July 3, a week before the press conference. He had a meeting with Golovin scheduled for two days later and had already sent over a preliminary typescript of the final, detailed mode comparison. He was sitting with one or two others in his office at the end of the day when, at about 6:15, the phone rang. It was Jim Webb. "Now look," he said to Shea, "Jerry Wiesner just called me and he's in a highly emotional state; he thinks L.O.R. is the worst mistake in the world, and we're risking these guys like mad—get over and see him."

Shea walked the few blocks to the Executive Office Building and found Wiesner in his office with Golovin. As Shea recalled it, Wiesner was furious. Lunar-orbit rendezvous was a "technological travesty." Even NASA's own analyses showed that it was. "Look at that," he said to Shea, pointing to a page in the draft report, "even your own numbers say that it's wrong!" Instead of the probability of mission success of .4 that M.S.C. had calculated, Golovin, in checking Shea's work, had come out with a probability of .3.*

"I said I didn't think the numbers were wrong," Shea recalled, and

---

* Both numbers were unacceptably low as true estimates of mission success, but this points to a distinction that needs to be made: In comparing two modes, the question was not whether the absolute reliabilities were correct, but whether the figures to compare reliabilities were consistent. To simplify for purposes of illustration, it didn't make any difference whether the real probability of a successful Saturn V launch was .9 or .95, as long as the same probability was used for the analysis of both modes. Shea and the others using these numbers felt all of the estimates of mission success were unrealistically low in an absolute sense. The position of Wiesner and Golovin was that even the comparative reliabilities were untrustworthy because each mode had some events that were unique to that mode, and the numbers representing the probability of success of those unique events were essentially made up.

looked at the draft for himself. Then he discovered what had upset Wiesner. In the typescript, which had yet to be proofread, there was an error in the calculations: The one that was wrong was the safety number summarizing the L.O.R. alternative.

Wiesner and Golovin did not remember it that way. "We pointed this out to them," Wiesner recalled later, "and in a week they came back with a new set of numbers which [still] showed that L.O.R. was better. . . . My recollection is not that they made a mistake. It was just that they added another nine to the cycle of things they multiplied to get [their reliability estimate], to get the number up where they wanted."

That evening in Wiesner's office, Shea went over for Wiesner and Golovin the reasoning that had led NASA to the L.O.R. decision. Yes, it was true that effecting a rendezvous in lunar orbit carried a risk, but so did everything else. Shea remembered that Wiesner kept saying, "No, it's got to be more dangerous than earth-orbit rendezvous," and that he, Shea, finally said, "Look, your background is electronics. You ought to understand it better than most guys in the program because they're not in electronics." But after an hour and a half, they were at an impasse. Shea could not understand why he could not convince Wiesner. Wiesner was convinced that NASA had done a shoddy analysis of the modes.*

Many factors made up the muddle—Golovin's history at NASA, Shea's cocky confidence in his own numbers, Wiesner's conviction that lunar-orbit rendezvous was intrinsically more dangerous no matter what the numbers said. But the core of the controversy continued to be the disagreement on how to estimate reliability, and on this point the divisions were deep and irreconcilable. Is reliability to be based on the statistical analysis of repeated trials? Or is it to be based on judgments about the hardware's design and the adequacy of the ground testing?

Because of PSAC's objections to L.O.R., NASA hedged at the press conference on July 11. The choice of lunar-orbit rendezvous was tentative, Webb said. More studies would be conducted.

Throughout the summer and into the fall, PSAC mounted a persistent campaign to have the decision reversed. The issue came to a head in September, at Huntsville, when John Kennedy was on a tour of NASA's facilities. Von Braun, Wiesner, and the President were looking at a mockup of the first stage of the Saturn V at Marshall. "I understand you and Jerry disagree about the right way to go to the moon," the President

---

* Wiesner remained convinced thereafter—years later, Seamans recalled, Wiesner would say "I still think you made the wrong decision" when he encountered Seamans on the M.I.T. campus.

said. Von Braun acknowledged that this was the case. "We were having an intelligent discussion," Wiesner recalled. "I was starting to tell Kennedy why I thought they were wrong when Jim Webb came up, saw us talking, thought we were arguing, and began hammering away at me for being on the wrong side of the issue. And then I began to argue with Webb."

The dispute about the mode, which the press had hitherto seen as a low-key technical issue, was suddenly in the open. There, in the middle of a Marshall hangar, with the press corps bunched behind the ropes a dozen yards away trying to make out the words, Wiesner and Webb were each saying heatedly that the other fellow was flat wrong.

On the way to his next destination, Kennedy predicted the eventual outcome. British prime minister Macmillan's science adviser, who had listened to the confrontation, was with Kennedy and Wiesner on Air Force One and asked how it was going to come out in the end.

"Jerry's going to lose, it's obvious," said Jack Kennedy.

"Why?" the Englishman asked.

"Webb's got all the money, and Jerry's only got me."

Nonetheless, Wiesner was not ready to give up. He continued to complain about the L.O.R. decision until, on October 24, Webb wrote him a tart letter, saying in effect that the L.O.R. decision was as sound now as it had been in July and they were going to go ahead with it. If Wiesner wanted to stop it, he was going to have to get the President to stop Webb. Wiesner couldn't very well do that right away, because on October 24, 1962, John Kennedy was in the midst of the Cuban missile crisis. But he did what he could. Wiesner asked to examine all the contractors' materials relating to the mode decision—and was refused, on grounds that much of the material was proprietary. Golovin and his staff prepared a brief for a two-man lunar mission using earth-orbit rendezvous; Shea and his staff blasted it for going back to techniques that had been found unacceptable months before, after exhaustive analysis. Finally, conceding that L.O.R. was a feasible mode for going to the moon—albeit an inferior one—Wiesner decided to let the issue die.

On November 7, 1962, Webb announced to the press that he was confirming the tentative decision of July, and that the Grumman Engineering Corporation had been chosen to build the lunar module. Eighteen months after the nation had decided to go to the moon, NASA had decided how.

10

"It aged me, I'm sure"

On June 28, 1962, the same day Jim Webb learned that the centers had agreed on lunar-orbit rendezvous, NASA's first F-1 engine destroyed itself on a test stand at Edwards Air Force Base in California.

The F-1 engine was the heart of the Saturn V and the individual technological achievement that more than any other made Kennedy's lunar commitment possible. At the time Kennedy was considering his decision to go to the moon, Low's task group was calculating that the spacecraft and rocket for a direct lunar ascent would weigh at least 10 million pounds. While this didn't necessarily mean that they had to build a rocket with more than 10 million pounds of thrust (if they decided to go to the moon by earth-orbit rendezvous instead of direct ascent, they could split the load up), it did demand an engine that was a quantum leap ahead of the largest one in the American inventory.*

Marshall had already let a contract with the Rocketdyne Corporation for such an engine, the F-1, when these calculations were being made. Engineered to produce 1.5 million pounds of thrust, eight F-1s in a single first stage (the contemplated Nova) could lift the required payload for lunar landing via direct ascent; four F-1s could achieve an earth-orbit

---

* In 1961, an uprated H-1 engine was the largest in operation, with 188,000 pounds of thrust.

rendezvous mission in two launches; and five F-1s, producing 7.5 million pounds of thrust, could permit a lunar landing in a single launch by using lunar-orbit rendezvous.

The question in the spring of 1961 was whether the F-1 was in fact feasible—not even the Russians with their large boosters had an engine half its size. On April 6, 1961, a week before Gagarin's flight, a prototype thrust chamber for the F-1 had achieved a peak thrust of 1,640,000 pounds. But this was a brief one-time test of a prototype, which was far from being a functioning engine. Kennedy's May 25 decision to go to the moon had to be based on the prediction that eventually the F-1 would work. If it didn't, the Apollo Program would have to come to a halt until an alternative could be developed.

This prediction, moreover, was by no means a sure thing. NASA's rocket engineers had chosen 1.5 million pounds of thrust as the goal for the F-1 not because they knew it was within their capabilities, but because that's what they needed for a launch vehicle that could be used to build space stations or go to the moon, the missions that von Braun had in mind. "They were cavaliers and they were pioneers," said an American engineer of the Germans. "They were willing to say, 'Well, okay, it's going to be one and a half million pounds.' They didn't go through mountains and mountains of computer programs to try to figure it out. A lot of personal judgment went into it." Despite the progress that had been made by the spring of 1961, the F-1 was still a gamble. The early development work that had produced the brief 1.64 million pounds of thrust had also been troubled by persistent problems. A number of senior advisers, including at least one member of the President's Science Advisory Committee, were arguing that the F-1 was just too big to work.

Rocketdyne, the contractor for the F-1, was located at Canoga Park in the San Fernando Valley north of Los Angeles. The company had a long and successful history of collaboration with von Braun's team, and during the remainder of 1961 and into the first half of 1962, it looked as if the F-1 would be another thoroughgoing success. But the initial tests of the engine were done in small increments, with little high-frequency instrumentation, and throughout this optimistic period the engine was shut down at the first hint of problems.

The Rocketdyne engineers tested their engine at Edwards Air Force Base, 120 miles east of Canoga Park, where a gigantic test stand for the F-1 had been anchored deep in the floor of the Mojave Desert. It was there, on June 28, 1962, that they let the F-1 run for a few seconds longer than before and it destroyed itself. "Destroy" in this case meant melting through thick, high-strength steel plate in a quarter of a second. But that wasn't the worst of it. The F-1 failed not because of a faulty weld or a flaw in an alloy. Those kinds of failures were comparatively easy to remedy. Instead, the F-1 destroyed itself because of "combustion instability."

1

In principle, liquid rocket engines are simple, far simpler than the internal combustion engine. Liquid fuel is pumped into a combustion chamber in the presence of liquid oxygen and a flame and is made to burn. That's all there is to it. There are no crankshafts to turn, no pistons to drive. The burning fuel produces energy in the form of gases that exit through the rocket's nozzle. The force the gases produce against the top of the engine is called thrust. The thrust is transmitted through the rocket's structure and, if it is greater than the weight of the rocket, the rocket lifts off. Put in its most basic terms, for any rocket to work there are two things that must be done extremely well: The propellants must be brought together, and then they must burn smoothly.

In the F-1, just pumping the propellants to the combustion chamber raised unprecedented demands. The F-1 used liquid oxygen (LOX) and R.P.-1, a form of kerosene. The pumps, one for the fuel and one for the LOX, had to deliver the kerosene from the tankage to the combustion chamber at the rate of 15,741 gallons per minute, and the LOX at the rate of 24,811 gallons per minute.* Driven by a 55,000-horsepower turbine, the pumps had to operate at drastically different temperatures: 60 degrees Fahrenheit for the fuel, −300 degrees for the LOX, while the turbine itself ran at 1,200 degrees. To complicate matters, the whole assembly had to be light and compact enough to fit on board the rocket and nonetheless sturdy enough to resist the pressures, vibrations, and other stresses of launch and flight.

Developing the pumps, however, was still not as hard as solving the second basic problem of rocket engines: making them burn smoothly once they had reached the combustion chamber. The pumps brought the kerosene and the LOX to a circular metal slab three feet in diameter and

---

* The LOX pump would have filled a swimming pool twenty-five feet long, ten feet wide, and six feet deep in about twenty-seven seconds.

about four inches thick, weighing 1,000 pounds, called the "injector plate."* The injector plate was pocked with 6,300 holes less than a quarter of an inch in diameter through which the kerosene and LOX entered the combustion chamber. Most of the propellant streams were arranged in groups of five. Two of the five, both kerosene, impinged on each other at a carefully defined distance below the top of the plate, forming a fan-shaped spray. The other three in each five-hole group were of LOX. These also impinged on one another, forming another fan. The two fans intersected. There, given the presence of a flame, they would combust.

In the F-1, the combustion chamber was a barrel about thirty-six inches wide and thirty inches long, closed at one end by the injection plate and opening into a nozzle at the other end. A few seconds before ignition, four small pre-burners in the combustion chamber—pilot lights, in effect—were lit, providing a flame at the point of impingement. As the pumps screamed up to speed, valves snapped open and more than a ton of kerosene and two tons of liquid oxygen burst into the combustion chamber. Per second. The gases produced by their ignition roared out through the "throat," the open bottom of the barrel, into the cone of the nozzle below. In the course of the few seconds from ignition to full power (mainstage), the interior of the combustion chamber went from ambient temperature to 5,000 degrees Fahrenheit. At the face of the injector plate, pressure went from zero to 1,150 pounds per square inch. Given that combination of propellants, pressures, and nozzle design, the force generated totaled 1.5 million pounds. In the first stage of a Saturn V, five F-1s were to ignite simultaneously and sustain mainstage combustion for 150 seconds.

By the early 1960s, creating an engine to withstand the temperatures and the pressures of the F-1 was, thanks to new metallurgical and engineering techniques, not in itself a formidable problem. The difficulty was to achieve what the engineers called a "smooth flame front," in which the kerosene and oxygen combined and burned at a uniform temperature across the face of the injector plate.

Achieving this stable combustion with an injector plate three feet in

---

* The kerosene arrived by a circuitous route, first acting as a coolant for the engine shell by passing through a labyrinth of tubing on the walls of the combustion chamber and nozzle. The kerosene carried off heat from the walls of the engine. It made for a much more complicated pumping system, but it also saved weight—an engine made of metal strong enough to withstand the temperatures of the F-1 throughout launch would have been prohibitively heavy.

diameter created unprecedented problems. If, for example, the holes in the plate were drilled so that one side of the flame front had a slightly higher oxygen content than the other side, the high-oxygen area would get hotter and produce higher pressures on that side. In a smaller combustion chamber, this imbalance might not create difficulties. But in the F-1, there was plenty of room for a "racetrack" effect to get started, in which a higher pressure on one side of the chamber would bounce, starting a wave front that would begin careening around the perimeter of the barrel. Within milliseconds, the heat fluxes inside the chamber would be bounding back and forth across the combustion chamber, reinforcing each other, going out of control, and destroying the engine.

"The slightest thing could trigger it," said one of F-1's engineers of combustion instability. This was a vexing situation, because the inside of an F-1 combustion chamber during launch was prone to develop a variety of "slightest things." If the pumps cavitated and failed to supply the propellants to the injector plate at an absolutely uniform rate, the streams of propellant and LOX impinged at the wrong points and could disrupt the burning process. Thermal shocks as the engine went from ambient temperature to 5,000 degrees could disrupt the burning process. Acoustical shocks that hit the chamber at the moment of ignition were the most troublesome of all. With the sole exception of a nuclear explosion, the noise of a Saturn launch was the loudest noise ever produced by man. The only sound in nature known to have exceeded the noise of a Saturn V was the fall of the Great Siberian Meteorite in 1883. Sound waves of such force tended to disrupt the burning process.

Combustion instability had been a problem in rocketry ever since Peenemünde days, and it had been solved for each generation of rockets through fixes that were more or less ad hoc. With the F-1, all the usual difficulties of fixing combustion instability were compounded by the enormous size of the engine. No one had ever tried to cure combustion instability under remotely comparable conditions.

On June 28, 1962, Jerry Thomson was working as chief of Marshall's Liquid Fuel Engines Systems Branch. Thomson, a soft-spoken, small-town boy from Bessemer, Alabama, had come to Marshall by way of the Marines, where he had been a seventeen-year-old rifleman in World War II, and Auburn University, where he had gotten his degree in mechanical engineering. Early in the 1950s, Thomson had chanced into the rocket engineering business and discovered that he had an aptitude for designing combustion chambers. By 1962, his aptitude was such that "somehow," as Thomson put it, he "got picked to resolve this combustion instability situation." NASA headquarters and Marshall considered the problem so critical that Thomson was told to turn over the operation of his branch to his deputy and move out to Canoga Park.

"It aged me, I'm sure," he later said with a sigh. While Thomson represented NASA, Paul Castenholz, a propulsion engineer, and Dan Klute, a Ph.D. in mechanical engineering research, headed up the contractor team for Rocketdyne. Like Thomson, Castenholz and Klute had a special talent for the half-science, half-art of combustion chamber design, and like Thomson they had been pulled off their management duties as senior engineers at Rocketdyne to work full time on the combustion instability problem. In all, about fifty engineers and technicians were assigned to what they called the Combustion Devices Team, augmented by technical support back at Marshall and consultants from universities, other NASA centers, and the Air Force. Within Rocketdyne, Castenholz remembered, they got whomever they wanted, to do whatever they asked, whenever they needed it—the Combustion Devices Team had the highest priority in the company.

As a process, the attack on the F-1's combustion instability was a model of government-contractor collaboration. Neither side blamed the other for the failures; both sides worked together as if for a single employer. The process was fine; its only defect was that for months thereafter it showed few signs of producing a solution.

At first, the team thought they could resolve the problem without having to redesign the whole system. They concentrated on the system's hydraulics—the flow rates of the fuel and LOX and the patterns of the holes. When they had completed a modified version of the combustion chamber, they tried it out at the Edwards test stand. After a few tests it too went unstable and did not recover. They redesigned again, tested again, and again the engine failed.

There was no pattern to the instability, and it would occur "for reasons we never quite understood," Thomson said. "It could initiate soon after you got the engine started, the engine would go unstable and destroy itself, or it might occur toward the end of the run. It just depended on the conditions that perpetuated the instability."

By January 1963, their tests had completely destroyed two more engines, and Brainerd Holmes called the NASA-Rocketdyne team up to Washington. Holmes was prepared to ask Congress for funds to begin a parallel development effort on another system, he informed them. This was not the time for false pride. The program depended on making the right decision. He went around the table, asking each man what he thought. Each in turn told Holmes that a parallel effort wouldn't be necessary—they would fix the F-1. And back they went to Canoga Park.

They set a new goal for themselves. It was useless, apparently, to try to design the F-1 so that it never began to go unstable. The engine was too big and subject to too many disturbances. Now they would consider it tolerable for the engine to initiate instability—all it had to do was achieve what they called "dynamic stability," which meant that it would correct itself. After instability began, the engine must (according to their goal) damp it out within 400 milliseconds.

In addition, they decided that for testing purposes they must "be able to initiate this instability at our command," as Castenholz put it. Nature, left to herself, wasn't producing instabilities frequently enough for the team to learn how to control them. Therefore, they decided to explode a bomb in the combustion chamber while the test was in progress—"If that didn't drive the engine into instability, then nothing would," they reasoned. Then they could concentrate on making the engine damp out the instabilities they had induced.

The Combustion Devices Team set out to master the peculiar art of exploding a bomb inside an inferno. It wasn't easy. The bombs they made were little cylinders about three inches long containing a black powder charge, heavily insulated to keep them from burning up before they were exploded. But even so, it was a problem to get a bomb inside an operating F-1 engine and detonate it at the right time. "We tried to pop 'em up through the throat, and that didn't work very well," recalled Castenholz, so they eventually devised a way to insert the bomb inside the combustion chamber ahead of time, running the detonating wires through a tube that extended from the injector plate down six or eight inches into the combustion chamber. They attached the bomb to the end of the tube, stood back, and started the test. The bomb would go off and the chamber pressure would suddenly jump from 1,150 to as much as 4,000 pounds per square inch. It was a terrific way to induce combustion instability.

Getting the F-1 to run under normal conditions had been an unprecedented challenge; now they were trying to design an F-1 that would run normally even if someone set off a bomb in it. It was an extraordinarily severe test to put to their new designs, and simply modifying the hydraulics wouldn't be enough. By the spring of 1963, the Combustion Devices Team was ready to acknowledge that they must completely redesign the injector plate.

They began by inserting baffles—copper plates extending from the injector plate into the combustion chamber that would interrupt the rebounding waves. They tried using thin, uncooled blades at first, "and it bent those baffles over like a tornado went through," said Thomson. They replaced the thin blades with massive "dams" of solid copper protruding about four inches from the injector plate, more than two inches wide at the base and tapering to about half an inch at the top, cooled by kerosene flowing through orifices in the baffles just as kerosene flowed through orifices in the injector plate. "Everyone thought that the baffles would be enough," said Thomson. But they weren't. They did help, and, if unperturbed, the new design worked quite well. But the bombs could still induce runaway instability.

"That's when we tried every trick that we could think up," said Thomson. He lost count of the design modifications they tested—forty or fifty of them, maybe. They would have one that they thought was going to be good, and in the first tests it would work. Then, unpredictably, it would fail. One of the Marshall engineers who specialized in sensors, Jim Mizell, was sent out to Canoga Park to help measure what was going on inside. He watched as the combustion instability team kept trying to find a fix that worked. "It got so bad that the engineers couldn't come up with a theory for the plate that they hadn't tried before," he recalled. "They turned it over to a bunch of craftsmen in back of the plant." Mizell was out there one night and saw them "boring holes like crazy." Mizell finally said to them, "What are you guys doing?" They replied, "Well, we've got this plate and we're supposed to bore holes in there until we get tired, and you guys are going to take it out to the test stand and fire it for us." They had code-named this particular injector plate the "Kitchen Sink," they told Mizell.

Jerry Thomson bristled when it was suggested that things ever went that far. "We resorted to prayer but not to anything quite that wild," he said. "We had engineering logic for all those things that we tried. Now you can say that the engineering logic wasn't very scientific, and that I do admit, but we weren't just punching holes to see what happened." As the months went on, though, and a solution continued to elude them, Thomson and Castenholz sometimes wondered whether it might come to that.

2

By now, it was the middle of 1963 and the Apollo Program was foundering. Or at least that's what *The New York Times* said, with a headline for its article of July 13, 1963, that read "LUNAR PROGRAM IN CRISIS." The trigger for the outcry wasn't the unsolved combustion instability. However crucial to the success of the program, that kind of problem was too esoteric to get much attention unless NASA publicized it, something that NASA was not about to do. Rather, the uproar in the press was over Brainerd Holmes's resignation in late June.

Holmes, head of the Office of Manned Space Flight, wanted authority to run Apollo with a free hand, the same kind of authority that Abe Silverstein had sought and been denied in 1961. Holmes and Webb had been sparring for months, with Webb continuing to refuse Holmes's key request, that the center directors report directly to him. Finally, Holmes took a stand on an issue (he wanted Webb to submit a $400 million supplemental budget request) on which Webb had the President on his side. Holmes tried to force the issue and lost. Isolated and demoralized, he left under circumstances that were officially described as Holmes's personal decision to return to private life for financial reasons.

Holmes was highly regarded in the engineering fraternity and had gotten a good press as well. *Time* had put him on its cover, billing him as the managerial mastermind behind the BMEWS early-warning radar network, as a "restless, dynamic worker . . . a scientist who [is] not afraid to work with his own hands," and as a man under whom "ambitious workers" were "clamoring to work." So when a politician like Webb ousted this eminent engineering manager, much of the press inferred that Webb was jeopardizing the best interests of Apollo out of personal pique. *Missiles and Rockets* entitled its editorial about Holmes's resignation "An American Tragedy."

But Holmes's loss to the program was only part of the story. The larger reality was that the nation's romance with Apollo had cooled. In the press, in the public opinion polls, and even in the White House, people were asking why all this money had to be spent so fast on such a distant goal with so little payoff. Congress had held a series of hearings that spring where a parade of scientists had derided the Apollo Program's scientific value. The competitive urge had faded as the Soviets failed to show signs that the United States was in a race. There were obviously other priorities that needed money (in mid-1963, cancer research and foreign aid topped the list; poverty would not become an issue for another year). Why not move at a more leisurely pace? Why not, perhaps, rethink whether the nation really wanted to do this at all? "It is probably too much to say, as some of NASA's more panicky partisans have, that the whole U.S. space program now stands in mortal peril," began an article in *Fortune* that fall of 1963. "Nevertheless, NASA and the space program have reached a critical stage. . . ."

It was in this context that an engineer named George Mueller moved into Holmes's office in September to take over the Office of Manned Space Flight. John Disher, one of the few who had been in O.M.S.F. through both the Silverstein and Holmes regimes, was still there, working in the Advanced Projects Section. A few days after Mueller arrived, he went to Disher's office.

Mueller—he pronounced it "Miller"—had an assignment for Disher. It was to be fast turnaround, utterly discreet. Disher and another old hand at headquarters, Del Tischler, were to conduct a private assessment of where the Apollo schedule stood. They were not to pay any attention to the official schedule. Mueller wanted realism. Disher and Tischler were to make their best estimate, based on everything they had learned from Mercury and Apollo so far, on how long things would really take.

For two intense weeks, Disher and Tischler worked on their commission. On September 28, they went into Mueller's office to brief him on their findings. For a program already under siege, those findings were devastating. After running through an elaborate analysis of tasks and schedules, costs and probabilities, they came to the last page of their briefing, the one headed "Conclusions and Recommendations." The first conclusion on their list was that "lunar landing cannot likely be attained within the decade with acceptable risk." The second one was that, in their best estimate, the first attempt to land men on the moon would probably take place in late 1971. The two engineers added their own personal guess that the odds of getting to the moon before 1970 were about one in ten.

They finished the briefing. Then "George took Del and me hand in hand and we went over to Bob Seamans's office," Disher remembered. "He wanted to give the report to Seamans, obviously just to say, 'Well here's the status of things as I'm taking over.' " Disher and Tischler went through their presentation again, and once again came to that stark last page of conclusions.

Seamans listened without comment. When it was over, he told Mueller he wanted to speak with him privately and Disher and Tischler left the room. In a few minutes, Disher recalled, Mueller came back alone and told them that Seamans wanted them to destroy the material. Seamans didn't remember asking anyone to destroy material, but he did remember thinking the results were unsatisfactory and telling Mueller to go back to the drawing board. Trying to hold NASA together in a time of troubles, Seamans needed something a lot more imaginative than a study saying that NASA couldn't get to the moon before 1971.

George Mueller, on the other hand, was content. Disher and Tischler had just given him the two-by-four he needed. He was about to ram through the most radical decision in the whole of the Apollo Program, a decision that was instrumental in getting the United States to the moon by the end of the decade.

CHAPTER

11

"It sounded reckless"

Engineering is rightly regarded as one of the most pragmatic of professions, but even engineers have their creeds and dogmas. Thus when Joe Shea once was asked to adjudicate a Marshall-Houston dispute over the correct way to do a certain type of soldering, he declined. "That's not technology," he said of the warring views. "That's theology." Now George Mueller was going to propose a plan for rescuing the Apollo schedule that would horrify Marshall and Houston in equal measure, for it would violate a taboo.

1

Looking back on the program as a whole, Shea would see Apollo as the story of three cultures. The differences among the groups he had in mind went far beyond just "points of view" or "schools of thought." There were the Germans from Peenemünde, the old N.A.C.A. hands from Langley and Lewis, and the systems engineers from the I.C.B.M. world, each band with its own tribal history and folkways and prejudices and dialect.

In the early days of the space program, the Germans and the N.A.C.A. hands lived together peacefully if suspiciously. True, the Germans had seniority and the N.A.C.A. people struggled to achieve parity, but they had in common not only their love of engineering and of things that flew, but also a tradition of craftsmanship. The Germans and the N.A.C.A. hands were essentially builders of fine machines.

Mueller, like Holmes and Shea, was part of the third tribe, the invaders from the world of systems engineering, socialized by the experience of building missile systems and early-warning radar systems for the Department of Defense. They, too, were craftsmen, but of a different sort. Their craft was not building hardware, but machining the managerial equipment for huge, highly technical, highly complex tasks.

Another difference that separated the systems engineers from the Germans and the N.A.C.A. hands grew out of the disparate circumstances under which their crafts had evolved. The Germans had grown up alongside the rocketry they were inventing. There was no such thing as rocket technology when the young Wernher von Braun teamed up with the young Arthur Rudolph and Walter Reidel and Bernhard Tessman. They had only embryonic ideas and grand ambitions. They had to inch their way forward. Failure was their tool for making progress. Furthermore, the Germans began at a time when the equipment for diagnosing failures was still primitive, and telemetry of data was almost unknown. When rockets failed, they did so spectacularly, in fireballs, obliterating recording equipment and sensors in their explosions. Often, the Germans found, you couldn't even be sure what had failed, let alone why.

One of their reactions to this experience was their fabled conservatism, giving each part and each system within the rocket a generous margin of extra strength or capacity. Another reaction was to compartmentalize their testing and development programs into the smallest possible packages, so that each new step involved testing just one new item. Each step was repeated many times before going on to the next. Their procedures were methodical, Germanic—and successful.

The N.A.C.A. hands grew up in the flight-test business. Unlike the Germans, N.A.C.A. engineers had the advantage of working with an already reliable machine, the airplane. Their chief shaping influence was the fact that human beings rode in their machines. Failures, however rare in statistical terms, were tallied in human lives, and so the N.A.C.A. engineers too learned to proceed on a painfully slow incremental schedule. They wouldn't think of taxiing the prototype of an untried design out to the field and taking off. First the pilot would run up the motor, and the engineers would gather test data to be taken back to the office for analysis; then, another day, the pilot would taxi the aircraft, and the engineers would look carefully at all those data; then, on still another day, he would reach near-takeoff speeds; until finally, cautiously, on a day with perfect weather conditions, the plane would lift into the air on its first flight. And this would be true even with a plane that had performed superbly in the wind tunnel, not all that different from other designs that had been flying safely for years. Out of their different experiences, the N.A.C.A. hands and the Germans came in the end to similar points of view on flight testing.

The systems engineers looked upon this philosophical alliance with a certain disdain. "The systems guys were given a very different chore," as Jim Elms once put it. "Their chore was, if we have a billion dollars and five years, what's the best way to get the most pounds of atomic bombs over some place like Moscow? They had a choice like, 'Let's see, we can build a thousand missiles with a reliability of seventy-five percent. That means we can get seven hundred and fifty of those missiles to actually go there for that billion dollars. Or if we want to make them perfect, we might be able to get twenty of them over there.' So their goal was to figure out where they wanted to be on that reliability scale."

When they came to the Apollo Program, the systems engineers accepted that they had one parameter—the safety of the crew—that was sacrosanct. But the mentality they brought to the program led them to approach the problem of safety in different ways than the Germans and the N.A.C.A. hands. Cast out all taboos, they said, and take a fresh, cold-eyed, analytic look at the problem, following the implications wherever the data lead. The results led them to fly in the face of everything that the Germans and the N.A.C.A. hands believed about flight testing.

The idea came out of experience on the Titan II and Minuteman programs. Shea had already broached it to Holmes, but Holmes had said no—they could never sell it to Huntsville and Houston, he thought. George Mueller, who like Shea had come out of the Air Force ballistic missile program, didn't worry about whether he could sell it or not. Huntsville and Houston would learn to like it or else. For this is what his analysis had led him to:

Point number one: They were never going to be able to fly enough Saturn Vs to be confident of the vehicle because of the number of times it had worked. So what if they tested the Saturn V six times instead of four? Or eight times instead of six? Statistically, the extra successes (assuming they were successes) would be meaningless. All they would have done is to use up two pieces of hardware that could have been used for real missions. The only way they were going to man-rate a Saturn V was by having confidence in the engineering and the ground testing that had gone into it. There was no other choice.

Point number two: This business of testing one stage of the rocket at a time, as Marshall always did, didn't accomplish what people thought it accomplished. You didn't really build brick by brick any more; all that did was to waste time. Whenever you added a new stage, the ground support equipment was different, the checkout procedures were different, the countdown was different, the hardware was different. You had to relearn everything anyhow.

Point number three: The step-by-step approach locked you into an assumption that you're going to fail. For example, Marshall had scheduled four test flights of the first stage of the Saturn I. The first stage had worked perfectly on the first flight, and as Rocco Petrone remembered, even they had asked themselves, "What the hell are we going to do with the next three?" Mueller wanted a testing program that put NASA in a position to "take advantage of success."

Therefore, what Mueller proposed to do was to scrap the plans for incremental testing of the stages and to condense drastically the testing schedule for the spacecraft. This was called "all-up" testing—"up" meaning that a stage is a flight-ready piece of hardware, "all-up" meaning that everything on the Saturn V would be a real, functioning stage the very first time they launched it.

George Mueller, the man who proposed this and then saw it through, is one of the most elusive figures in the Apollo story. For the most part, Apollo was led by men of great ability, often colorful and occasionally eccentric; but always they were men who plainly bled when they were pricked. With Mueller, it was harder to be sure.

In background and intellect, Mueller was like Shea—or the other way around, since Mueller was Shea's senior. Mueller was a newcomer to the manned space program, as Shea had been. Like Shea, he had worked in the I.C.B.M. world. Like Shea, he was a systems engineer. Brilliant like Shea. Intellectually arrogant like Shea. They had even worked together before, first at Bell Labs and then at T.R.W., where Mueller had been Shea's boss. But whereas Shea had an open complement of human vulnerabilities, Mueller kept his well hidden. Some of his colleagues weren't certain he had any.

No one disputed Mueller's technical capabilities. John Disher, who remained in O.M.S.F. with Mueller for many years, put him a notch above even Silverstein, Disher's other hero, as the "only bona-fide genius I've ever worked with." Sam Phillips, who ran the Apollo Program Office under Mueller, could never understand why the histories of Apollo paid so little attention to Mueller—he hadn't ever "gotten the credit that he really deserves for the success of Apollo."* It wasn't just the big things, like the all-up decision, Phillips emphasized. Mueller could work the Hill one day, then go off and supervise a highly technical task force to deal with problems in the flight operations software. People like George Low and Chris Kraft, who would clash repeatedly with Mueller, readily conceded his strengths. "I've never dealt with a more capable man, in terms of his technical ability," said Chris Kraft, who dealt with everybody. Nor was Mueller an office tyrant. On the contrary, he was considerate with his own staff—getting along with him was "a piece of cake" for John Disher, who found Mueller to be always pleasant, a pleasure to work for. But when Mueller was dealing with senior people from the centers, it was a different story.

It wasn't that he would try to humiliate people, as another senior manager in NASA had been known to do. Rather, Mueller appeared to be almost inhumanly rational, unmoved by any argument that was not scrupulously logical and grounded in data, indifferent to the human content that was so much a part of even the most technically abstruse struggles during Apollo. "If you saw a meeting and were not tuned into the content," said one man who knew him well, "he would seem to be reasonably charming and affable. But there's a coldness there." Disher made the same point: Mueller was "always the epitome of politeness, but you know down deep he's just as hard as steel."

You had to remain on your toes, no matter where you were in the NASA hinterlands. "He never hesitates to pick up a phone and call Joe Schmo and challenge him in the greatest technical detail on a technical problem, and ninety-nine times out of a hundred he'll wrestle the guy to the ground in his specialty," remarked one of his colleagues. "Once in a while one of the supervisors who has been bypassed gets real disturbed. He might bring it up to George and George will say, 'I'm so sorry.' He smiles and says, 'So sorry about that.' But completely unperturbed, you know."

Mueller's way of putting people down wasn't obvious, another

* Others made the same point about Mueller but usually added Sam Phillips's name as the other most underappreciated person in the success of Apollo, in terms of public credit.

engineer observed. He didn't shout at people, didn't call them names, but he was a master of the intellectual put-down. It's like the joke about the knife fight, the engineer said, where the guy says "Hah! You missed!" and then discovers he can't shake his head because it's been severed at the neck. Mueller cut you like that.

Let there be provisos and caveats to that description, for Mueller was a complex man. When he was once asked what the highlight of the Apollo Program had been for him, his answer had nothing to do with honors or awards or the moment of his own most personal triumph, the successful flight of the first Saturn V. The most memorable moment of all his years on Apollo, George Mueller said, had been the surprise birthday party that his staff gave him after Apollo 11. Or there would be the night after the Apollo fire, when the supposedly implacable Mueller would twice put aside technical arguments in order to help a troubled Joe Shea. The people who saw Mueller at closest hand seem to have admired him the most, which is one of the ultimate compliments. But at least this much may be said without qualification: He was the undisputed boss of manned space flight from the day he walked into the office in 1963 until the day he left six years later.

2

Mueller didn't so much sell the all-up idea as dictate it. He had arrived at NASA at the beginning of September. At the end of September, he got the Disher-Tischler briefing. Two weeks later, he canceled four manned flights of the Saturn I. On the 29th of October, at a Manned Space Flight Management Council meeting, he told a dumbfounded audience from Houston and Huntsville about the all-up concept. Lest there be any doubt, he followed that up two days later with a priority teletype message spelling out the new flight schedules. He asked for responses by November 11, and announced his intention to post "an official schedule reflecting the philosophy outlined here by November 25, 1963." It was a way of putting things that didn't leave a lot of room for argument.

The people at Marshall were incredulous when they first got wind of the all-up decision. John Disher was the first to tell Willy Mrazek, the head of von Braun's Structures and Propulsion Laboratory: Instead of testing each of the stages, they were going to fly the first stage, using the

F-1 engine (which was still experiencing combustion instability), the second stage (which used hydrogen as a fuel, involving new and tricky technology), and the third stage, the S-IVB, all together, on their first flight. Mrazek snorted. It was impossible. He asked sarcastically, "And how many launches are you going to have before you put the man on?" Disher said, "One successful." Mrazek said, "You're out of your mind."

On November 16, 1963, President Kennedy flew on Air Force One to the Launch Operations Center at Cape Canaveral, where he inspected the construction sites for the V.A.B. and Pad 39, saw mockups of the hardware to come, and was briefed by senior NASA officials on their progress toward the goal he had set two and a half years earlier. On the following Wednesday, the Senate cut $612 million from his budget request for NASA, another sign of public disaffection with the space program which, *The New York Times* observed, "raises a serious question of whether the Administration can count on the budgetary support necessary to achieve a lunar landing by the 1969 deadline." Kennedy responded the next day, warmly commending the space program in a speech at San Antonio, Texas, where he had begun a three-day political fence-mending trip. That same day, he flew to Houston and attended a testimonial dinner for Albert Thomas, the man who did most to bring M.S.C. to Houston. Kennedy spent the night in Fort Worth. The next morning, Friday, November 22, he flew to Dallas. Six days later, less than two weeks after Kurt Debus had taken President Kennedy on a helicopter tour of the Launch Operations Center, it was renamed the John F. Kennedy Space Center in his memory. Work on the Apollo Program continued without a perceptible break.

Despite Marshall's consternation, Mueller was immovable. Arthur Rudolph found that out when Bob Seamans and Mueller visited Marshall in early December. Rudolph showed Seamans a model of the Saturn V next to a model of the Minuteman missile, which the systems engineers had successfully tested all-up. Rudolph carefully walked Seamans through the enormous differences between a small solid-fuel rocket like the Minuteman and the complex, state-of-the-art Saturn V. Use all-up for the Saturn V? Rudolph looked at Seamans: "Now really, Bob!" he said, pleading for understanding. "I see what you mean, Arthur," Seamans said. Rudolph, thinking that he was making progress, got Mueller over beside the models and went through the same pitch. The complexities of the Saturn. The size. The unknowns. He finished with a flourish and looked at Mueller hopefully. "So what?" said Mueller.

Marshall never really did "agree." Von Braun and Rees granted Mueller's central point: Without all-up, they couldn't possibly get to the moon by the end of the decade. Marshall could not dispute, except emotionally, Mueller's contention that the Saturn V would be as safe for manned flight (as far as anyone could tell) under all-up as under the more extended—but still inevitably limited—testing process. Left without a good objection, von Braun went on record supporting all-up testing, overriding the vocal objections of many of his senior staff. One of them, Dieter Grau, later wrote, "I'm not aware that a consensus was obtained on this subject in favor of the all-up concept." What happened, he said, was that "just as Dr. Mueller could not guarantee that this concept would succeed, the opponents could not guarantee that it would fail." Von Braun, in Grau's view, decided that Marshall "should share the risk with him."

In later years, von Braun would write good-humoredly about how "George Mueller visited Marshall and casually introduced us to his philosophy of 'all-up testing.' . . . It sounded reckless," von Braun said, but "Mueller's reasoning was impeccable." Moreover, the leader of the German rocket team continued, "In retrospect it is clear that without all-up testing the first manned lunar landing could not have taken place as early as 1969."

3

Mueller also moved quickly to accomplish what had eluded Silverstein and Holmes before him: Webb finally gave in to a reorganization of NASA that would have the center directors reporting directly to Mueller on matters affecting manned space flight. And it was not to be merely a paper triumph. For better or worse, Mueller was the man from headquarters who finally pacified the independent duchies that had come together to make up NASA. Brainerd Holmes had made a few inroads, but when Mueller took over, the centers were just about as feisty and independent as they had ever been. Mueller brought them to heel, even Marshall, the proudest and most independent of them all. Under Mueller, manned space flight became much like other federal bureaucracies, run from the headquarters in Washington with strict management controls or, depending on your point of view, stifling interference.

Of more immediate importance in those early months of Mueller's reign, however, were two key management moves. Walt Williams, deputy director at Houston and one of the chief reasons for the success of Mercury, was brought to Washington, where he was effectively removed from power. Joe Shea took over the Apollo Spacecraft Program Office (ASPO), the office at Houston responsible for producing the spacecraft. The two events were not unrelated.

The preceding spring, a few months before Brainerd Holmes lost his bureaucratic duel with Jim Webb, Charlie Frick, head of ASPO in Houston, departed. The spacecraft contract with North American was falling badly behind and was beset with other troubles that were only beginning to surface. On top of that, there was a big new contract with Grumman for the lunar module that had to be managed. The whole effort had acquired a slapdash quality that was worrisome to Gilruth and headquarters alike. In March, ASPO was reorganized in one more of the attempts to bring Apollo under control. Frick left in April, half of his own accord, half pushed, physically exhausted. In May, when ASPO was reorganized still again, Bob Piland, one of the early members of the Space Task Group, agreed to serve as acting manager.

When it had become obvious that Frick wasn't going to be around much longer, Jim Elms had come up with an idea for Bob Gilruth to consider. Elms, Gilruth's deputy at Houston, had known Shea since his Titan days, had worked with him at headquarters, and had been impressed with the way that Shea had taken hold of the mode decision. Elms suggested to Gilruth: Why not bring Joe Shea down to Houston to run ASPO? It sounded good to Gilruth. He broached the idea to Brainerd Holmes, who quickly squelched it. It turned out that Shea himself had already come to Holmes, asking for the ASPO job, and had been refused. Didn't Shea realize that it would be a demotion? That in the ASPO job he'd be two or three layers farther down in the organization than he was in his present position as associate director of O.M.S.F.? Holmes wouldn't hear of it.

By fall, Holmes was gone. Bob Piland, still serving as acting manager of ASPO, had already told Gilruth that he didn't want to continue in the job. When Mueller was making his first tour of the centers, Gilruth asked Mueller about sending Shea to Houston. Displeased with what he was hearing about the progress of the spacecraft, Mueller phoned Shea from

Houston and asked whether he wanted it. Sure, Shea said. There were, however, certain things that Shea wanted changed. In a memorandum dated September 17, 1963, he wrote to Mueller, "I think you know that I would be enthusiastic about the assignment" if certain conditions could be met. And he proceeded to set out a mighty set of ifs.

There was the problem with Walt Williams, deputy director for Operations at Houston, who, according to a senior person close to the situation, "wanted so badly to be the head of this organization [the Manned Spacecraft Center] that it overcame his better judgment." There were reasons for Williams to think he deserved the job; to many of the people who saw Project Mercury in its early days, Walt Williams was the man who saved it from disaster. But Williams was not an easy man to get along with; he made enemies in a way that the easygoing, more diplomatic Gilruth did not, and key people at headquarters couldn't see Williams in the center director's chair. To Joe Shea, the point of all this was that he wasn't going to go to Houston with Williams there. "Joe didn't want to be in the middle of it," George Low said. Thus Shea wrote to Mueller in the confidential memorandum that he would accept the ASPO job in Houston only if Williams were shifted "to a position in Washington in which he can be effective but not dominant." Shea added that Elms's authority should be augmented so that Elms could pull together "all elements of Houston to support the program."

There were other conditions as well, aimed at centralizing the authority of ASPO at the factories, at Houston, at the Cape—all in all, a pretty cocky set of conditions. It didn't faze Mueller, however. His mind and Shea's were working in synchrony—the things Shea wanted done were things that Mueller wanted done anyway. And he also was aware, as Shea would admit later, that Shea's memo wasn't really much of an ultimatum. Shea would have taken the job no matter what, even if Mueller had done none of the things Shea had asked. He wanted it that much.

On October 8, NASA announced that Joseph F. Shea was being reassigned from the Office of Manned Space Flight to the Manned Spacecraft Center in Houston, where he would manage the Apollo Spacecraft Program Office. Two weeks later, George Mueller had another personnel announcement to make: Walter C. Williams was being reassigned to Washington and promoted to the position of deputy associate administrator for Manned Space Flight in O.M.S.F.

Two more important changes followed in close order. On the last day of 1963, Sam Phillips, a brigadier general in the Air Force, became director of the Apollo Program Office within O.M.S.F., working under

Mueller. Six weeks later, George Low left O.M.S.F. for Houston, where he took over the deputy director's position vacated by Elms, who wanted to return to private industry.

These changes marked an end to the two years of personnel ins and outs and revised organization charts that had beset Washington and Houston. The dramatis personae for the Apollo Program from then through the spring of 1967 was now set: in Washington, Mueller, running all of Manned Space Flight with a disciplined hand; also Phillips, running a centralized Apollo Program Office under Mueller. In Houston, Low, deputy to Gilruth; and Shea, ramrod for the spacecraft.

12

"Hey, it isn't that complicated"

"I went to Houston in November '63," Shea recalled, "and we'd made the decision to go to the moon L.O.R. in November '62. And there was still no design for a command and service module compatible with lunar-orbit rendezvous. *No* design. I found it hard to understand." And so Shea began to see firsthand what came to be the closest thing to an out-and-out scandal in the Apollo Program, the origins and first years of North American's contract to build the spacecraft.

1

In awarding large contracts, the federal government uses a standard procedure. First it publishes a Request for Proposals in which it describes the work to be done and names a time and date by which proposals for this work must be received. Companies that wish to bid submit proposals by that date. The government evaluates the proposals. The contract is awarded to the proposal scoring highest on a prearranged formula that weighs the technical approach, the personnel the bidder will use, the bidder's corporate expertise in this area, and so on.

At the time the R.F.P. for the Apollo command and service module (C.S.M.) was distributed, in late July 1961, North American Aviation was one of the leading aerospace companies in the United States. It boasted an especially strong record on military and experimental aircraft: North American had designed and produced both the F-100 and F-86, and more recently had built the experimental X-15, which in its initial flights was proving to be a surpassingly fine aircraft.

The man who had managed the X-15 and who ran North American's other space-related projects was Harrison Storms, known universally as Stormy—a nickname his friends thought to be descriptive of the man. Storms was not enthusiastic about bidding for the C.S.M. North American had a grab bag of subcontracts on the Mercury and Gemini programs and was front-runner to win the contract for the second stage of the Saturn—a contract which was in fact awarded to North American in early September. It is an understood thing in the government contracting world that the goodies are spread around, and it seemed to Storms that their chances of winning the C.S.M. contract as well were "pretty much impossible." But he figured that "since we were training these people to be a Space Division, and since we needed the exercise and we could always go subcontractor, then it was a very worthwhile exercise for us to go through the motions of bidding," Storms said later. "So we went ahead and bid."

Five companies submitted proposals: Convair, General Dynamics, Martin, McDonnell, and North American. On October 9, NASA's Source Evaluation Board, heavily loaded with Space Task Group people, met at the Chamberlain Hotel at Old Point Comfort, near Langley.

Tom Markley was secretary for the subcommittee evaluating "administrative capacity." His subcommittee rated Convair first, Martin second, and North American last. A few days after his subcommittee had finished its work, he got a call from Bob Piland. "The scores on your Management Subcommittee look a little skewed," Piland told him. "We'd like to have you evaluate some other criteria, in addition to what you guys evaluated already."

This seemed a little odd to Markley. Why hadn't they thought of these other criteria beforehand? Well, he was told, the Source Evaluation Board thought it ought to give a little more weight to the bidder's experience in producing experimental aircraft. The other subcommittees got similar instructions: Rescore the proposals, giving more weight to experience with experimental aircraft. And so everybody did.

The result of the rescoring helped North American considerably, since the leading example of an experimental aircraft was the X-15. North American was still ranked fifth on the "business" criteria. It was ranked fifth on "technical approach." But in "technical qualification" it had risen to first by a thin margin over Martin. Overall, the summary score put Martin first, with Convair and North American tied for second. The Source Evaluation Board was unequivocal when it turned in its recommendation: "The Martin Company is considered the outstanding source for the Apollo prime contractor," it wrote in its report to Webb.

The rescoring wasn't the only funny thing that had been happening with the C.S.M. contract. One Space Task Group engineer was assigned to a review panel along with astronaut Al Shepard. During the panel meetings, Shepard was openly contemptuous of the rigmarole they were being asked to go through. "This is all a waste of time," he kept saying. "It doesn't make any difference what the score is. North American is going to win." Another man who had been with the Space Task Group for about two weeks remembered walking to lunch during the oral presentations, before North American had even given its pitch yet, and hearing another astronaut, Gus Grissom, say to his companions, "By God, I'm going to do everything I can do to make sure North American gets this contract." The man was shocked. What kind of outfit am I in, anyway? he asked himself.

Others recall the ways in which, in those years, North American seemed to act as if it had the government in its pocket. One senior NASA official remembered being present in Brainerd Holmes's office when Holmes told a senior executive from North American that he would not be permitted to bid on the lunar module because of North American's backlog of work. The executive is supposed to have said, explicitly, "If you don't let me bid, I'll have your job"—which seemed to the observer to be a strange thing for a corporation executive to think he could say to an official of the government.

Then there were curious coincidences involving Bobby Baker, still operating under the unofficial patronage of Lyndon Johnson and retained as North American's Washington lobbyist, and the powerful senator Robert Kerr of Oklahoma. Baker had begun a company (Serv-U) with the help of a bank closely connected to Kerr. Serv-U then got a big contract to place vending machines in the North American plants. Shortly after being awarded the C.S.M. contract, North American moved a large part of the Apollo work to a plant in Tulsa. No one ever proved that anything shady had gone on, but the coincidences were noted and the rumors persisted.

After finishing his work with the subcommittee, Tom Markley was offered the job of resident manager at the C.S.M. contractor's plant (he would be resident for ASPO, hence RASPO). Markley agreed and, since he knew that the board had recommended the Martin Company, Markley went to Baltimore, where Martin had its main plant, to start looking for a house.

In a few days, Markley got another call from Piland, who now asked Tom whether he would be willing to take the RASPO job if the place "turned out to be a warmer climate." Markley was perplexed.

"There's no place in Florida they're going to build this," he said.

"What about Southern California?" Piland asked.

"Bob, General Dynamics doesn't have a chance," Markley said. "And there's no one else in California who could win this thing."

"Well, I just wanted to know your opinion," said Piland. "What if it *is* California?"

Markley finally got the drift. He paused, and then he told Piland that in that case he didn't think he wanted to go. He just wouldn't be able to start off with the right attitude.

Webb had set aside the recommendation of the Source Evaluation Board and given the C.S.M. contract to North American. But he had done so not at the behest of North American's powerful political allies; rather, he had followed the advice of Bob Gilruth, George Low, and Walt Williams. Piecing together various accounts, it seems to have happened more or less like this:

After the Evaluation Board had finished briefing Webb on its assessment, Webb took Seamans and Dryden into his office to talk. Probably Brainerd Holmes was called in too. Then after a while Webb called in Gilruth, Williams, and Low, and asked them whether there were any factors other than those presented by the Evaluation Board that he should take into account. Gilruth and Williams promptly said that they felt North American was more qualified than Martin, and Low indicated that he agreed.

And therein lies the complication that keeps this from being a simple story of favoritism and rumors of shady dealings. A lot of people within NASA wanted North American to win because they thought North

American would do the best job. George Low said later that when they saw the Source Evaluation Board's report recommending Martin, "we didn't believe it." He and the others told Webb and Seamans "to think of this spacecraft as an outgrowth of more conventional flying machines." They worried that Martin could not do what North American had already proved it could do with planes like the F-86, the F-100, the B-70, and the X-15. The X-15, the closest analogue to the command and service module in an aircraft, loomed especially large. "It flew like a bird; it was more or less within cost, within plan, on time," Caldwell Johnson would say later. "They weren't just talking about it, they had done it and there it was." That sort of thing meant a lot to the old N.A.C.A. hands who, above all, were persuaded by evidence that someone could make a fine flying machine.

Webb listened carefully to all these arguments. By this time it was late Friday evening. He said he wanted to sleep on it over the weekend. Monday morning, he told them his decision. North American won the contract, and with it trials and tribulations that would not fully end until 1968.

The tensions between NASA and North American began the night of the award, when someone at North American made up some baseball caps with "NA$A" embroidered on the front. That infuriated a number of NASA engineers: The moon program was not supposed to be business as usual. There was also the problem of staffing: North American was already strained to the limit with its new contract for the Saturn S-II stage. Simultaneously, North American was involved in trying to develop the paraglider that, it was then thought, would be used for the Gemini capsule's landing system. So during the early months of the award, North American's Apollo spacecraft team was often understaffed and studded with inexperienced people.

Time would not soften Joe Shea's judgment about the situation he found when he went to ASPO in Houston two years after North American began work. "I do not have a high opinion of North American and their motives in the early days," he said many years later. "I think they were more interested in the financial aspects of the program than in the technical content of the program. I think Storms was a very bad general manager, I think Atwood [North American's president] had dollar signs in his eyes. Their first program manager was a first-class jerk. . . . There were spots of good guys, but it was just an ineffective organization. They had no discipline, no concept of change control. If anything, they were interested in pumping the program up rather than in what the program really was.'' Shea was not alone in his opinion. During the first years of the C.S.M. contract, North American acquired a very bad reputation around NASA.

It wasn't all North American's fault. The Apollo spacecraft was a completely new machine and more complex than any aircraft ever built. The Mercury capsule had been much simpler—yet McDonnell Aircraft Company had run into the same kinds of problems that North American was stumbling over now. For its first year, many at NASA had thought McDonnell was a money-grubbing, incompetent contractor that was shipping half-finished, misbegotten capsules to the Cape where they had to be rebuilt from scratch. By the time McDonnell was making Gemini capsules, it had gotten so good at its job that the Geminis routinely arrived at the Cape in launch-ready condition, and McDonnell had in NASA's eyes become a paragon of corporate virtue.

Nor should it be forgotten that many of the NASA people who were managing the hardware contracts had come out of Langley. When Sam Phillips came to the Apollo Program at the end of 1963, at the same time Shea went to Houston, he was startled to find such a complete contrast between ''a tremendous technical capability [and] a very, very thin capability to manage a large-scale program activity,'' and this was nowhere truer than at Houston. No one had learned how to manage contracts at Langley, because there weren't any. The men of the old Space Task Group were engineers in bone and spirit, and suddenly they had become managers by decree. Some of them became good managers, but it took them a while. Many never were comfortable in that role. Was there ever a less plausible manager than Caldwell Johnson—without a shred of diplomacy, a loner, a natural designer who by his own admission lost interest in his designs once they were ''pretty well patted into shape''? Yet when Shea arrived to take over ASPO, there was Johnson, managing NASA's C.S.M. contract with North American, poring over work schedules and time charges and management systems. (Shea shifted him back to a design job forthwith.)

Lee Atwood thought that much of the criticism of North American was really part of an inevitable charade to which contractors must reconcile themselves. ''The contractor is always the underdog when something comes up,'' he said later. ''He has to absorb the criticism.'' NASA's technical people love to think they invent things: ''That's kind of a handicap for the contractor's technical people, because they are being second-guessed by a large gathering of technical people in NASA. If the work goes wrong, the contractor is wrong. If it goes right, the NASA technical man gets his medal or whatever.''

But however the blame is parceled out, the fact remains that the first years of North American's effort to build the command and service module were rife with wasted effort, lack of coordination, and intermittent warfare between contractor and customer.

2

To get control of the program, Shea called in his new assistant, Tom Markley, and said to him that somehow they had to keep this thing on track. "Go out and fix me a management tool," he told Markley.

Markley sent his men out to look at the major aerospace companies and their sophisticated management control centers. Then Markley himself went to visit what everyone said was the best of them all, at the Martin Company. It was beautiful, Markley thought, sort of a management war room. He figured he would set something like that up for Joe—which would be a big relief to Markley because, until he did, Shea was expecting him to know all the answers. Whenever Markley didn't know something, he had to live with Joe riding the living hell out of him until he found out—"so it was best if I got a good system going."

Shea turned the Martin model down flat. Markley submitted a revised plan, scaled down a little. Shea turned that down flat too. This is ridiculous, Markley thought, so finally he walked in and asked Shea what kind of management system he wanted. Shea told him. "What he wanted," Markley said, "was to get a notebook every Thursday night, and in it he wanted the entire program structured, meaning all the interfaces with Marshall, everything going on with the command module, everything going on with the service module, everything going on with the SLA, everything going on with guidance and navigation, everything going on with the LEM, everything going on with the launch pad, the ground support equipment, with General Electric, everything Boeing was doing, everything everyone was doing.''* All this was to be in one

---

* LEM refers to the lunar excursion module, the lunar lander. The word "excursion" was insisted upon by Owen Maynard and others at Langley who wanted to emphasize its limitations. Eventually headquarters shortened the official title to lunar module, and the official abbreviation was shortened to LM. Everyone continued to call it the "lem," so we have compromised with purity by spelling it "LEM" throughout the book. The "SLA" was the spacecraft lunar adapter, the slanted section just below the C.S.M. that contained the lunar module during the launch. SLA is pronounced "slaw," as in cole.

notebook, and should include not only what had been done in the last seven days but also comparisons against the programmed schedule and costs.

And so Markley went away to produce the first edition of the loose-leaf notebook that he and his staff would prepare for Shea every week for the next 165 weeks. Each notebook ran to more than a hundred pages and was on Shea's desk by close of business Thursday. Shea would get up at 4 A.M. Friday and start to annotate it, usually with technical comments and instructions, occasionally with remarks such as "You've gotta be kidding." He would work on the notebook intermittently from Friday morning through the weekend and return it to Markley on Monday morning. By Monday afternoon, everyone in the ASPO network would have gotten his response from Shea.

With the annotated notebook in his hands, Markley became the action officer. "I'd take this notebook, and then I'd fan it back out very quickly and say, 'Hey, we've got these questions and I want the answers in the notebook by this Thursday.' We'd get the answers back and then we'd take Joe's marked-up notebook and start off in the front, that would be the first part of the notebook, anything he had marks on. We'd have the answers there. And then we'd have the brand-new section with the newest status too." The scrawled notations to and fro were a way of keeping up "a running communication," in Shea's words, without having to type up long memos. "Even if I didn't see [the project officers] directly, I'd at least been in touch with them, and they'd been back in touch with me."

"I'd never seen it done before and I haven't seen it done since," recalled the project manager for the LEM. "Shea came into the organization and he said, 'I want only those things that you want me to read and that you want some kind of answer on. Just don't tell me things are going along great, but if you want some decision, do it through your weekly activities report.' So we'd send them in on Thursday and on Monday morning you'd have your copy back with his comments on it." He could never figure out where Shea got the time.

By now, in 1964, the program had exploded in size. Faget's engineering division alone employed 1,400 men, serving (among other duties) as ASPO's technical resource. At North American's Downey, California, plant, 4,000 engineers were working on the command and service module. Grumman was going full speed on the lunar module. Other contractors were making the space suits, the guidance and navigation systems, and a dozen other major systems, not to mention the hundreds of smaller contracts and subcontracts.* All of these systems were being designed, developed, and assembled simultaneously. The most minute change in one system was likely to require a cascade of changes in others. The complexity of the program and the coordination required to keep all the pieces growing in harmony increased exponentially.

The coordinating power for the spacecraft was in Shea's hands, and he exercised it without a blink. As huge as the program was becoming, with its growing paraphernalia of collective decision making, any decision about the C.S.M. or the lunar module ultimately came to Shea, and Shea didn't worry about getting a consensus. In the fall of 1964, for example, he established a Configuration Control Board (known as "the Change Board") that would systematize and limit the changes being made to the spacecraft. But when it came to the final decision, the one man whose opinion really counted was Joe Shea. An engineer who had to deal with Shea remembered what it was like. If you had a change you wanted to propose in the spacecraft, and if it got by the Configuration Control Board, the last step in the process was to go into Shea's office on the seventh floor over in Building 2 at M.S.C. You'd walk in, and there Joe would be, by himself, behind his desk, and you had to explain to him what it was you wanted and why it was absolutely essential that this change be made in the spacecraft. And then Joe would begin to grill you. There was no point in trying to sidestep or fudge an answer. It made no difference if Shea wasn't a specialist in your area. "If you understand it, you can make me understand it," Shea kept saying, and you had better understand it absolutely cold or Shea would dissect you.

Also, you had to keep ahead of him, which could be difficult. "He'd stop you in the middle of a presentation and say, 'You're going to have to include that you need four batteries in the circuit,' " one engineer recalled. "And you'd say, 'I'm going to get around to that.' And he'd say, 'Well, tell me right now why the hell we need four batteries. And goddamn it, if you can't tell me why we need four, then you ought to come back and see me next week. You're wasting my time.' "

At the end of the meeting, Shea would say yes or no. And that was it. It took people a while to get used to that, and sometimes during his first

---

* This was just Joe Shea's shop, having nothing to do with the equally large and complex organization being run out of Marshall to build the Saturn V, and a third one at the Cape to build the V.A.B. and Complex 39.

months at Houston, an item that Shea had decided against would be brought up at another meeting. "I want to tell you that an adverse decision is not a decision delayed," Shea icily informed an engineer one day, and after that the practice stopped. "I never ran the Change Board as a democratic process," Shea admitted.

"The better is the enemy of the good," Shea told them again and again.* The biggest problem with a new product in its developmental phase, Shea thought, was that a good engineer could always think of ways to make it better. This was fine, except that they couldn't keep changing the spacecraft forever. Sooner or later they had to lock it into one configuration, so that they could make that configuration work. Keep the changes down.

Keep it simple—that was the other half of the Shea doctrine. To Shea, it seemed as if everyone were saying, "Apollo is the most complex job in the world; therefore, every part has to be as complicated as possible." And that was exactly backwards. What they should be trying to do was figure out how to do things as simply as possible, with as few ways as possible to go wrong.

Fuel gauges, for example, in the tanks in the reaction and control system on the spacecraft. Okay, Shea agreed, it's tough to measure how much fuel you've got in a tank when you're in zero gravity. The usual devices don't work. And furthermore, it was vital that they keep enough fuel in those tanks right up to the end of the flight, because the R.C.S. was what put the spacecraft in the right attitude to enter the atmosphere. But the people who wrote the specifications for the fuel gauges had reasoned that, because it was so important, they needed a highly accurate fuel gauge system. The way they would do it was to build a Geiger-counter arrangement that would infer the bulk of fuel left in the tank by the attenuation of radiation sources that would be transmitted through the fuel. It was a challenging bit of technology, no doubt about it—so challenging that the gauge they had developed so far was accurate only to about 10 percent and it kept failing quality tests for reliability.

Shea let this process run for a couple of months and then stepped in. "This thing doesn't make any sense to me," he said. "We don't need

* There are variants on the phrase, which has been popular among engineers for many years. Bill Donovan, head of the O.S.S. during World War II, liked to say that "the perfect is the enemy of the good." Shea's version enjoyed a vogue during the Apollo Program. Jim Webb and Robert McNamara both used it, always referring to it as an old Russian proverb. One historian of the Apollo Program, Robert Sherrod, was able to track a variation back to Voltaire (*"le mieux est l'ennemi du bien"*). Voltaire in turn seemed to think it was an old Italian adage.

anything that complicated.'' Shea had a Karmann Ghia which didn't even need a fuel gauge, because it had a little reserve fuel tank that he could switch to. And that's what they ended up doing for the R.C.S. fuel tanks. They had a big tank that they used during the mission, and a little tank, not to be touched unless the big tank ran out, but that could get the crew home safely. No more nuclear gas gauge.

There was the problem of the heat shield. The specification the North American engineers were trying to meet required, logically enough, that the heat shield maintain its integrity under the extreme shifts from cold to heat that it would experience on its way to the moon and back, depending on whether it was in sun or shadow. But when they began testing the heat shield material under entry conditions, they found to their consternation that after exposure to severe cold it began to crack and craze and flake. They were going to have to create a new heat shield material, which would take untold quantities of money and time. Shea asked how long it took for the heat shield to cool down to the point where the problems began. The answer was about thirteen hours. So why did the spacecraft have to stay in the same attitude for that long? Why couldn't it rotate, so the heat shield would remain nice and warm all the time? And that was the origin of what came to be known as the ''barbecue'' mode, or passive thermal control (P.T.C.), in which the spacecraft rotated once an hour all the way out to the moon and back. Keep it simple.

In Shea's mind, nothing was sacred about the specs for the individual components of the spacecraft. There were only three sacred specs, and they were man, moon, decade. ''If those are the real three things you've gotta do, then everything else can be traded off underneath,'' he would tell them, and so he kept going back to the why of things: Why these numbers and not others on the spec? Was the product good enough to do the job it was supposed to do? Was the job it was supposed to do in the spec the job it would have to do on the flight? ''Because if we fail a qual test and if I have to send it back to redesign,'' Shea explained, ''then I have a less mature product when it comes back than I had already gotten up to that point.'' The better is the enemy of the good.

Again and again he preached: ''Hey, it isn't that complicated. It is very understandable. The engines work this way, the guidance system works that way, the transistors work this way, so don't get yourself in a state of mind thinking that it's too complex. It really is very simple. It's piece by piece.'' It's awfully big, that's all, Shea kept saying. Piece by piece, it's simple.

Sometimes it seemed a losing battle, as contractors became so obsessed with the specifications that common sense got lost in the shuffle. Honeywell, for example, was late in delivering the auto-pilot for spacecrafts 012 and 014; the reason, they told ASPO, was that they were redesigning some connectors and printed circuits to pass Shea's humidity spec. How badly had it failed? Shea asked. Well, actually, they hadn't even tested it against the humidity spec. What they were doing was immersing the connector board in water, so they would be sure of passing the humidity test when they got to it, and they were having trouble getting it to survive the immersion.

Keep it simple. Keep to the schedule. Do your job or get out of the way. Shea's methods could be abrasive and tough, but they could be exhilarating. It all depended on your point of view, and the people down in Houston had all sorts of opinions of Joe Shea.

He was too hard on people, too hard-driving. "Reminds you of a military guy," said one engineer: "Take that hill and don't worry too much about the casualties." No, said another. "If you came in and had the right story, and if you didn't try to bullshit him—because he was too darned smart to be fooled—he treated you pretty well." That was the key, said another. "Shea had a big ego, but he was extremely sharp and he was fair. He just wanted to know that you knew what you were talking about." Wait a minute, another said: "Shea was an egotist, very much so. If you buttered him, you got the gravy jobs." But he was watching from afar, he added, and "other people that I got to know real well after I transferred over to ASPO thought he was great." Yes, but he was pushing too fast on the hardware, said others. He'd be out at Downey during the graveyard shift, signing off on things just to keep to the schedule. That's right, said one of the inspectors: "He was out there telling North American, 'Hey, hurry up, don't worry about that little stuff, those little things.' Joe was pushing kind of hard." But then, of course, he added, "You had to. They'd have never got anything out of that plant if he wasn't pushing." At least, said still another, there was no doubt about who was running ASPO: "Shea wanted complete control, and he had it."

And what of Joe Shea and Max Faget? "Well," said Shea later, a little abashedly, "Max grew on me." Despite their rocky beginning, Shea and Faget forged a friendship. Professionally, they worked out a system whereby Shea in ASPO tapped into Faget's expertise over in Engineering without either of them stepping on the other's toes. And it turned out that Nancy Faget and Berry Shea liked each other too, so as time went on the families became friends.

It was always to be a friendship with an edge, as could hardly have been otherwise with two men as competitive as Faget and Shea. One time, after Faget had started jogging in his spare time, they were on a plane to Washington and Faget announced to Shea that, although Shea might once have been a runner in school, Faget was getting pretty fast himself. In fact, Faget thought that he could beat Shea. Shea doubted that. Later that day, pedestrians walking near the Georgetown Inn could see two men in early middle age, one tall and one short, racing up the crowded sidewalks of Wisconsin Avenue in a hundred-yard dash. Shea won. "It was that kind of relationship," Shea said, and they had a lot of fun with it.

So Joe Shea became part of Houston, and as time went on he thought of himself less and less as a headquarters man. He found himself liking the N.A.C.A. hands, even though he insisted on continuing to call them the "Fly-Boys," and liking the organization. "Houston was the only place I've ever seen," he recalled, "where every guy was comfortable in his job. He didn't want a promotion. There was no jockeying for position, there was real rapport."

Shea never became fully part of the Space Task Group family—that was probably impossible for anyone who hadn't been at Langley or Lewis—but it wasn't necessary. These were Joe Shea's glory days, and whatever the swirl of opinions about this gifted, enigmatic man, he was taking an effort that had been foundering and driving it forward. This, said Tom Markley later, was the reason Shea walked into a staff meeting one Monday morning to find everybody sitting there, feet up on the table, all with red socks on. "What that showed was our love for the guy," Markley said; and he then paused to reconsider his choice of words. Not "love" exactly—Joe didn't let many people get that close to him. They wore the socks out of deep respect and admiration for Shea, he said. "He was a walking encyclopedia, a design guy, a systems integration guy, and not only that, but he could understand contract negotiations. And he could handle senators and congressmen when they came down, just as well as he could handle his own design people." Joe Shea could do it all.

And when the pressure got too high, they had one other thing going for them. "Down at Houston we could argue like hell," Shea recalled. "But afterwards we would be out in the parking lot and look up at the moon and say, 'You really want to go there?' " And the answer was always yes.

13

"We want you to go fix it"

In the first half of 1964, while Joe Shea was taking hold of ASPO, the Combustion Devices Team out at Canoga Park doggedly continued to work on the injector plate for the F-1. Management at both NASA and Rocketdyne began to wonder whether some sort of compromise might be possible. The combustion chamber with the baffle usually worked—not on every test, but on a high proportion of them. Jerry Thomson began to hear things like, "Why not just accept that once in a while one will go unstable?" This school of thought held that a failure was unlikely in the first place, and that even if an engine did fail, the first stage of the Saturn V had an "engine-out capability"—it could continue to fly on four of its five engines. The mission might be jeopardized, but the crew would be safe.

But the vehicle would continue to fly and the crew would be safe only if a runaway engine shut itself down before it caused an explosion—which it might or might not be able to do, depending on the nature of the instability. Thomson thought the danger was substantial—"If you had gone unstable on the F-1 engine, you'd have lost the bird." He and Paul Castenholz continued to insist that they would deliver a dynamically stable injector that would damp in no more than 400 milliseconds.

They were now pursuing another line of attack, borrowing from some test results on another engine, the H-1, which had been developed for earlier versions of the Saturn. Though much smaller than the F-1, the H-1 had also been troubled by combustion instability. The problem had finally been solved by changing the impingement angle of the propellant streams. The angle of the holes through the injector plate had been altered so that the fans of LOX and of kerosene were formed farther down in the combustion chamber.

There were some disadvantages to this modification, mainly involving the efficiency of the engine. The farther down the streams impinged, the less completely they burned before being expelled through the throat. It was not a trivial loss—when the Combustion Devices Team tried this fix with an appropriately modified injector, they did indeed find they had cut the efficiency of the F-1 by a few percent. But some decrease in efficiency could be tolerated (that German conservatism once more gave them precious leeway), and with the new angle of impingement, plus the redesigned baffles, they found that they had reduced the occurrence of instability substantially.

More months went by, and much fine-tuning. They modified the angle again and they increased the orifice size by a fraction of an inch. They fiddled and nudged, and the incidence of instability decreased still further until, by late 1964, they weren't getting it anymore. The bombs would explode, the pressure in the combustion chamber would skyrocket—and then the engine would be running smoothly again, not just within the 400-millisecond goal the team had set for itself, but within 100 milliseconds.

The men who had worked so singlemindedly and so long on solving the combustion instability problem in the F-1 had no moment of triumph, no equivalent of the splashdown party that other people in the Apollo Program enjoyed. They never knew for sure that they had finally won. "There was an apprehension that something would happen that you didn't know about," Castenholz said. "I think you'll find that with almost all rocket engineers." For Castenholz, it lingered throughout Apollo, even after the lunar landing. It wasn't as if they had ever managed to write the equations that explained exactly how combustion instability could be done away with. All they had done was redesign until it was gone. But though they could not know it then, it truly was gone, never to reappear in any F-1 that ever flew.

In January 1965, the injector for the F-1 was rated flight ready by Marshall Space Flight Center. On April 16, 1965, a full complement of five F-1 engines, mounted as they would be in flight, were ignited on the Huntsville test stand. During the course of the 6.5 seconds of ignition, they generated 7.5 million pounds of thrust.

1

At the Cape, the last half of 1965 and the beginning of 1966 saw the pieces of the launch complex finally begin to come together. In October, the first portions of the High Bay in the V.A.B. were occupied.* In January 1966, the crawler, which had been giving Don Buchanan fits for almost four years, successfully returned the mobile launcher to the V.A.B. (for six months, the launcher had been stranded a mile and a half away, where the crawler had taken it and then broken down). In March, the first stage of a full-sized Saturn V test article, called the Saturn 500-F, with the same tankage and umbilical connections as a real Saturn V, was lifted to a vertical attitude in the V.A.B. Ten days later, the second stage of the 500-F was lifted 200 feet into the air and mated to the first stage. Five days after that, on March 30, 1966, the crews in the V.A.B. mated the third stage, then a test-article version of the C.S.M., and for the first time the people of Apollo saw what a Saturn V really looked like when assembled—all 363 gleaming white feet of it, standing under the harsh lights of the V.A.B. On May 8, the last of the swing arms which had caused so much trouble was installed on the umbilical tower.

On May 25, 1966, five years to the day after John Kennedy had made his speech promising the moon within the decade, the giant doors of the V.A.B. slid open and the crawler emerged, bearing on its back a launcher, an umbilical tower, and a full-sized, full-weight mockup of the Saturn V. The crawler took its load out to Launch Complex 39, reaching Pad A at dusk. Five weeks later, Rocco Petrone, leaving his job as program manager for the Apollo facilities at the Cape, moved over to become director of Launch Operations. If Marshall could give them a vehicle and Houston could give them a spacecraft, Stage Zero was ready to fly.

* The V.A.B. is not really a box. The "High Bay," 525 feet high, is the tall portion of the V.A.B. Less noticeable is a wing of the V.A.B. called the "Low Bay," which is a mere 210 feet high. First-time visitors to the V.A.B. often do not realize they are entering through the Low Bay. They marvel at how big the building is, only to walk into the High Bay and realize that what came before was only the little shed on the side.

2

By August 1966, Joe Shea's ASPO was nearing achievement of Houston's part of the bargain. During 1964 and 1965, eight boilerplate spacecraft had flown on Saturn Is and Big Joes. Only one was a failure, that one because the launch vehicle blew up. In 1966, three unmanned command and service modules had been launched so far, their flights including a high-altitude abort test, a test of the compatibility of the spacecraft and the launch vehicle, and an orbital test of the S-IVB third stage. All three had been unmanned versions of what were called "Block I" spacecraft, meaning that they couldn't link up with a LEM.*

ASPO still did not have a lunar module ready to fly, nor did it expect to have one for another year. There had been a multitude of problems with the LEM; indeed, after the F-1 problem was resolved, the LEM had become the "pacing item" in the program, the one that would hold back all the rest unless it made up time. But Grumman was by common consent the finest prime contractor that NASA had. When problems arose with the LEM—no surprise in a vehicle so unprecedented in its function and design—ASPO found that Grumman got on top of them as fast and as energetically as anyone could wish. Within NASA there was confidence that, one way or another, the LEM would be arriving pretty much when Grumman said it would, and that when it arrived, it would work.†

The relationship between ASPO and North American had reached a modus vivendi. At the engineering level, some close working relationships had been established. Shea found that Dale Myers, who had replaced John Paup as the contractor's program manager for the spacecraft, brought the same frame of mind to Downey that Shea brought to Houston. They had joined forces during the first year to define the Block II spacecraft—"ganged up on the rest of the world," Myers recalled, "and said we're going to have some specs [for Block IIs] that are right and meaningful." They had pushed them through, and subsequently had seen the Block II spacecraft come close to completion.

* Originally, the Apollo Program had planned to fly four manned Block I Apollo spacecraft in low earth orbit, using the Saturn I and IB as the launch vehicles. These were canceled as part of Mueller's all-up decision in November 1963. As of 1966, only one Block I spacecraft was scheduled for a manned flight, A.S.-204, with Grissom commanding.

† One NASA engineer wondered whether the differences were geographic—"You could look across the country from the East Coast to the West Coast and watch the personality change" among the contractors, he thought. Contractors on the East Coast (Grumman was headquartered on Long Island) had a go-get-'em, do-it-right attitude; the people in the Midwest (McDonnell was in St. Louis) were somewhere in the middle; "and the folks in California—well, we had to push them a lot in California."

The tensions between NASA and North American were not over, however. Late in 1965, Sam Phillips and his staff, prompted by continuing problems and delays, had conducted an extensive investigation of the status of NASA's contracts with North American for the C.S.M. and the S-II stage of the Saturn V and prepared a highly critical report that later would become famous as the "Phillips Report." "I am definitely not satisfied with the progress and outlook of either program," Phillips had written to Lee Atwood, and listed a long set of needed changes. Eberhard Rees, von Braun's deputy center director at Marshall, was near despair over the state of the S-II. In a confidential memorandum with only three copies—for Phillips, von Braun, and his own files—he wrote, "I believe NASA has to resort to very drastic measures," including the possibility of shifting to a new contractor. "For me, it is just unbearable to deal further with a non-performing contractor who has the government 'tightly over a barrel' when it comes to a multibillion-dollar venture of such national importance."

But Rees was referring to the S-II. The C.S.M. was doing much better, as North American began funneling some of its best people into the Downey plant. North American's director of engineering, George Jeffs, a rising star with a professional reputation matching any in NASA, was sent over to reinforce program manager Dale Myers as his deputy manager and chief engineer. Jeffs brought with him a small cadre of his own people and cut the number of engineers in the program from 4,000 to 2,000. "We had lots of areas to clean up," Jeffs recalled, "lots of areas to get down to the meat and potatoes, and get rid of some of the ice cream," and he proceeded briskly. Houston, remembering the early years of the space program, hoped that the experience with North American was going to turn out like McDonnell on the Mercury and Gemini capsules—a slow start, then accelerating progress as they moved over the hump of the learning curve.

On August 19, 1966, spacecraft number 012, the first Apollo designated for a manned mission, was ready for the Contractor's Acceptance Readiness Review (CARR). This was a major event in the life of every spacecraft, the moment when it was officially judged to have met the specifications of the contract and became the property of the United States government.

The CARR for spacecraft 012 was held at the North American facility in Downey, a sprawling complex of hangars and manufacturing plants ten miles east of Los Angeles International Airport. Joe Shea presided over the meeting, which was held in the low fieldstone building of 1940s vintage that served as office space for the North American executives. The astronauts who would fly 012 were there: Gus Grissom, one of the original seven astronauts and the second American to fly in space; Ed White, the first man to walk in space; and a rookie astronaut, Roger Chaffee. Also present were Faget, representing Engineering, Chris Kraft, representing Flight Operations, and a dozen other senior people from Houston, plus all the senior North American executives responsible for the C.S.M. contract.

Shea opened the meeting by saying, "This is not a meeting to bring up old bitches. It's a meeting specifically concerned with spacecraft 012 and its suitability to leave the plant and begin the checkout procedures and booster mating procedures down at the Cape." He cautioned everyone not to get "ourselves all tangled up between design changes and the specific checkout of this spacecraft as it is presently configured." Spacecraft 012 had been moving slowly, Shea acknowledged. "It's still not completely through all its tests and the CARR is in that sense somewhat provisional but I intend to go ahead with it anyway." He was in high spirits that day, clicking through the items, punning, arbitrating disagreements. The atmosphere was relaxed, with wisecracks and laughter punctuating the discussion.

Most of the problems raised at the meeting were minor, or had been hashed over already in previous sessions. Even so, the list of items to be reviewed was so long that the CARR went on for six hours. As the meeting was winding down, Gus Grissom asked for the floor and pulled out two photographs from a large envelope. They were identical, but with different inscriptions. The picture showed Grissom, White, and Chaffee seated behind a table on which a small model of the Apollo capsule rested. Their heads were bowed, hands steepled in a caricature of prayer. Grissom gave the first copy to Stormy Storms, general manager of the North American Space Division. "We've got one for Joe Shea also," Grissom said, and passed the second photograph down the table to him. "Joe advised us to practice our backup procedures religiously, so here we are practicing." There was loud laughter. The picture, signed by all three astronauts, was inscribed, "It isn't that we don't trust you, Joe, but this time we've decided to go over your head."

At one point during the CARR, the question of flammable materials in the cabin came up. Since Mercury days, the astronauts had used Raschel netting, attached with Velcro, for rigging up pouches to store the pens, flight plans, and other paraphernalia they had nowhere else to keep. The astronauts had also gotten into the habit of customizing their spacecraft, putting a pouch in exactly such-and-such a place (and having it made of such-and-such a color—some of the astronauts became quite fussy about these details of interior decoration) for a particular purpose. The problem was that Raschel netting and Velcro were flammable. They talked about it at the CARR for several minutes. Shea ended the discussion by noting tersely that the fire rules for the spacecraft prohibited anything that was potentially flammable from being closer than four inches to anything that could create a spark. He told North American to see that the cabin was cleaned up.

A month later, on September 26, North American submitted a memo listing its CARR Action Responses for C.S.M. 012. Under Section 14, Problem 14.7.2, "PROBLEM—Flammable Materials in CM," North American reported that a "walk-through" inspection of 012 had been performed and the results were documented in NAA IL 693-300-040-66-1009, dated August 22, 1966. It added that "specific NASA direction . . . on the findings must be made to N.A.A." Shea told his staff to prepare a direction. They did, and put it into the pipeline. But there were lots of directions going out, and Shea didn't try to keep track of this particular one.

In early October, Shea received a two-page letter from Hilliard W. Paige, vice-president and general manager of the Missile and Space Division at General Electric. Paige was writing to express his concerns about the possibility of a fire in the spacecraft. It was a friendly "Dear Joe" letter; still, Paige was worried. "I do not think it technically prudent to be unduly influenced by the ground and flight success history of Mercury and Gemini under a 100 percent oxygen environment," he wrote. "The first fire in a spacecraft may well be fatal." Paige suggested testing the ability of man's senses to detect a fire or an incipient fire condition. Also, ASPO might consider putting some sort of fire extinguisher in the cabin. He added a handwritten postscript saying that he hoped Joe would personally review the matter.

Shea called in ASPO's Reliability, Quality, and Test Division and asked for a review of the nonmetallic materials control program. Seven weeks later the chief of the division, Bill Bland, submitted his report to Shea. They had not been able to schedule the review that Shea had requested because of "our usual press of business with more significant problems." But Bland pointed out that they had just completed technical assessments of the fire hazard in the crew compartment in the command

and lunar modules. He attached a copy of the two assessments to his memo, noting the conclusion that "our inherent hazards from fire in the spacecraft are low." They should of course continue to review their fire hazard potential. Shea wrote back a "Dear Hilly" letter to Paige, sending copies of the evaluations of the command and lunar modules "so you may see how secure we are." Shea, too, added a handwritten postscript: "The problem is sticky—we think we have enough margin to keep fire from starting—if one ever does, we do have problems. Suitable extinguishing agents are not yet developed."

The exchange of letters was duly filed away in the rows of file cabinets of letters, analyses, and memoranda involving the spacecraft.

3

As 1967 began, Joe Shea was beginning to be a celebrity outside of the confines of NASA. When Shea took a few hours off to watch his favorite team, the Green Bay Packers, beat the Kansas City Chiefs in the first Super Bowl, he sat with Walter Cronkite. He was in demand as a speaker and was increasingly looked upon as a spokesman for the Apollo Program. With the red socks and the puns, the quicksilver mind and the intense, Black Irish good looks, Shea finally gave Houston a plausible competitor to von Braun for a place in the public imagination.

Time magazine was about to stamp its imprimatur on Shea's emerging status by putting him on the cover. The story was to coincide with the launch of Apollo 1, now scheduled for late February. Shea had already undergone the obligatory lunch with the *Time* editors, and by the last week in January there remained only some wrap-up reporting. When Shea flew to the Cape on Wednesday, the 25th of January, Ben Cate, *Time*'s Houston correspondent, tagged along to grab interview time on the plane and at meals.

Shea spent Wednesday and Thursday huddling with Rocco Petrone, trying to work out the problems that had plagued spacecraft 012 since it had arrived at the Cape.* Shea and Petrone had never been the best of

---

* Later, the story would spread that Grissom had hung a lemon on 012 as a sign of his disgust with it. Apparently he in fact hung the lemon on Riley McCafferty's simulator at the Cape, not the spacecraft. The spacecraft was changing so fast that McCafferty, chief of Flight Crew Operations at MILA, always had at least 150 or 200 new modifications that hadn't yet been incorporated into the simulator. The ones that did get patched into the software usually had bugs at first, so the simulator was in a constant state of incipient breakdown.

friends, and each was inclined to see the problems as being the other's fault. Why had Shea let North American ship a half-finished bucket of bolts to K.S.C.? Why couldn't Petrone get his launch teams to get the spacecraft checked out on schedule? It was, Shea remembered, "a knockdown and drag-out." In Petrone's opinion, Shea hadn't spent enough of his career with hardware to understand the nature of the problems the Cape faced.

On Thursday afternoon, Wally Schirra, the backup for Grissom, was chatting with Shea when he came up with an idea. Why didn't Shea get into the spacecraft himself and go through the countdown test with the crew? That way he could find out what it was like from their point of view. Shea had never done such a thing, but he liked the idea and asked the K.S.C. technicians to wire up a fourth communications loop into 012. Shea went off to have dinner with Cate and then finished the evening by having a couple of drinks with Schirra. He changed his airplane reservations so that he could stay through the test.

The next morning, Friday the 27th, Shea was at breakfast with the crew when the K.S.C. communications people reported that they couldn't Rube Goldberg a fourth communications loop in time. There wasn't a fourth jack in the spacecraft, an extra loop would have to go through the hatch. . . . It was too complicated.

Grissom still wanted Shea to be with them in the spacecraft. The practice countdowns just hadn't been smooth enough. Things didn't flow, and when they said something into their headsets, the spacecraft didn't get an answer right away. The subsystems didn't check out as well as they should. "It's really messy," Grissom told him. "We want you to go fix it." But if he couldn't have a headset, Shea didn't see the point. "You think I'm going to sit at your feet for four hours and not be able to communicate?" Shea said to Grissom. "You're nuts. You go through the test; I'll go back to Houston and I'll come back Monday and do it in the simulator with you."

At about eleven o'clock, Shea and Petrone went out to Pad 34 with the crew and took a look at 012 for themselves. Both of them noticed that some polyurethane pads were still in the spacecraft—the technicians used them when they were in the spacecraft to avoid damaging the wiring. Petrone directed that they be removed, except for one that Ed White had asked them to leave—White, sitting in the center seat, was the one who opened the hatch at the end of the test, and he wanted to be able to stand on the pad while he did it so that he wouldn't step on the wiring.

Shea went back to the blockhouse while the crew was sealed into 012, and the countdown began. He called Ben Cate and told him he was going back to Houston with him after all—there was no point in hanging around if he wasn't going to be in the spacecraft. They caught the 2:30 National Airlines flight from Melbourne and arrived a little after five in the afternoon, Houston time. Cate headed for home after arranging for the Sheas to come to dinner on Saturday. Shea went to his office at M.S.C., arriving there at about 5:30. With the one-hour time difference, that made it 6:30 P.M. back at the Cape.

CHAPTER

14

"Did he say 'fire'?"

Among the people of Apollo, it is known simply as the Fire, needing no other label. Twenty years later, after a visitor had questioned him about it, the Fire would still keep Don Arabian from getting to sleep until three in the morning, turning the data over in his mind, trying to make all the pieces fit. Marty Cioffoletti's voice would still tighten when he remembered having to listen to the tape of the crew's voices, again and again. Some of the men who had been on the consoles would still be unable to talk about it, stopping in mid-sentence. The history of Apollo is divided into two time periods, Before the Fire and After the Fire.

1

Scott Simpkinson was the engineer in charge of disassembling the burnt-out spacecraft. He also acted as editor of the official NASA account of what happened, *Report of the Apollo 204 Review Board*. Sitting in the card room at the Baywood Country Club where Simpkinson spent his days after he retired, he settled back to discuss the fire once again, after twenty years. The question was put to him: Did they ever find out exactly what caused it?

"There was no positive proof," Simpkinson replied. "It was the 'most probable cause,' I believe we stated in the report. Of course, I have my own opinion, which was shared by many of the people on the board."

He pulled on his cigarette, trying to decide how to explain it to an outsider, then took out a pen and sketched the floor plan of the Apollo spacecraft on a napkin. "The commander was on the left," he began. "On the pad he would be lying on his back. And right here"—the pen etched a delicate X—"in the side of the spacecraft beside him was a compartment for lithium hydroxide, which takes out the CO_2 so the crew can rebreathe the oxygen. It had a metal door with a sharp edge . . ."

The door opened into the Environmental Control Unit (E.C.U.). Just underneath it was a cable, part of the E.C.U.'s instrumentation harness. The cable was wedged against the bottom of the door by other bundles of wires beneath it. Each time the door was pushed shut, the edge of the metal door scraped against the cable. The slight but repeated abrasion had exposed two tiny sections of wire in the cable. Something—maybe Grissom opened the door—caused a brief electrical arc between the exposed portions of wire.

Even with the pure oxygen environment, the spark from the short would not have caused a conflagration. But it also happened that, just below the two scuffs in the cable, a length of aluminum tubing took a ninety-degree turn. This particular ninety-degree corner joint, one of hundreds in the tubing that interlaced the spacecraft, probably had sprung a leak. The joint had passed redundant inspections, but it had been subject to "creep" from stress at remote points in the system—the spacecraft was constantly being worked on, and things got bumped around. Certainly even a small leak would help explain what happened, because the tubing carried a glycol cooling fluid which, when exposed to air, turned into fumes. The liquid was not flammable, but the fumes were.

The scuffed cable wedged against the bottom of the door was not originally intended to cross the tubing at the joint nor was it originally intended to lie quite as close to the tubing as it did. But there had been thousands of changes in the Block I spacecraft since the plans had been drawn, and the changes had put many cables, this one among them, in different positions from the ones that had originally been thought out at the design tables.

There was also some Raschel netting near the scuffed cable, closer than it should have been—ASPO's directive about enforcing the fire rules had not yet been acted upon. Besides, astronauts had been customizing their spacecraft ever since the first Mercury flight. And there were so many rules. And this was just a practice countdown, not a real flight. And no one had focused on how highly flammable the netting was in a pure-oxygen environment.

Simpkinson surmised that the hot spot in the netting moved horizontally at first. Or it could have just hung there, slowly growing larger. Possibly it created an acrid smell. For about thirty seconds just before the crisis the crew wasn't saying anything but White and Grissom were moving around, doing something. One explanation is that they were trying to determine where the smell was coming from.

Heat rises. The glowing spot reached a vertical strand of the netting and began to climb, rising in temperature as it fed upon itself, finally bursting into an open flame. It was at this moment, 6:31:04 P.M. (as later determined by the Medical Analysis Panel), that the sensors attached to Ed White registered "a marked change in the senior pilot's respiratory and heart rates." A second later, the first message came over Black 3, the crew's radio channel: "Fire," Grissom said, or perhaps it was "Hey." Two seconds later, Chaffee said clearly, "We've got a fire in the cockpit." His tone was businesslike.

One hundred and eighty-two feet below, R.C.A. technician Gary Propst was seated before a bank of television monitors. He was trying to adjust the brightness on Camera 29, one of the dozens of cameras that continuously fed pictures into the Operational Television (O.T.V.) control racks during any simulation or launch. Some of the cameras were focused on critical components of the launch process—the swing arms on the mobile launch tower, the umbilical connections on the Saturn. Others covered the exterior of the entire vehicle, not because anything particularly interesting was expected to occur, but for the same reason that cameras record every public appearance of a president of the United States—just in case. There was no camera inside the cabin.

The monitor immediately above and to the left of the one that Propst was working with was fed by Camera 24, located in the White Room—the enclosed end of the swing arm, immediately adjoining the spacecraft—on Level A8 of the tower. Operated by remote control, it showed activity in the White Room; or, if zoomed in on the porthole in the hatch, it could provide a partial, shadowy view of the interior of the spacecraft as it looked from behind and above the center couch—in effect, showing the view over Ed White's shoulder. Propst's primary communications loop was Black 7, but he also monitored Black 3 to keep in touch with events in the spacecraft. He heard the call of "fire in the cockpit" and shifted his attention to the monitor for Camera 24.

At first, all Propst could see was a bright glow within the normally dim spacecraft. Then he could see flame flickering across the porthole and Ed White's hands reaching above his head toward the bolts that secured the hatch. There was a lot of motion, Propst observed, as White seemed to fumble with something and then quickly pull his arms back, then reach out again. Another pair of arms came into view from the left—those would have been Grissom's—as the flames spread from the far left-hand corner of the spacecraft toward the porthole.

It seemed to Gary Propst that the crew's agony was endless. Four days later, testifying before the Review Board, he would estimate that he watched movement within the spacecraft for about two minutes before the spreading flames blocked the view. Bitterly angry, he wondered why no one was getting them out.

Up on Level A8, Don Babbitt, North American Aviation's pad leader on the second shift, was standing at the pad leader's desk when he heard Grissom's first message over the comm box. The spacecraft's hatch was only twelve feet away, across a swing arm and through the White Room. For Babbitt, time felt speeded up. He couldn't make out the whole message, but he registered the word "fire." He told Jim Gleaves, his lead technician, to get the hatch off and then turned from the desk toward the comm box to call George Page down in the blockhouse. As he turned, there was a venting sound, a sort of *WHOOOOOSH!,* followed by what seemed to Babbitt to be a sheet of flame shooting from the spacecraft and arching over his head, charring the papers on the desk.

Gleaves was already moving toward the White Room when he heard the noise. To Gleaves, it sounded as if Grissom had dumped the cabin pressure. Then there was a flash, indicating to Gleaves that "something was fixing to happen," and he turned away to run just as the heat shield ruptured, spattering him with fire and debris. "It was like when you were a kid and you put a firecracker in a tin can," Gleaves said in his testimony to the Review Board, "and it blew the whole side out of the tin can with the flames shooting out." The force of the blast knocked Gleaves up against the room's orange door ("which," he would pointedly tell the Review Board, "I might say opens the wrong direction").

Gary Propst's recollection notwithstanding, the rending of the space- craft took place with merciful speed. From the first call of fire to the final scream on the communication tape, from the sudden rise in the senior pilot's pulse to the explosion that slammed Jim Gleaves up against the orange door, just eighteen seconds had passed.

2

A plugs-out test was a major event in the preparation for a manned launch and as such it required the combined ministrations of three widely separated control centers. One was Mission Control in Houston, which oversaw the simulated post-ignition phase from the time the vehicle cleared the top of the launch tower until insertion into orbit, when the simulation ended. But the main purpose of a plugs-out test was to demonstrate that the systems of the launch vehicle and the spacecraft worked—that a given switch turned from "Off" to "On" produced the desired consequences in the innards of the machines. So for a plugs-out test, the two control centers at the Cape had most of the action. Their duties were split at the joint that divided the people at the Cape into one of two primary groups, the yin and yang of manned spaceflight—the world of the launch vehicle and the world of the spacecraft.

The checkout of the launch vehicle, ensuring that the hundreds of valves and pumps and tons of propellants would produce a controlled explosion of precisely the required magnitude and characteristics at precisely the right moment, was entrusted to the Firing Room. In 1967, at Pad 34, the Firing Room was in a concrete blockhouse located only a few hundred yards from the launch tower. Meanwhile, the tests of the spacecraft's systems were conducted by the Automatic Checkout Equipment (ACE) Control Room, located in the Operations and Checkout (O&C) Building a few miles from the launch complex.

Only one person in each control room was authorized to talk directly to the astronauts. In the ACE Control Room, that person was Skip Chauvin, the spacecraft test conductor. At 6:31:04, the test was in a scheduled hold at T − 10 minutes and Chauvin was examining the next page in the test protocol, thinking about what came next, his pen poised to make another note. Like Babbitt up on the tower, he didn't really catch the first message, just something about a fire. The voice was calm and matter-of-fact. Chauvin looked up at the TV monitor, but the picture wasn't much help—something was burning, Chauvin could tell, but his first thought was that the fire was outside the spacecraft.

John Tribe, North American's lead engineer on the C.S.M.'s reaction and control system, was sitting at a console a few rows in front of Chauvin. He was taking advantage of the hold to scribble out a procedure he wanted to run the next day when he too heard the word "fire" on his loop. Tribe wasn't sure that he had heard right. "Did he say 'fire'?" he asked the man next to him. Yes, that's what he had heard too. "What the hell are they talking about?" Tribe wondered.

Then Chaffee's voice again, agitated now. "We've got a bad fire— let's get out. . . . We're burning up!" There was a short scream, then silence.

Chauvin was calling to the crew on Black 3, getting no response. He shouted across the room to the people at the electrical power consoles to power down the spacecraft—startling many, for the composed Chauvin never shouted. To Chauvin, as to Propst, everything during those moments seemed to take forever. Unlike Propst, he recognized how illusory his reactions had been—when he testified before the Review Board a week later, he would refuse to make time estimates for specific events, understanding that he was trying to remember a nightmare.

At the time, the significance of what they had seen and heard took a while to sink in. Mostly, people were bewildered. Only a few had been watching the television monitor—Tribe couldn't even see it from his console—and those who were in a position to see couldn't make out much. After a few minutes, Tribe was sure at least that he was going to be working late that night. He took off his headset and went to phone his wife, telling her he wouldn't be home for a while. She wasn't to worry if she heard about some flap out at the Cape. No, he couldn't say any more, but she shouldn't worry, he was okay. After he hung up, one of Tribe's friends came over and picked up the telephone. "Hey," he asked, "what's the matter with this phone?" Nothing, Tribe replied, he had just used it. The man shook his head. "It's dead." Tribe turned around and saw a guard at the door of the ACE Control Room. Rocco Petrone, watching the test from the blockhouse, had already moved to seal off Kennedy Space Center.

Petrone was over in the blockhouse beside the pad, sitting beside Deke Slayton, one of the original seven astronauts chosen for Mercury, now the director of Flight Crew Operations. Nearby was Stu Roosa, acting as Stony, the person in the blockhouse who communicated with the crew (the position called CapCom in Mission Control).

Petrone and Slayton had been talking idly about the erratic communications when they heard a muffled comment from the crew. Petrone, looking at one of the three television monitors at his console, saw what looked like a shadow moving in the spacecraft. Then, to his puzzlement, he noticed that another monitor, focused on the service module beneath the spacecraft, showed that the cables attached to the service module were swaying.

Slayton saw flames around the hatch area, but it was hard to make out what was going on. The television picture added to the confusion—it changed quickly from flames into a blinding white glare, then subsided into flames and smoke again. When Slayton heard the frantic "We're burning up," his first impression was that the pad leader was crying for help, that a fire had broken out on Level A8 in the White Room.

"And then things just went crazy," recalled Ed Fannin, sitting at one of the Environmental Control Unit's consoles. One of the spacecraft people called him on the loop asking for him to increase the flow of air into the service module. Fannin didn't think it was a good idea, since they didn't know yet what was going on up there, but he did what they asked. It was pretty quiet in the blockhouse after that, he remembered, "except for the communications going over the loop. You could hear guys out there saying 'It's too hot, it's too hot, you can't touch it, it's too hot.' "

As they listened to Skip Chauvin calling helplessly to the crew and saw the image on the monitor cloud over with greasy smoke, Slayton thought he knew what they were likely to find when they opened the hatch. He arranged for the medics to get up the tower, then picked up the phone to get things cranked up back in Houston in case this was as bad as he was beginning to fear.

Up on Level A8, the situation was slowly stabilizing. When spacecraft 012 ruptured and flame and debris spewed into the White Room, Babbitt and the men with him scrambled across the swing arm into the main tower where a crew was standing by the elevator, waiting to participate in the final stage of the plugs-out test. Babbitt shouted to the elevator technician to tell the blockhouse that the spacecraft was on fire, and to get some help. Meanwhile, they looked for gas masks so they could get back to the spacecraft.

"I became scared," L. D. Reece told the Review Board, the only person in all the Review Board testimony who openly acknowledged that Level A8 was a frightening place to be at that moment. Smoke was billowing out of the umbilical arm and secondary fires were burning not only on Level A8 but on Level A7 as well. Though the Saturn's propellant tanks were empty, the spacecraft was still a bundle of pyrotechnics, with explosive materials packed into all kinds of places. And fastened to the top of the command module was the escape tower assembly, powered by a solid-fuel escape rocket with greater initial thrust than a Redstone. To be on Level A8 was not unlike being in a fire at a dynamite factory, with the added thrill of being 200 feet high on an open tower. Nonetheless, Babbitt and his men returned to the swing arm.

Jim Gleaves tried first, heading back toward the White Room almost immediately after reaching the umbilical tower, but he was stopped by the smoke and flames before he could get to the spacecraft. In the meantime, Reece had found the gas masks. Some didn't work at all and the ones that did weren't much help—they had been designed to screen toxic gases, not thick smoke—but the flames had died down enough so that Babbitt, Gleaves, Jerry Hawkins, and Steve Clemmons took the only fire extinguisher they could find and went back into the White Room. Hawkins managed to put out the fire around the hatch area of the spacecraft before the hose of his gas mask came off and he had to retreat again. Presently, Gleaves too had to flee the White Room, coughing badly.

To get the astronauts out, Babbitt and his men had to remove three separate hatches: the boost protective cover (B.P.C.) that would shield the command module during launch, the ablative hatch, and the inner hatch. After one of the early trips, Gleaves returned, again choking from the smoke and fumes. "I got the B.P.C. hatch," he told Hawkins. "Get the others." It took several more trips—they could stay beside the spacecraft for only seconds at a time. On what he thinks was his third trip, the smoke had cleared enough so that Babbitt could see his headset where it had fallen to the floor. Miraculously, it still worked. Thereafter, each time that Babbitt left the side of the spacecraft to get some air, he paused to report to George Page, giving the blockhouse and the ACE Control Room their first reliable information about what was going on.

The smoke was still so thick that they were working mostly by feel. The ablative hatch finally came off. The face of the inner hatch remained too hot to touch, but the handles were cool enough to hold on to. Even so, they couldn't rotate it to get it out and they couldn't seem to push it straight into the cabin. To make matters worse, Babbitt and his men were near collapse. Gleaves, almost unconscious from the smoke, had to be ordered to leave. Hawkins and Clemmons weren't in much better shape. Dodging back into the umbilical tower on one of his trips for air, Babbitt asked Reece and Henry Rogers, a NASA quality control inspector who had happened to be on his way up to Level A8 when the fire occurred, to go in and help get the hatch off. The five of them—Babbitt, Hawkins, Clemmons, Reece, and Rogers—continued their frantic relay, working for as many seconds as they could in the choking smoke, running back to the umbilical tower to breathe, running back to the hatch to wrestle with it some more. Finally they managed to push it a little way into the spacecraft and to one side.

A wave of heat and smoke poured from the opened spacecraft as the hatch gave way. Hawkins and Clemmons peered into the blackness. The small fluorescent lights near the astronauts' headrests were still lit, glowing dimly in the roiling smoke and ash, but the two men could make out nothing within. They had time only to lean into the spacecraft and try to feel with their hands for the astronauts before being driven outside for more air.

L. D. Reece was working his way toward the spacecraft as Hawkins and Clemmons staggered by. As he approached, he thought he could hear the astronauts calling out for help. He got to the hatch and leaned in as far as he could, feeling around the center couch. He felt nothing. Still convinced he had heard the crew calling, he took his mask off and yelled several times into the blackness. He was choking now, so he dropped his mask on the center couch and fled back to the swing arm.

Finally the smoke thinned and the ash settled enough to see. Two days later, citing an unnamed "official source," *The New York Times* would report a charnel house of horrors, with incinerated bodies consisting of little more than bones, with fingernail scratches and shreds of flesh embedded in the metal of the hatch where Ed White had scrambled frantically to get out. The reality was not quite so dreadful as that. There was Ed White's hand print on the hatch, outlined in ash, but no scratches in the metal, no torn flesh. The astronauts were not incinerated. The flames had extinguished themselves quickly and the suits had provided substantial protection. Grissom had sustained burns to his leg and Chaffee to his back (from the pad that White had asked be left in the capsule), but they were survivable. The crew had died of asphyxiation, not burns, and they had not suffered long. "It is estimated that consciousness was lost between fifteen and thirty seconds after the first [astronaut] suit failed," the Medical Analysis Panel concluded, and the first suit had failed no later than the rupture of the spacecraft—the Medical Panel knew that, because a portion of Grissom's suit had been blown through the breech in the spacecraft, landing five feet outside the command module on the White Room floor.

Still, it had been a terrible few seconds. Grissom was lying on his back on the floor of the command module, where he had crawled to try to escape the fire. White had in the last moments given up his attempt to open the hatch and was lying transversely across the spacecraft just below the hatch. Chaffee was still on his couch.

Babbitt went back to his headset and called to the test conductor. In pain from smoke inhalation and flash burns, his eyes red and puffing shut, his voice down to a rasp, Babbitt tried to concentrate on how he should handle this. Some part of him said he shouldn't be saying that the crew was dead over an open loop. So when he got to the headset he said simply that the hatches were open and "I cannot describe what I saw."

From the time when the fire was first reported to the time when Babbitt first reported that the hatch had been opened, five minutes and twenty-seven seconds had elapsed.

Petrone watched the televised picture of the open hatch on his monitor. There was nothing to be seen, just a featureless black hole in the side of the spacecraft. He couldn't make himself take his eyes off the monitor, couldn't bear to look at it any longer. He turned to Paul Donnelly, his test supervisor.

"Paul, have 'em cut that camera off." The black hole was everywhere in the blockhouse, on perhaps thirty screens.

"Roc, that's life," Donnelly said.

"Yeah, I know that's life," Petrone said, "but I want that camera off." The screens went blank.

Sam Beddingfield, driving up the narrow access road to Pad 34, saw exactly what he expected to see as he returned from his dinner in Cocoa Beach: the umbilical tower rising beyond the blockhouse alongside the Saturn I, the conical spacecraft gleaming in the spotlights against the last remnants of twilight. He was returning to the blockhouse to conduct the final item in the checklist. There was no hurry. He had phoned from the restaurant and learned that countdown was far behind schedule, just as Beddingfield had expected it to be.

So why were the guards at the gate waving at him like that? Beddingfield wondered. Then he saw in his rearview mirror that he was being overtaken by an ambulance. He pulled aside and the ambulance raced past him toward the launch tower. Beddingfield pushed his car back into gear and drove up to the blockhouse, wondering what had happened and whether it would interfere with that last item on the checklist, the one he was going to supervise. They were going to simulate an emergency to see how fast the astronauts could get out of the spacecraft.

3

To John Hodge, the flight director in Houston's Mission Control that night, the fire will always be something that happened early in the morning. Flight simulations were so realistic that Hodge would find his palms starting to sweat at $T-3$ minutes, and according to this one's schedule they were nearing the moment of an early-morning launch. But in fact it was the end of a long day. This was not one of the hair-raising flight simulations, it had been going on for a long time, there wasn't that much for Houston to do in the pre-launch phase, and at 5:31:04 Houston time, Hodge was presiding over a team of flight controllers who were about as somnolent as flight controllers ever get.

John Aaron, sitting at the EECOM console, knew immediately that something was terribly wrong.* Aaron wasn't even supposed to be the EECOM that night; his buddy Rod Loe was. But January 27 was Loe's wedding anniversary and his wife, Tina, had arranged for some friends to come over and play poker. So Loe and Aaron had traded off, and it was Aaron who was sitting at the EECOM console when the data flickering on his screen abruptly showed a bulge in the cabin pressure, then disappeared.

John Frere, sitting at a console in one of the computer rooms at the Control Center, was heartsick. The Control Center was still working out the bugs in its communications system with the Cape, and recently they had been having a mysterious problem. For no apparent reason, a computer at Houston would occasionally send a command to the Cape—a valid, correctly formatted command that no one at Houston had intended to send. He telephoned his boss, Pete Clements, who was out at Ellington Air Force Base. Clements raced his car to the Control Center, thinking to himself that somehow his computer had caused this thing, whatever it was. Frere ran out to meet Clements's car as he pulled into the parking lot. "It's not us! It's not us!" he shouted.

Hodge was getting calls from his controllers that they had lost telemetry, but he couldn't make out what was going on. One minute they'd had the loop, then came "a great jumble of talk, difficult to

* The EECOM position monitored the environmental and electrical systems on the spacecraft.

understand.'' Then nothing but silence. Hodge called to the Cape on his loop to tell them that Houston had lost its telemetry. ''Stand by one, Houston,'' the Cape said—that's what the Cape would say whenever Houston called and they were busy. That's about all they got for a while, said Hodge. ''We could surmise that something serious had happened, but we didn't think in terms of all three of them being killed.'' He called Chris Kraft, who came over immediately. Shortly thereafter, George Low and Joe Shea walked into Mission Control.

Normally, Shea would have gone straight home from the airport, but he'd been away from the office for three days and wanted to leaf through the mail. He was at his desk for only a few minutes when Low called on the squawk box that linked the senior staff in Building 2, saying that there had been a fire in the spacecraft and that they had better get over to the Control Center. Shea swore briefly. It was one damned thing after another, and this accident, no matter how trivial it turned out to be, was going to mean a delay.

He and Low walked quickly across the M.S.C. grounds to Mission Control in Building 30. The spacecraft's hatch was still closed when they got there—or at least, that was the last word they had had from the Cape. They watched, and waited, and occasionally asked a question. ''Stand by one, Houston.'' Finally the word came through that the astronauts were dead.

Shea arranged for a NASA plane to fly him to the Cape that night, and then he drove home. By the time he arrived, his wife and daughters had heard the first bulletins over the radio. Shea took his wife to their bedroom to explain what had happened and momentarily allowed the anguish that was to consume him to burst into the open. But he stopped it there and packed quickly and went back to his office long enough to give Tom Markley some instructions. Shea would be working out of the Cape for a while. Tom would have to protect the program in Houston. It was going to be a difficult time. Shea left for Ellington Air Force Base to catch the plane.

It was the beginning of a long trip for Joe Shea. Some months earlier, speaking to a U.P.I. correspondents' dinner in Dallas, Shea had speculated on what this moment would be like. Sooner or later an astronaut would die, he had told them. That was inevitable. The odd thing was that in a way the effects might be worse for the people on the ground. The astronauts would be dead, whereas the people on the ground would have to live with the knowledge that they might have done things differently. This could be more painful than dying, Shea had said. Presumably, most of his audience thought Shea had been exaggerating to make a point.

4

By a strange coincidence—the fire was strewn with strange coincidences—the top executives of the whole Apollo Program were in the same room when the fire occurred. Webb, von Braun, Gilruth, Debus, Phillips, Mueller, plus the chief executives of the companies that held the prime contracts for Apollo—North American, Grumman, Boeing, Martin, Douglas, McDonnell—were all together, in one room, celebrating a successful day.

Some years earlier, Mueller had established what was known first as the Gemini Executives Group, later as the Gemini-Apollo Executives Group. At this meeting, with the Gemini flights successfully completed, it would become the Apollo Executives Group. It was an elite club that Mueller liked to bring together four times a year. Friday was the first day of a two-day meeting, and as of dinnertime it had been especially glittering. Sixty ambassadors had signed an international "space treaty" in the East Room of the White House that afternoon, and Webb had obtained President Johnson's permission to bring the Executives Group along for the reception.* It had been quite impressive—"a spine-tingling affair," Lady Bird Johnson wrote in her diary. Among other details, she recalled "Jim Webb's happy face."

The Executives Group had left the White House for dinner at the International Club on 19th Street a few blocks away. Vice-President Hubert Humphrey attended, along with members of the Senate and House who held key positions on the Space and Appropriations Committees. They were still standing around with drinks in their hands when the calls began coming in. No one is sure who got the first one—probably Kurt Debus, from Rocco Petrone in the blockhouse. Lee Atwood heard via a phone call from Stormy Storms at Downey.

Storms said abruptly that there had been a bad fire at the Cape. The capsule had burned. Atwood's first thought was that the whole notion was incredible—how could a capsule *burn?* But even as he turned to hand the phone to Gilruth, the waiter was paging Gilruth for another call, this one

* The space treaty forbade use of space for military purposes.

from Kraft. Atwood felt someone grabbing his arm. It was Webb, who wanted to introduce Atwood to one of the senators. Atwood told him the news and Webb left to call the White House.

The word spread slowly among the rest of the guests. Lew Evans of Grumman knew that something awful had happened—just watching Bob Gilruth's and Lee Atwood's faces told him that. But the rumors were confused—a fire in the stack, that's all that they knew for sure. Webb finally called for quiet and broke the news officially. After that, Evans recalled, "all the conversation was deep from the heart." The contractors who didn't have hardware that was involved remained at the International Club. The others hurriedly left to find airplanes to take them to the Cape. Gilruth, Mueller, and Phillips caught a ride with Debus.

5

It had taken an hour for NASA to put out the first bulletins, until 7:40 in the evening Eastern time, 4:40 in the afternoon in Downey. Scott Simpkinson was sitting in his room at the Tahitian Village, a gaudy motel with a go-go club decorated in a South Sea island motif. It was a great favorite of the NASA engineers working at the North American plant.

In the years since Simpkinson had led the pioneer team of Space Task Group engineers down to Hangar S at the Cape, he had become a troubleshooter for the manned space program. During Gemini, Simpkinson had managed the Office of Test Operations, a job that required him to prepare the post-flight technical report on each flight. Gemini's last flight had been two months earlier, and Simpkinson had finished the report just the week before. Upon his transfer to ASPO, Shea had sent him directly to Downey to assess the production situation for the Block I command modules.

At 4:40, Simpkinson and Andy Hoboken were going over the memo they had prepared for Shea. It was Friday, they figured they could get the final version of the memo in the mail that night, and they were talking about renting a car and driving up to the ski resort at Big Bear Lake for the weekend.

The television was on in the background as the two chatted; it took a moment before "fire" and "spacecraft" registered when the first bulletin was read. They looked at each other, wondering if they had heard right. Then Hoboken ran to the TV to turn up the volume, but the bulletin was over. He started switching channels while Simpkinson turned on the radio and scanned the dial until finally he found another newscaster quoting the first bulletin. It was unspecific, almost cryptic: "There has been an accidental fire at Launch Complex 34. There is fatality. More will be announced after next-of-kin are notified. The prime crew was in the spacecraft." There was no other information.

Simpkinson picked up the phone to dial a closely held number that would connect him directly with Mission Control in Houston. But it was as if the phone had gone dead—there was no operator, not even a dial tone. Finally Simpkinson said hello into the silent receiver and found that Robert Gilruth was on the other end—he had rung Simpkinson's motel room at the precise moment when Simpkinson had picked up the receiver to call Mission Control. Years earlier at Lewis, Simpkinson had studied the ways in which airplane fires propagate after a crash. Now, they had urgent need of his expertise. Gilruth told him to get to the Cape as soon as possible.

6

And so the word spread, and NASA changed forever. It was not only three astronauts who died on that evening in January 1967: Some of the space program's lightheartedness and exuberance died too. Things had been changing all along, of course, as Apollo swelled to thousands, then tens of thousands of workers. Things had had to change. Already it had been a long time since Abe Silverstein could name Apollo as he would name his baby, a long time since Max Faget had thrown paper plates from a balcony at Langley or Jack Kinzler had lashed a spacecraft onto the back of a flatbed truck. But the fire was a demarcation of the loss of innocence. Never again would individuals be allowed to take so much responsibility onto themselves, to place so much faith in their own experience and judgment. By the time of the fire, there were already many books of rules to go by, and thenceforth the people who had written the rules would themselves have to be governed by them. Nor would there ever again be so many new things to invent from scratch. Before the fire, there had been a sense in which Marshall was still Wernher von Braun's little band of Germans grown larger, the Cape still Kurt Debus's launch team grown larger, Houston still Bob Gilruth's Space Task Group grown larger. No more.

For many of the Apollo people, the night of the fire was also when they stopped believing in their invulnerability. "With a young guy, you don't realize sometimes you can't do something, so you go ahead and do it," said one Cape engineer. "As you get older, you get more cautious, and you have a fear of how you would look if you failed." The night of the fire, the young men of the space program suddenly got much older. After that night, there would always be a heightened sense not just that things could go wrong (they had always known that), but that things might actually go wrong, something that had been hard for the young ones to recognize.

But these were changes to be discovered in retrospect. That night the people of Apollo grieved. It made no difference how many thousands of people were in the program, it made no difference that only a few of them had known the astronauts personally. That night, the astronauts were mourned as family.

At the Cape, it was the numbed grief of people who have to carry on—hardly anyone on the vast Merritt Island reservation got home before the early hours of Saturday morning, and many didn't get home for days. The reaction was worst for the North American people: It was their spacecraft and they were the people who had certified that it was ready to fly. Marty Cioffoletti was by now working on the reaction and control system for the command module. He and a half dozen of the other North American engineers, people who had been working together for years, close friends, gathered in one of the North American offices. Emotions ran high. "People started saying, 'It had to be your system, because I know my system was clean,' " Cioffoletti recalled. "Those guys knew each other, and they would have staked their lives that the other guy would do the right thing in an emergency; and here they were, almost coming to a fistfight that night over whose system was to blame. And nobody knew whose system was to blame—Jesus, it was only an hour after the thing happened!"

At Houston, far from the scene, with nothing for most of them to do, the grief was more uncomplicated. When Jerry Bostick drove into the parking lot outside the Control Center—he knew only that "something pretty bad had happened"—the first thing he saw was his friend Dutch von Ehrenfried crying uncontrollably in the parking lot, saying over and over, "It's horrible! It's horrible!" Bostick was mystified. He wondered whether Dutch had experienced some strange kind of breakdown. John Aaron, who would be so dispassionately in control during moments of crisis in Apollo 12 and 13, was so shaken that Rod Loe went to the

Control Center to drive him home. Rod and Tina Loe would have no celebratory poker party that night.

Some of the other controllers who had been on shift in the Control Center went to the Singing Wheel, a nearby roadhouse that was a favorite hangout. They usually went there to wind down after simulations or to celebrate splashdowns. But they discovered that night that the Singing Wheel was also a good place for a wake. They sat at the bar, drinking and talking quietly until the small hours of the morning.

CHAPTER

15

"The Crucible"

The first of the small planes touched down at the K.S.C. airstrip shortly before midnight after the two-hour flight from Washington. Gilruth, Mueller, Debus, Phillips, and Atwood were aboard. The plane from Houston carrying Faget, Shea, and a half dozen of the key technical people in ASPO arrived an hour later.

They gathered first in the ACE Control Room, listening on headsets as the ACE controllers talked to the men on the tower who were removing the astronauts' bodies. For some of them, it was a time of belated revelation. Lee Atwood, president of North American, found himself "staggered" when he found out what the test conditions were—"locking those men in a spacecraft, pumping sixteen pounds of oxygen on them, and telling them to operate complicated electrical equipment . . ." It all seemed so obviously wrong to him—now. And yet they had been doing countdown tests in pure-oxygen environments since Mercury. They had sealed astronauts into spacecraft for pure-oxygen ground tests throughout Gemini. The designers of the Apollo spacecraft had weighed the pros and cons of changing to a two-gas system. There seemed to be too many cons, and finally they decided to leave it as it was.*

* As in many aspects of Apollo, all of the alternatives had dangers associated with them. Pure oxygen was essential in orbit (and accordingly was retained in orbit even after the fire) because cabin

Shea decided later that night to move into the astronauts' quarters. This was not normal procedure. The astronauts' quarters on the third floor of the Operations and Checkout Building were treated by everyone as the astronauts' private property—"a sacristy usually inviolate," in Shea's words. But now Shea felt that's precisely where he had to stay. He was already shifting into the mode he had used for the Titan project when it had fallen behind, for the Apollo spacecraft when it was stalled out at Downey: Move onto the site, work all three shifts, understand what has gone wrong, fix it. That's what he intended to do now at the Cape.

Shea then went down to a meeting with Gilruth and Mueller and Low. First they decided who was going to be on the investigative panel. They moved quickly through most of the names, but got into an argument when it came to choosing an astronaut for the panel. Gilruth wanted Wally Schirra, Grissom's backup for Apollo 1. Shea was opposed—Schirra was a good astronaut, Shea said, but he was not as immersed in the details of the spacecraft's design as were Jim McDivitt and Frank Borman, the prime candidates for the first lunar landing. The way Deke Slayton's rotation system was set up, it looked as if McDivitt might now be tied up preparing for the first Apollo flight. That left Borman. The rest were persuaded, and Borman was put on the panel. Shea finally got to sleep at 4 A.M.

1

Scott Simpkinson caught a red-eye from Los Angeles and got to Melbourne in the pre-dawn darkness. He rented a car and drove to the large, hangar-like areas in the O&C Building where the spacecraft were checked out. He arrived at about eight o'clock to find a collection of people beginning to set up the investigation.

---

pressure in orbit was only about one-third the pressure at sea level. Normal air at that pressure doesn't have enough oxygen to sustain life. For sea-level operations, the original Mercury capsule had pure oxygen in the suit loop and normal sea-level atmosphere in the cabin. But a test pilot for McDonnell (and brother of Warren North, who was in charge of astronaut training at Houston) almost died during a test involving the Mercury capsule precisely because (for a complicated chain of technical reasons) the cabin was not flooded with pure oxygen. This led to the initial decision to keep the cabin at pure oxygen on the ground as well as in space. That history, plus weight and reliability disadvantages of a two-gas system, affected subsequent design decisions during Gemini and Apollo. A similar balancing of dangers had to attend the decision after the fire to convert the hatch into an outward-opening one. The new design was safer than an inward-opening hatch while the astronauts were on the ground. It was less safe when the astronauts were in space, and enough heavier that it required a redesign of the parachute system, which in turn involved a host of competing considerations to be weighed and balanced.

Nobody was really running the meeting. Tommy Thompson, center director at Langley, had already been named as head of the Review Board, but he was still on his way down from Virginia. Shea was there, awake again after a few hours of restless sleep. Faget was there, and the others who had come in from Houston. The assignment of tasks could be roughed out without waiting for those who hadn't arrived. They told Simpkinson that he would be in charge of disassembling the spacecraft, and that he was to take it apart in such a way that every component of every system could be exonerated or implicated as a cause. It was the kind of meticulous job Simpkinson had made his reputation on, but this morning, with the world watching over their shoulders, everyone was on edge. They kept emphasizing to Simpkinson that he must find a way to disassemble 012 without destroying even a fragment of potential evidence. Simpkinson had never liked people telling him his job, and after putting up with this for a while, getting more and more irritated, Simpkinson left for Pad 34.

By the Saturday morning after the fire, a thick blanket of security had already been pulled over the Cape. Entry into the Merritt Island reservation was open only to people with high-priority official business. Launch Complex 34 was off-limits to anyone who was not directly involved in the investigation. Nobody got to Level A8 of the umbilical tower except someone like Simpkinson. He was alone as he took his first look at 012.

He touched nothing, of course. No one had touched anything, except for the medical team who had removed the bodies of the astronauts in the early hours of the morning (a grisly job made worse because the heat had melted the material in the suits and fused it to the couches and the cabin floor). Even looking into the spacecraft from outside the hatch, however, revealed the problem he faced. Everything had to be examined first without being touched. But the interior of the spacecraft had been baked into a friable crust. Simpkinson's dilemma was that he couldn't get into the spacecraft without messing it up; and he couldn't examine it without getting into it. Simpkinson shivered in the chill January morning, waiting for an idea. Finally he got back into the elevator and went looking for Sam Beddingfield.

Like many of the staff at Kennedy that morning, Beddingfield had worked through the night, grabbing a catnap on his desk as dawn broke. He was a pyrotechnics expert, and sitting atop the ruined spacecraft at Pad 34 was a fully operational Apollo escape rocket. No one knew how badly the support structure of the escape rocket had been weakened by the fire and no one wanted to find out what would happen if the structure collapsed. Beddingfield's job during the night had been to rig the crane on the launch tower so that it took the weight of the escape rocket off the spacecraft. He still had to decide how to detach the struts of the escape rocket from the body of 012 without jostling anything.

Simpkinson had known Beddingfield from the old days. In fact, Simpkinson had interviewed the young Air Force flight-test engineer when he had applied for a job at the Cape back in 1959. Later, he had sent Beddingfield to Picatinny Arsenal to study pyrotechnics. They hadn't worked together since Mercury, but Simpkinson remembered Beddingfield as "one of the few guys I could give a job to and it got done," and so it came about that Beddingfield found himself drafted for membership on the Disassembly Panel.

Simpkinson and Beddingfield returned to the O&C Building to figure out how to get into the spacecraft. By the end of the day, they had come up with a solution. They would use the four vertical struts on the outside of the spacecraft that supported the launch escape tower as the suspension points for a cantilevered Lucite floor. The floor would be hinged, like an end table, so that it could fit inside the hatch of 012 and then be unfolded to cover the interior of the spacecraft. That way the investigators could crawl on hands and knees around the interior, looking and photographing but not touching anything.

"If you wanted something made, it got made before you could blink an eye," recalled Simpkinson. All the resources and every employee of the Kennedy Space Center—as well as any of the other centers and the contractors—were at the disposal of the investigating teams, day or night. Simpkinson wanted a Lucite floor? Cantilevered but strong enough to hold the weight of a few men at a time? Sculpted to fit the inside of the spacecraft? Hinged? The contraption was designed, fabricated, and assembled within the next two days.

2

Joe Shea went from the meeting in the O&C Building to the offices in the front of the building to brief Bob Seamans, who had just arrived. By that time, Shea was able to recite to Seamans a large number of things that probably hadn't caused the fire; the actual cause remained mysterious. Shea also ordered that spacecraft 014 out at Downey be shipped to the

Cape—014 was nearly a duplicate of 012. It would be used for practice in taking apart the burned spacecraft.

Those tasks out of the way, Shea went back to the working area in the O&C where endless strips recording the data from Friday's Countdown Demonstration Test (C.D.D.T.) now papered the walls. It was there that Shea, with Slayton and a handful of others, first listened to the recording of the astronauts' voices in the last moments. Everyone in the room was carefully clinical. Had the astronauts had any earlier indications of problems? No evidence of that on the tape. Perhaps more sophisticated analysis might be able to tease it out.

Shea spent the rest of that Saturday arguing about the schedule with Mueller and Phillips and Gilruth. The fire was going to set them back. Mueller wanted to make up the time. The easiest way to do it would be to put the all-up philosophy to work again. Instead of flying a repeat of the Grissom mission—Grissom's crew was to have orbited the earth in a command module, without a lunar module aboard—Mueller argued for skipping it and going directly to an earth-orbit test of both the command module and the lunar module. Gilruth was opposed. If Apollo 1 had been a useful mission (and if it wasn't, why risk three men?), then they owed it to the program and to the memory of the lost crew to do the mission. It was ultimately an emotional argument, Shea thought, but Gilruth's strategy was right for the wrong reasons. Shea had been part of the trauma at Houston just twenty-four hours ago. The next flight was going to have everybody on edge anyway. Adding a LEM to the first mission after the fire would be an unnecessary stress on an organization that was already scraped raw. Make it as simple as possible, Shea thought: Use a Block II C.S.M. on the first flight, but leave the LEM behind.

Shea found himself taking Gilruth's side against Mueller, which didn't happen often, and "the arguments on my side were all qualitative, nonanalytic, the worst kind to try to win with Mueller," Shea said. Mueller was intransigent at first, but Gilruth and Shea kept at it. At two o'clock on Sunday morning, Mueller went to bed, saying perhaps it could be worked out. Shea, who had now been functioning almost continuously since he had had breakfast with the Apollo 1 crew on Friday morning, still couldn't sleep. He stayed up again until four o'clock, talking with Sam Phillips.

By eight o'clock Sunday morning, Shea, Mueller, Gilruth, and Phillips were conferring again, pinning down the details of their decision of the night before. It was a three-hour job, finally ending with agreement that the next flight would be a Block II C.S.M., Grumman would get some breathing space in its schedule, but that the all-up test for the Saturn V—Mueller's central concern—would go ahead as planned.

That resolved, the next question came up: Who would represent the Office of Manned Space Flight during the investigation of the fire? Mueller was head of manned space flight for all of NASA, but he couldn't do it; he would have to stay in Washington to represent O.M.S.F. with the administration and Congress. Sam Phillips, Mueller's deputy and head of the Apollo Program Office at headquarters, was the logical alternative. Phillips had already said so to Mueller. But Shea wanted the job for himself. He proposed a break and took Mueller aside.

ASPO was his responsibility, Shea told Mueller, and ASPO was going to take responsibility for the fire. But if ASPO was going to take responsibility, then the guys from ASPO had to be "the catalysts to solve the problem." That was the only way for them to work through this thing. So Shea had to be O.M.S.F.'s lead man. Shea added that he wasn't "going to sleep anyway until I understood what had happened." For the second time in twenty-four hours, Mueller accepted an argument based on emotional considerations. Shea had the job.

At mid-afternoon on Sunday, Shea left for Houston and still, after almost forty-eight hours, his work for that weekend wasn't done. He went home, changed, and spent the rest of Sunday with his wife, driving from home to home in Clear Lake, visiting the widows of Gus Grissom, Ed White, and Roger Chaffee. Then back to the Cape.

Early Monday morning, he threw himself into the investigation with an intensity that eclipsed even his previous episodes on the Titan and Apollo—he was "on high blower," he would say later, a jet on afterburner. He broke off only once during the first week, to return once more to Houston to brief a meeting of the Apollo contractors. "What is done is done," Shea began, and he clicked off a brisk summary of where the program stood. He discussed how they must guard against overreaction. He concluded with a call to do better as fulfillment of their obligation to the dead crew.

The speech was balanced and disciplined—dispassionate without being cold, optimistic without being sanguine. Seamans watched him and thought that the words sounded fine, but that Shea looked exhausted. Something in Shea's tone bothered Deke Slayton, too. He telephoned Shea at home that night, just to say he had heard the speech and was worried that Joe might be taking the accident too personally. He shouldn't do that, Slayton said. Everyone had confidence in him. Shea thanked him for his concern, and went back to the meeting he was having in his living room with Charles Berry, the chief of the Medical Division. Shea got to bed about midnight. He was up before dawn to catch a plane back to the Cape.

On the way to the Cape after the speech to the contractors, he wrote down in his technical diary an eleven-point agenda for getting the program back on track. At the end of it, he wrote a note to himself: ''I may be over optimistic (as usual) but the above seems like it is so obvious now that it has to go. Get memo dictated in detail today & assignments going. Begin to focus Board redesign suggestions so good guys can concentrate.''

3

Up on Level A8, and later in the cavernous Pyrotechnic Installation Building over near the O&C Building, Simpkinson's disassembly of spacecraft 012 was under way, an investigation that would be remembered by many as the most intense single effort during the Apollo Program.

It was a tortuous experience. Marty Cioffoletti, the young North American engineer, remembered writing Test Preparation Sheets (T.P.S.s) for taking out the engines from the command module and disassembling the propulsion system. He explained the process. Suppose the disassembly was approaching the point where the team would be ready to take out a screw in one of Cioffoletti's systems. Before Cioffoletti or any of his technicians could touch it, Cioffoletti had to write a T.P.S. The T.P.S. would specify the physical action (unscrew the screw). The part number of that particular screw. The torque that was supposed to be necessary to break the screw loose.

With the T.P.S. in hand, Cioffoletti, the presiding engineer, would read an instruction. A North American quality inspector would move into place. One of NASA's inspectors would move into place. A photographer would be called over. A North American technician would get into the burned-out spacecraft (observing whatever specific precautions were detailed for getting to that particular part of the spacecraft at that particular stage of the disassembly process) and then, using the specified mechanical device, take the screw out. The engineer would record the torque necessary to break it free. The technician would hand the screw to the North American quality inspector, who would make sure that it was the right part and the right part number, recording his results on his copy of the T.P.S. The North American inspector would hand it to the NASA inspector, who would record his observations. The photographer would take a picture of the part. The engineer would then put the screw in a plastic bag, label it, and take it to the appropriate repository.

If in the course of this process you found that an unanticipated step had to be added, or a planned step changed, everything came to a halt while the revised T.P.S. was sent to the Review Board for approval of the deviation—"even though you were the one who wrote the procedure in the first place." This, for a screw. It was, Cioffoletti said, "the most excruciating technical dissection of a machine I could ever imagine happening."

It drove some of the astronauts to distraction as well. They knew the interior of that spacecraft. It was a second home. They knew what it looked like and felt like, how you moved around in it, the things that got joggled and bumped and scraped. If only they could get to 012—but there was Scotty Simpkinson, the Cerberus of the disassembly hangar, growling at anyone who tried to get close without following procedures. In retrospect, Simpkinson was sympathetic. "They just got disgusted with the rigmarole you had to go through to take a look at an idea you had, you know. 'What if this happened? Well, if we could just see that thing, we could tell if it did.' Then they'd have to go through badges, and certain times of day, and all that kind of stuff. And it kind of got them upset." But there wasn't any choice, Simpkinson said. "You just cannot go haphazard at an accident review, that's all. Just can't do it."

And through it all, the North American engineers kept saying to themselves, Please don't let it be my system, please let my system be clean, please let it be someone else's responsibility. "I hate to say that," Cioffoletti said, "but you really thanked God it wasn't your system, that it was somebody else's, and you could breathe easy again."

Once the data had been pulled together and one could be fairly sure about what events had occurred, in what order—Simpkinson's Disassembly Panel was just one of twenty-one panels working on the overall investigation—there remained the problem of deciding the cause. A fire has three requirements: oxygen, flammable materials, and an ignition source. The spacecraft had oxygen in abundance, pure and under pressure. The spacecraft also contained flammable materials, more than anyone had intended. All that the investigators needed to do was discover the ignition source.

The panels on the investigation kept weeding out possibilities until finally two things became clear. First, they would never be able to prove exactly what had happened, though a short in the cable bundle near the E.C.U. was the prime suspect. Second, this had not been a one-in-a-million freak accident that could never happen again. There were lots of ways a fire could have gotten started in that spacecraft. Combustibles? There were 5,000 square inches of Velcro in a cabin that was supposed to have contained 500. A source of ignition? Unless the elements of the electrical system were installed immaculately, the tangle of wiring in the spacecraft could be a scattering of match heads waiting for something to scratch them.

North American quickly came under scrutiny. The next spacecraft in line, C.M. 017, intended for the first all-up flight of the Saturn V, had arrived at the Cape two weeks before the fire and been subjected to the standard inspection before being taken to the V.A.B. for mating with the launch vehicle. After the fire, it was removed and taken back to the O&C Building's White Room for a new and more thorough examination.

Like everyone else at the Cape, the inspectors waited for the 204 Board to complete its work, fearful that the fire had been caused by something they had missed. Now, with that possibility still on everyone's mind, the inspectors went after 017 in a state of mind that bordered on obsessive. They began finding skinned wires and called Joe Bobik, the Cape's chief inspector for the spacecraft. Bobik went to the White Room and looked for himself. "The more I looked, the worse it looked. The wires weren't routed neat, they were crisscrossed in a junky way. That kind of stuff. I went to Rocco [Petrone] and I said, 'Hey, Rocco, we've got a problem. I want to inspect that whole vehicle.' " They were finding too many skinned wires, he told Petrone—twenty or thirty already. Petrone was incredulous. "Yeah," Bobik told Petrone, "don't look too great, either; workmanship's not the best."

"We can't take out any instrument panels," he remembered Petrone saying. "You take them down, and you've got to redo all the tests. Do you know what you're asking?"

"Yeah, but we gotta take 'em down," Bobik said. "Just too many problems. Can't fly that way."

Rocco Petrone came down to the White Room in the O&C to see the vehicle. He looked, and shook his head, and swore. The more Joe Bobik thought about it, the more he began to think that even taking down the instrument panels wasn't enough. He had been bothered for some time by the evidence of people "pushing too hard" out at Downey—skinned wires, sloppy work. "Send it back," he told Petrone. "Send the damn thing back to the factory. It's that bad."

Petrone couldn't do that. This was late February, only weeks after the fire, and for NASA to let it be known that the next spacecraft was being sent back to the factory was tantamount to killing the North American contract—Congress, gearing up for its own investigation, was unlikely to accept anything less. While few at NASA would grieve for North American, to switch the spacecraft contract to a new company would mean a long delay in the moon landing—two years at least, probably more, pushing it into the 1970s. More than that, the program itself would be in jeopardy.

A flurry of visits to spacecraft 017 followed. Shea came over from the disassembly area to examine it. Bobik would remember seeing tears in Shea's eyes as he left. Gilruth flew in from Houston. North American people flew in from Downey. And finally, Sam Phillips, chief of the Apollo Program Office at NASA headquarters, flew in from Washington. "I took him in the spacecraft, I started showin' him all this stuff," Bobik said. "He didn't know what to say. Very quiet. Came out, didn't say anything. Not one word or nothin'. I knew he didn't like it, but I couldn't figure out what he was going to do or say."

On the 2nd of March, ASPO told headquarters that because of many wiring discrepancies found in Apollo spacecraft 017, "a more thorough inspection was required," and all twelve main display control panels were to be removed. The "more thorough inspection" eventually revealed 1,407 errors.

4

The collective grief that had stricken the Apollo Program on the night of the fire hung on for weeks. "The more we probed for answers, the more depressed [the people in the investigation] got," Frank Borman wrote later. "They'd take downers to ease the pain of guilt and uppers so they could face the next day." The depression affected the Cioffolettis down near the bottom of the system and it affected those at the top. One of the NASA public-relations men remembered joining a gathering of the senior people in the program in a Cocoa Beach bar less than two weeks after the fire. People who hardly ever drank were drinking martinis, steadily. To Frank Borman, one of the most straight-arrow of the astronauts, "getting drunk seemed like a good idea." Max Faget, abstemious by temperament and usually too wrapped up in his latest idea to have time for a drink anyway, ended up doing a handstand on a chair as a go-go dancer continued her act a few feet away. They ended the evening throwing glasses against the wall—it reminded Borman of an old war movie.

It was the gaiety of a wake, and lasted only as long as the party. Later that night, Shea (who had also been part of the group) went back to Borman's room at the Holiday Inn, where they talked somberly until 3:30 in the morning. Shea was up again at 7:30.

That was his schedule during the first three weeks—work for sixteen hours or so, sleep for four or five, and use the rest to play handball or go out with friends or drive at high speeds on Merritt Island's narrow roads, using physical depletion as a drug. Even exhaustion wasn't always enough, and Shea too found himself using Seconals and scotch to help out. But the hours paid off in progress, and four weeks after the fire he moved back to Houston to pick up his work at ASPO, confident that the investigation had gotten on top of the situation at the Cape.

To some of the others in NASA, Shea seemed to be fraying at the edges—he was too exhausted, too driven. Shea himself didn't feel exhausted at all; if anything, he felt as if he had almost too much energy. His colleagues' worries were compounded when without warning a senior executive for one of the contractors, a man greatly admired within NASA, suffered a nervous breakdown in Shea's office during the middle of a briefing. Shea accompanied the executive home on his company's plane.

Shea's relationship with Gilruth, which had been cordial but never close, was also deteriorating. Just a few days after the contractor had broken down in Shea's office, Gilruth asked Shea why a certain set of data wasn't ready yet. Shea snapped that it wasn't ready because (a) he wanted it really right, and (b) he took an extra hour's sleep one night. Gilruth was disturbed. As the investigation continued, Gilruth decided that Shea hadn't kept him informed about what was going on before the fire, and began to wonder whether Shea himself was heading for a breakdown.

In mid-March, Shea was scheduled to deliver the annual Goddard Lecture, part of a prestigious two-day event within the space fraternity, in Washington. Webb and Seamans saw a copy of his draft and were dismayed. Shea had entitled the lecture "The Crucible of Development"— "crucible," with its connotations of fire and molten metal. It had been a deliberate choice of words. Shea had been thinking about the speech for weeks, ever since the accident, jotting down phrases and topics, and the concept of "crucible" had been one of the first to occur to him. Indeed, it was precisely because of the connotations of "crucible" that he had chosen it. Webb and Gilruth had forbidden him to talk about Apollo (the investigation was still ongoing), so Shea had tried to build a talk using historical analogies for what he wanted to say: that technological development is "a severe, searching test"—a dictionary definition of "crucible"—and the fire had been just such a test. Webb called Shea to his office and persuaded him to change the title to "Research and Development in Perspective." Shea consented, reluctantly. He in his turn was frustrated with Webb and Gilruth and the rest—in his view, he at least was trying to confront the larger truths of what had happened. The others were trying to back away from it.

Read in cold print, Shea's speech is a densely textured discussion of the ways in which human beings go about advancing their technology, interspersed with quotations from Montaigne and Milton, with historical references to Galileo's *The Starry Messenger* and to the anthropological discoveries of Louis Leakey. On top of everything else, Shea added some puns to the speech, trying for humor. He concluded with a passage from Shelley's *Prometheus Unbound*. It was not a run-of-the-mill engineering lecture.

Listening to it that evening in Washington, some of Shea's colleagues stirred uneasily in their seats. The puns didn't work, not now. The delivery, by a driven man, had an unnerving edge to it. This was not the Joe Shea they knew.

Webb played a fatherly role with Shea during February and March. When Shea came to Washington for the Goddard Lecture, Webb personally picked him up at the airport and persuaded Shea to stay at his home. Shea had come without his topcoat and it was still cold in Washington. Webb insisted that he take an old topcoat from Webb's State Department days, an elegant black one. Shea was touched, and he wore it for the next month, thinking of it as a symbol of Webb's caring and protection.

Webb's sympathy was authentic, but he also had a tough problem facing him: Both houses of Congress were about to convene hearings on the fire, and the fate of the Apollo Program hung on their outcome. Senator Walter Mondale of Minnesota, who had become a prominent critic of NASA, had already let it be known what they would be like, charging that NASA's engineers—presumably meaning chiefly Shea—were guilty of "criminal negligence." In Webb's opinion, Shea was under too much stress to testify. Webb and Berry, the physician, insisted that Shea take a vacation. Shortly after the Goddard Lecture, Shea took his wife and children up to a place in central Texas, planning to stay for two weeks.

After only a week, Berry came up to visit Shea at his vacation retreat. The news he brought was that Seamans had flown down to Houston from Washington and had sent Berry to tell Shea that they wanted him to take an extended leave. An announcement to that effect would be released tomorrow morning. Shea told him that in that case they could announce his resignation instead. "You've been working on half-truths and rumors and things you don't understand," he said. He and his guys down at Houston were doing fine. They were responding to the crisis, coming up with fixes, and weren't going into a state of shock. It was the people above them who were in shock, not the troops. However, Shea told Berry, he would abide by the decision of any competent psychiatrist. "If he thinks I need an R&R, then we'll discuss it. But I don't like amateurs judging my psyche."

They set up a session for that very evening at the Houston Medical Center, with two Houston psychiatrists chosen by NASA doing the evaluation. Shea spent two and a half hours with them. Their conclusion was that Shea was under strain, yes, but without a psychosis. In fact, the psychiatrists told Shea, he had so thoroughly analyzed the experience for himself that he was probably psychologically stronger than before the accident. Gilruth had been worried that Shea would come apart if another crisis occurred, but the psychiatrists disagreed. Shea did react strongly to such events, leading to a purely physiological problem of high blood pressure, but that could easily be treated with medication if another such crisis occurred. These findings were fed back to the waiting NASA officials. Shea declined to take extended leave and the press release was canceled. He thought "the thing was absolutely put to bed."

A week or so later, George Mueller called. He wanted Joe to move up to Washington and be his deputy. He made it sound enticing. Seamans called Shea, and told him how much he wanted him to accept—Seamans and Mueller had been drawing apart, Seamans said, and they needed Joe to serve as a bridge. Jim Webb joined the chorus. Joe had spent enough time on the detail level, Webb told him. It was time for him to be working at the policy level. When Webb and Shea happened to be on the same plane with agriculture secretary Orville Freeman, Webb introduced Shea to Freeman by saying that he wanted Shea to interact "with all you cabinet people." Webb knew how to take people up on the mountaintop, Shea decided later, after he had already accepted the job in Washington.

The reality was that the top management of NASA had determined to get Shea out of ASPO and out of the line of fire. Seamans's earlier trip to Houston, when Shea had been examined by the psychiatrists, had been undertaken with two purposes in mind, one open and one covert. The open one was to announce Shea's extended leave of absence. That plan had to be scrapped when Shea balked and then the results of the psychiatric examination thwarted them. The covert purpose was to come to an agreement with Gilruth and George Low, now the deputy director at Houston, that Shea would have to be replaced. The covert purpose was accomplished. Perhaps the psychiatrists were right that Shea had no psychosis. But the question for Seamans and Gilruth and Low wasn't whether Shea was sane; it was whether he should continue to run ASPO. They saw that as a judgment that the psychiatrists weren't competent to make. For that matter, some of the people who knew Shea had a feeling that the psychiatrists hadn't quite realized what they were taking on when they interviewed him. One of Shea's friends talked to the psychiatrists just after they had finished the examination. He would laugh ruefully when he recounted the story later. "The psychiatrists came back saying, 'He's so smart! He's so intelligent!' Here Joe was, ready to kill himself, but he could still outsmart the psychiatrists."

Shea felt he had passed that stage weeks earlier. Then and in later years, Shea's own assessment was that the psychiatrists were right. He had been deeply saddened by the fire, he acknowledged, but he believed that after a difficult beginning he had come to terms with it. To Shea, another factor was at issue. Shea felt strongly, and was saying so at the time, that it would be a mistake to go back into the Apollo spacecraft and make sweeping changes. The appropriate fixes were straightforward and limited; do more than that, and the program risked coming out of the process with a less mature spacecraft in which new problems might be hiding.

Shea had few allies. The prevailing view was that a sweeping review of the entire spacecraft was in order, and that fixes should not be limited in the way Shea wanted. These diametrically opposed views of the appropriate strategy made keeping Shea complicated for a variety of reasons. At the same time, Shea's senior colleagues in NASA were convinced that Shea had been too deeply affected by the fire to manage its aftermath, and believed that it would be best for all—including Shea—if he left ASPO. Even as Shea was assuming that the problem had been "put to bed," the others were deciding on the man to replace him.

On Monday, April 7, 1967, a House space subcommittee opened hearings into the causes and impact of the Apollo fire. On the same day, Shea left ASPO to become deputy associate administrator for Manned Space Flight at NASA headquarters, three years and five months after he had gone to Houston.

The next day, the lead editorial in *The New York Times* was devoted to a commentary on the report of the Apollo Review Board which had been released a few days earlier. The *Times* was appalled at "the incredible complacency" of the NASA engineers. "Even a high school chemistry student" should have known not to use a pure-oxygen environment. The editors' judgment was austere and unrelenting: "The dry technical prose of the report convicts those in charge of Project Appollo [sic] of incompetence and negligence."

During his first few days in Washington, Shea was excited about his new job, eager to work as Mueller's deputy. But then Mueller told Shea that of course Shea must realize that Sam Phillips, head of the Apollo Program at O.M.S.F., would not be reporting to Shea, and Shea decided he'd been had: Mueller's office was a good place to keep him tucked away during the congressional postmortems on Apollo.

Shea became uncomfortable at headquarters, and trying to work on a few pet projects there didn't help. His wife and daughters were still in Houston. He found himself sitting alone in his office for hours on end, reading the newspapers. Sometimes he would walk out into the streets of Washington, always alone, prowling the museums, sitting on a bench in the gardens of Dumbarton Oaks. The Phillips Gallery became a favorite haven in those days, quiet and serene. Sometimes, he found himself wishing he had been in the spacecraft on that Friday evening in January. Finally, in July 1967, six months after the fire, Joe Shea left NASA and took a job as head of engineering for a company up in Boston.

Another six months passed. One winter's day in early 1968, as Shea talked to someone on the telephone, Berry Shea heard her husband laugh. It was the first time she'd heard him do that in almost a year, she remarked when he hung up. Shea thought that it seemed longer.

5

So passed the fire. For those who weren't in the program, it was soon forgotten. Pad 34, far to the south of the huge complexes where they launched the Saturn Vs, is not a stop on the bus tour of Kennedy Space Center, and there isn't much to see anyway. The squat domed blockhouse remains, padlocked. The great slab of the launch pad is still there, grass growing through the cracks in the concrete, as are the rusty railroad tracks once used to bring up the gantry. The concrete flame deflectors, too massive to move, are off to one side, with faded painted stenciling— ABANDON IN PLACE. Otherwise, there is just scrub and sand and the sea.

The ironies surrounding the fire make it the stuff of fable. The astronauts, a few weeks from confronting the perils of outer space, died on top of an unfueled booster sitting on the pad—that is the most obvious irony. But there are others.

Consider the case of Gus Grissom and his appointment in Samarra. The first spacecraft, the Mercury capsule, had an explosive device on the hatch so that the astronaut could escape quickly in an emergency. If there had been such a device on the Apollo spacecraft, the crew would have blown off the hatch as soon as the fire started to spread, the pure oxygen under pressure in the spacecraft would instantaneously have been replaced by ordinary Florida air, and the fire would have guttered. The crew would have gotten a scare and Grissom might have gotten a singed leg.*

But at the end of his suborbital Mercury flight in *Liberty Bell 7,* Gus Grissom's hatch blew and Grissom nearly drowned. Grissom insisted that the hatch had misfired on its own. Because of that, the explosive hatch was reckoned to be too dangerous, and the designers of the hatch for Apollo focused on making sure that the hatch was safe from accidental opening, not on ensuring that it could be opened quickly. Thus, when the fire broke out, Grissom was sitting inside a spacecraft whose hatch was secured by six bolts.

Consider the strangely intertwined sequence of events that Sam Beddingfield was left to contemplate on quiet evenings. Sam Bedding- field was an old friend of Grissom's. He had been disturbed after *Liberty*

---

* Reflecting on this, Rocco Petrone pointed out that he probably wouldn't have permitted an explosive hatch to be armed for a ground test, because of the potential danger to the ground crew in the White Room—which opens a whole new set of what-ifs (one may imagine Petrone's unenviable position after the fire, in that case). The Apollo hatch as redesigned after the fire was not explosive, but could be opened in three seconds.

Bell 7 by how ready the astronauts and the engineers were to blame Grissom for the blown hatch. He believed in Grissom's composure and honesty, so he had approached his job as one of the investigators of the *Liberty Bell* accident with special zeal. It was Beddingfield who had identified two specific, documentable sequences of events that indeed could have blown the hatch in the manner that Grissom described. It was largely because of that work that the hatch was discarded for Gemini and Apollo. If Beddingfield had been less motivated by his friendship with Grissom, would those alternative explanations have been found and demonstrated? If they hadn't been found, would Grissom still have been on flying status during Apollo?

There were two smaller, bitter ironies for Beddingfield as well: The plugs-out countdown on January 27 was to have ended with a test of the emergency egress system—as it certainly did. And Sam Beddingfield himself was to have been the supervising engineer.

Consider this elusive overarching irony: Might it not, after all, have been a good thing for Apollo that the fire occurred? Many in NASA thought so. Shea himself decided that the fire provided a better and more "gentle"—his word—ambience at the top. "The fire could have been, and for a lot of people it was, a unifying force," he said. "It brought home how the things we used to bitch about, the trivia, the nitpicking, any of those itty-bitty things could cause a problem." After the fire, there was more of a "we're all in this together" feeling, Shea thought.

A senior ASPO engineer once talked about the personal loss the fire represented to him—as Houston's project officer for the LEM, he had gotten to know Grissom and White well—but, still, "from an overall standpoint of the program, it might have been one of the best things that could have happened." It is a tough judgment, he recognized, but "I think we got too complacent in the manned program. . . . The fire really woke people up." Another engineer thought that the fire gave the Apollo Program some time it was unwilling to give itself: "The Apollo design had progressed to a point where a lot of things were put up on the shelf as being the kind of thing—'Well, let's not worry about that right now, we'll pick it up later.' " He said it wasn't so much that Apollo 1 would have been dangerous if they'd flown it, but rather that "once the fire occurred, the flight schedule came to a screeching halt and everybody stopped and took stock." People then had the time to go back and work in a less pressured fashion on "all of these things that everyone had in their back pocket that they should have worked on, and hadn't had a chance to."

There were other considerations as well. If the fire hadn't delayed the flight schedule, there would have been delays anyway. The lunar module wasn't going to be ready for its first scheduled flight. The fire gave Grumman a chance to catch up. The Flight Operations Division was still in the process of re-equipping the Control Center to support Apollo missions; they would have been hard pressed to keep to the schedule that NASA was using before the fire.

And there were the Velcro and the Raschel netting: If it hadn't been for the fire on the ground, there might well have been a fire in space. Tom Markley grieved for what the fire did to Shea, but he remembered too the day a week after the fire when Caldwell Johnson came running into his office and set up a movie projector. "Let me show you how this damn stuff explodes in pure oxygen," Johnson said, and turned on the projector. Markley was "totally aghast" himself. "It just went *ZOOP!* It was unbelievable. The stuff burned like you couldn't imagine." It was Markley's opinion that "if we had had the Apollo fire in orbit or going to the moon, we wouldn't have flown for another decade."

For Joe Shea, the accident's ironies achieved the dimension of tragedy. Consider the paradox in Shea's narrow escape, when only a technical hitch kept him from being in the spacecraft during the fire. So lucky for him, it first seems, that they couldn't hook up the headset in time. So morbid of him, it first seems, to wish occasionally that he had been in the spacecraft after all. And yet if Shea had been in the spacecraft, he would have been sitting in the bay beneath the astronauts' couches—down beside the E.C.U. and the netting that probably first caught fire. Furthermore, the film that Caldwell Johnson showed to Markley also showed that the hot spot would have remained there like an ember for ten seconds before it went *ZOOP*—ten seconds for Shea, lying with his head a few feet from the E.C.U., to see the glow and smother the fire with his hand or shoulder. In the months after the fire, Shea, brooding about that possibility, even came to an estimate. It was better than an even chance, he decided, that he would have seen it and have been able to react in time to smother it. "And I really don't want to think about it any more," he said, "because you can't get any more data."

Consider how differently the dice fell for the men who followed Shea. Eighteen months after the fire, George Low persuaded NASA to take the audacious step of sending Apollo 8 to the moon on a circumlunar flight, arriving at the moon on Christmas Eve, 1968. Shea would have fought the decision—too big a risk. But it was a brilliant success. And yet, suppose that the accident which would occur on Apollo 13, when the oxygen tank exploded, had occurred on Apollo 8 instead. Apollo 13 used its lunar module as a lifeboat to keep the crew alive during the return to earth. Apollo 8 had no lunar module. The crew of Apollo 8 would have been dead when the spacecraft reached the moon on Christmas Eve, and George Low would be remembered as the man who took a crazy chance just to get to the moon by an arbitrary deadline.

Or there was Apollo 12, two and a half years after the fire, when the spacecraft was struck by lightning a few seconds after launch. Every system in the spacecraft was battered by a gigantic charge of electricity. But the flight controllers checked out the systems while the spacecraft remained in earth orbit, and, when all seemed to be in order, Chris Kraft and Rocco Petrone approved their recommendation to send Apollo 12 to the moon. It was another courageous decision, and the rest of the mission was nearly flawless. Yet, if the electrical surge had weakened one of the systems in an unsuspected way and a catastrophe had resulted—for that matter, even if something unrelated to the lightning had caused a catastrophe—then Kraft and Petrone would be remembered as the men who insisted on sending men to the moon in a spacecraft that had been hit by lightning.

For Apollo 1, a spark became an ember, the ember became a blaze, and Shea held himself responsible. Dispassionately, he could review the events leading up to the fire and conclude that he had not been pushing the program at the expense of safety; that, strictly speaking, the fire was not his fault. Nonetheless, he refused to grant himself absolution. Years later, the question was put to him: What then was the mistake for which you wouldn't forgive yourself? "For me, it isn't as simple as 'the mistake'— it goes back to a concept that when you are responsible for a project, there are no shades of gray," he wrote in response. "You may not know every detail, but you should construct an environment where *nothing* can fall through the cracks. For Apollo, that was rigorous ground test and closing out all action items before the real, irreversible danger begins at launch. Ironically, even that philosophy worked. The fire did indeed occur in ground test. The flaw was not recognizing the danger of the plugs-out test in pure oxygen at atmospheric pressure, and closing out the critical actions before that test."

"It's another lesson in life," Glynn Lunney once said, thinking of such things. "Boy, you can think you're the smartest sonofabitch anybody ever saw, but there's so many events that occur that can affect one's performance or one's role in life. You still can't stop people from stuffing rags in pipes and things like that, all of which make somebody look dumb in retrospect. The obvious question is 'Jesus, why didn't you guys see that?' '' That is the question Shea refused to quit asking of himself. For years after, as he neared the top of one of the nation's largest high-tech corporations, he kept the photograph that the crew of Apollo 1 had given him displayed prominently near the front entrance of his house. Shea would not let himself so much as enter or leave his own home without passing by the inscription: "It isn't that we don't trust you, Joe, but this time we've decided to go over your head."

John Hodge, the Houston flight director on duty during the fire, once reflected on how the fire brought out a difference between engineers like Shea who came out of the world of missiles and systems engineering, and those who had come out of the flight-test business. For a Joe Shea, there must always be a way to do it right, so that nobody gets hurt; it's just a matter of being smart enough and careful enough, and if you aren't, it's your fault. Hodge shook his head at that thought, comparing it with the attitude of the people who had come out of flight-test. "You understand that that risk is there," Hodge said, "and when it happens it's terrible, you wish it didn't happen, you wish you were smarter, but you know it's going to happen and so you learn to live with that. You worry about it a lot and you think about it a lot but when it does occur it doesn't kill you." There is a truth that his friend Joe Shea couldn't accept, Hodge said, that few people outside the business can: "You lose crew. Pilots die flying experimental aircraft."

16

"You've got to start
biting somewhere"

During the years when the spacecraft was being designed and built, the man who managed ASPO was the fulcrum of the spacecraft side of the Apollo Program, with an influence far beyond his place in the organization chart. This had been apparent in the fall of 1963, when Joe Shea went to Houston, and it remained true in the spring of 1967. On Sunday morning, April 2, Robert Gilruth and George Low flew to Washington to recruit Joe Shea's replacement.

Their choice was Chuck Mathews, a founding father of manned space flight—one of the Space Task Group's original forty-five, the first director of Flight Operations for Mercury, the man who rescued Gemini when it encountered managerial troubles. With the successful completion of Gemini, Mathews had recently been promoted out of Houston into Mueller's office at headquarters. Gilruth and Low checked into the Georgetown Inn and invited him over for a talk.

They ran into a stone wall. Mathews had just moved to Washington and promised his family that this was it—no more moves for a while, no more weeks on end when he hardly saw the children. The ASPO job was tempting, but he just couldn't accept. Anyway, Mathews and George

Mueller had been talking it over themselves and had decided on another person for the job. The person they had decided upon was George Low.

Low was nonplussed. Thinking back on it later, Mathews, who was one of Low's closest friends, reflected that he was putting Low into an unusual position. George Low always preferred that things proceed in an orderly process and that he, George Low, be in control of that orderly process. Usually, he was. Chuck Mathews was supposed to take the job, that's the way Low had planned it all out. "He had never, I guess, experienced a situation like this one," Mathews said, "where someone put the finger on him and he didn't have control." Finally Low said, "Chuck, I've got the worst headache I've ever had in my life." Mathews suggested they go for a walk.

They strolled in the spring twilight along the quiet residential streets of Georgetown, talking intently. The new job would be a demotion, of course, from number-two man at Houston to manager of a program office. But Joe Shea had taken a theoretical demotion when he went to ASPO; so had Low, when he went from headquarters to the deputy director's job at M.S.C. Apollo managers bounced around the hierarchy without paying much attention to that sort of thing. Low in particular had never been a man who "worried about the color of the carpet on the floor," in Mathews's words, and he knew that ASPO was the critical job right then no matter what the organization charts said. Still, Low had carefully decided what ought to happen—Mathews would manage ASPO—and it took a long time for him to reconcile himself to a different plan. They walked for hours, Mathews remembered, getting back to the Inn late at night. But finally Low agreed to take the job.*

1

By good luck as much as anything else, the Apollo Program seemed to get the right men at the right time. In 1963, they needed an impatient, charismatic man who would seize the spacecraft and pull it through North

---

* There followed the next morning a curious charade that has gone into the history books as the reality. It was considered impolitic to present Webb with a *fait accompli*. So Mathews's refusal was presented to Webb along with the suggestion that what they ought to do now was persuade Low to take the job. Webb thought it was a terrific idea, and nothing would do but that they call Page Terminal at National Airport and have them hold Gilruth's plane (which was about to return Gilruth and Low to Houston). Webb, Seamans, Phillips, and Mueller all piled into Webb's official car and they went out to the airport and "persuaded" George Low. In recounting the story later, Low never wavered from the approved version.

American. Because much of Houston was preoccupied with Gemini during this period, they also needed a man who liked to do everything himself. They got Joe Shea. In 1967, they needed a "knitter of people," in Glynn Lunney's graceful phrase, someone who could win the trust of the disparate elements in Houston and bring them together as a working whole. They got George Low.

Except in ability, Low and Shea were mirror images. They were both born in 1926, but instead of Bronx working-class Irish, Low was Viennese Austrian, the son of a well-to-do manufacturer, educated in private schools in Switzerland and England before his widowed mother brought Georg Wilhelm to the United States in 1938. Instead of a systems engineer working on nuclear missiles, Low was a Lewis man, with the N.A.C.A. from 1949 on, who wrote papers on esoteric problems in boundary layer theory and heat transfer. While Shea reveled in pressure and complexity and the clash of personalities, Low sought order and harmony.

In personal manner, Low was unassuming to the point of diffidence. Back in the late 1950s, Caldwell Johnson reminisced, when they were fabricating the first Mercury capsule in the Langley shops, they roped off the area immediately around the capsule so that visitors wouldn't get in the way of the engineers and technicians. One day Johnson noticed that a fellow had ducked under the ropes and was right up next to the capsule, even running his hand over the metal shingles. Johnson went over and reprimanded him. "You know you're supposed to stay behind ropes, don't you?" Johnson said belligerently. The man apologized and obediently went back behind the ropes.

Johnson returned to his work. Another engineer came over and whispered to him. "You know who you were talking to?" Johnson didn't. "That's George Low!" Johnson looked over again, and he was still there, peering intently at the spacecraft—a young man with a kindly face, thin, a little stooped. He was NASA's Program Manager for Manned Space Flight, and he was standing quietly behind the ropes as he had been told. He never lost that quietness. Later, in the days when he ran ASPO, subordinates would sometimes call the Low home late in the evening for advice on some new problem, and they would hear a soft voice answer the phone. "May I speak to your father?" they would ask George Low.

And yet this same man took some of the boldest initiatives in the manned space program, succeeding in them because he was a leader, classically defined—men followed him, as they followed Shea before him, Kraft in Flight Operations, Debus at the Cape, and von Braun at

Marshall. But he was specifically suited for a particular kind of leadership, the leadership of engineers.

He was, first of all, unwaveringly faithful to good engineering when he made decisions. ("To be friends with somebody is one thing, but when you're in the technical arena, it's technical. I don't give a damn whether I play golf with you or whatever, that's a different world. Low knew that.") He added to that an infinite capacity for detail. ("George was the kind of guy that if you gave him a job emptying wastebaskets, he would stretch it into overtime, not because he was loafing, but because he'd find more to emptying wastebaskets than you ever imagined could be there.") He had a memory that was by all accounts phenomenal. ("He'll tell you, 'You wrote me a memo a year and a half ago and you told me that the oxygen flow rate was going to be 7.6 pounds per hour, and why is it 7.3 now?' What can you say to a guy like that?") And he had a gift for organization and management that verged on genius.

Low had an effect on his fellow engineers in Apollo that might have been hard for an outsider to appreciate. About a year after Low took over ASPO, one of Webb's deputies in Washington got a call from a friend who had just arrived in Houston. The man in Washington was sufficiently amused by the call to jot down the exchange verbatim immediately afterward. "I hope you don't think I'm nuts, but I want to tell you something," the caller said. "We had an Apollo briefing in the ninth-floor conference room and George Low spoke to us. God! It was the most exciting, lucid, thrilling, dynamic thing I've ever heard. . . . Honest to God, it was just fantastic. He didn't even have a note. He held us all spellbound. . . . So help me, if he had said 'Let's go' I'd have followed him right off a cliff!" An observer who wasn't an engineer would have seen just a courteous man, speaking quietly.

2

Taking up the ASPO job in early April, Low began dictating a daily memorandum to Gilruth, usually two or three typewritten pages, summarizing the day's activities.* A few weeks later, returning from a trip to

* Following the fire, Gilruth made it known to others that Shea had not kept him informed about what was going on. The "Apollo Notes," as he called them, were part of Low's way of avoiding that problem. In the process, he also provided historians with a unique day-by-day narrative. The "Apollo Notes" are now part of the George M. Low Papers at Rensselaer Polytechnic Institute.

the North American plant in Downey, he wrote: "My general impression after this week's visit is that Dale Myers, Charlie Feltz, and George Jeffs are trying extremely hard to do the right things. . . . The next level below them, however, disturbs me." Frank Borman, writing of that period in his autobiography, put it more bluntly: "North American was positively schizophrenic, populated by conscientious men who knew what they were doing and at least an equal number who didn't know their butts from third base." After four years of prodding and pushing from ASPO and headquarters, even after the embarrassment of the Phillips Report, North American remained a problem.

Borman was in a position to know. Gilruth had appointed him as head of a "tiger team" (an Air Force term for a small group of troubleshooters) to go out to Downey and be in charge of redefining the Block II spacecraft with North American. Borman took Aaron Cohen as his design engineer and Simpkinson as his quality and reliability expert. For the next four months, they spent every day in the North American offices, with virtual carte blanche from Gilruth and Low to carry out the redefinition strategy that ASPO had decided upon.

Houston also kept trying, as it had for years, to get some changes in the way that Downey operated. After a few days there, Borman noticed that many of the technicians were going across the street and having a few beers at lunch. Outrageous, when you're working on a spacecraft, Borman told North American; and everyone, management and unions and the technicians themselves, agreed. The practice stopped. Jack Kinzler, the craftsman from Langley who had been heading the shops for the manned space program since the earliest days of the Space Task Group, was dispatched to Downey to look around the shops. Kinzler identified thirty or forty things that offended him—sloppy handling of materials, confused demarcation of work spaces on the shop floor, lots of wasted time—and wrote them up. Gilruth sent a copy to the president of North American. The offensive practices got fixed.

But other problems of scheduling and efficiency that had persisted for years at first seemed as impervious to solution as they ever had. In another "Apollo Note" to Gilruth, with "No other copies to anyone except RRG" scrawled in big letters across the top, Low listed the kinds of things he was finding when he went to North American. Item: Up to five people could work productively in a command module at any one time, but out at the Downey plant, Low and his managers rarely saw more than two or three; often only one. Item: The industry average for a wire termination was six minutes. At North American, they were taking up to an hour and a half. Item: Low's people had seen it take as long as four hours to install one washer. Item: The wire harness for spacecraft 101 was not even started until five weeks after it was supposed to be completed. Someone at North American had to start doing something about this.

3

The man most obviously on the spot at North American was Stormy Storms, head of North American's Space Division. He was a proud man with a proud history, the engineer who had guided the development and production of the X-15, but the fire had happened on his watch and the pressure was relentless. While Gilruth continued to think highly of Storms, Low, Phillips, and Mueller were agreed that Storms had to go. Storms himself thought he was being "unjustifiably crucified," in the words of one of his close associates. "There's not a goddamn thing wrong with those spacecraft," he remembered Storms saying shortly after the fire. "If they want to fly one this December, just fly what we've got."

Few others agreed, and the fact that Storms could think nothing needed fixing was persuasive evidence to them that he wasn't the man to rescue the contract. Webb called in Lee Atwood, now North American's chairman, and told him that Storms had to go or NASA would turn to another contractor. However unrealistic the threat—if carried out, it would have hurt NASA as badly as North American—the word around Downey was that Webb had already held preliminary meetings with Boeing for just that purpose. Storms was retired from his position as president of the Space Division.

The man whom Lee Atwood chose to rescue the situation was Bill Bergen. For thirty years, Bergen had been one of the star engineers at the Martin Company—the company that had originally been recommended for the spacecraft contract back in 1961. When the Apollo fire occurred, he had just become corporate vice-president at North American for the Space and Propulsion Group, meaning that Stormy Storms's Space Division came under him.

Bergen was at an air show in Phoenix when he got a call from Lee Atwood. Bergen volunteered to give up the Space and Propulsion Group and take over the Space Division—to demote himself, just as Low and

Shea had demoted themselves before him. "God bless you," said Atwood. Bergen took over his new job on April Fool's Day.

"People think I came in here and turned this place around like crazy," said Bergen later. But he claimed that he hadn't. "I think if you look back at it, soberly, there were very few changes made, but some very key ones." One of the first of them was to send Bastian Hello, known as Buzz, to take over North American's Cape operation.

Buzz Hello arrived at the Cape on May 7, 1967, to find a place where, in his words, "morale had sagged like a clothesline with ice on it." The North American people at the Cape were the people who had somehow burned up three astronauts. "They had no way of knowing where they had failed, what they had done wrong," Hello recalled. "They had lost very close friends of theirs in the spacecraft. . . . The whole world had turned against them." There was no one particular thing that had to be fixed. "You just sort of wade into it," Hello said. "It's like a gigantic piece of cheese—you've got to start biting somewhere."

Hello saw that somehow there had to be a "welding together" of his people. It could take the form of small psychological things, like a campaign to keep all the workplaces immaculate. Or of organizational things, like designating a separate room for tracking the status of each spacecraft, a place where that spacecraft's schedule, test preparation sheets, engineering orders, and test results could be posted and tracked—"so it's perfectly clear who's doing what to who and who's holding the bucket of water." Or of training things, like "Crew Qualification," in which each person who worked on the spacecraft had to undergo a "stand board," appearing before a board of three or four of his peers to be cross-examined on the workings of his system. But the most important thing that Buzz Hello did to turn the situation at the Cape around was to hire Tom O'Malley.

By 1967, Thomas J. O'Malley had been working at the Cape for ten years. First he had run Convair's Cape office for the Atlas program. In Mercury days, the man with his finger on the launch button for the Atlas launches was Tom O'Malley. During Gemini, O'Malley had been with General Dynamics on the Atlas-Agena, then was promoted away from the Cape to the foggy precincts of New London, Connecticut, to be a senior manager in the General Dynamics Electric Boat Division. Now, Hello wanted him for North American and O'Malley wanted to come home to the space program.

"A bellerin' type," said one of the spacecraft inspectors, who were perpetually at war with him. "He was a driver, a real driver," said Buzz Hello. "There are very few people down here that work as hard as Tom O'Malley, and woe be unto the man who tries to pass some short story off on him, because he has probably been there himself a little earlier in the morning and stayed a little later at night. Feeding him a short story is. . ." "Death?" asked Hello's interviewer. "It's worse than that," said Hello. "It's torture and death."

There was not a whole lot of this "tough but fair" business about Tom O'Malley, either. On the contrary, "he tended to form opinions [about people] on limited data points sometimes, which bothered me," said a Cape engineer who worked closely with him. "He got on fine with me because we'd been together a lot in the past, so he had me calibrated and I had him calibrated. But sometimes if a guy didn't say the right thing the first time he met him, that guy was doomed." What O'Malley had going for him was a ruthless integrity about getting the job done, as well as an absolute indifference about whom he offended in the course of doing it. And that's why Hello hired him.

O'Malley soon decided that North American had been running what he called a "country club" down at the Cape. One of the first days after he took the job, an engineer came to him to get a routine signature on a travel order—he was going out to a convention of retired Air Force officers in San Francisco. O'Malley told him that they weren't doing business that way any more. The engineer was outraged—those conventions were an important way for a company like North American to tie into the old-boy network, and such trips had always been one of the perks. He got hold of an ex-general who was a senior executive at Downey. The ex-general complained to Bill Bergen. Bill Bergen chewed out the ex-general for interfering with O'Malley, and said that anyway O'Malley was right. The word quickly got out that the old days were over.

O'Malley was out at the pad one day, watching liquid oxygen being pumped from a tank up onto the umbilical tower. He asked the engineer where the LOX was being pumped to. "It beats the hell out of me," the engineer replied. "Once it gets up there, I don't know what happens to it." Shortly after that, O'Malley called a meeting. Thenceforth, every engineer was expected to learn everything about the system he was running—where the stuff came from, where it was going, and all the things that might go wrong in between. And they did. "You understood it from the very beginning of the tank farm to the last nut and connector in that system," said one engineer, "and you could just about close your eyes and draw a schematic of the electrical and mechanical systems by heart. Everybody had a thing called a 'Smarts Book,' and in that Smarts Book each system engineer had every fact that you could possibly collect about that system that you worked on. . . . You understood that system from womb to tomb and there was nothing in that system you couldn't recite by heart, including torque values, safety wire specs—you name it and a system engineer down there could tell you what it was. Those guys were terribly intense.''

If O'Malley was not always the most judicious and fairest of men, he was nonetheless the most loyal when he respected someone. ''If I'd asked my guys to push that V.A.B. building over, they'd be out there trying to do it,'' O'Malley said. ''You couldn't get a bunch like I had.'' And those who worked for O'Malley assumed that, if he had asked them to push over the V.A.B., somehow that's what they would do. ''O'Malley is unique in this program,'' said John Tribe, who worked for him from Mercury days through Apollo. ''He's rough, bluff, crude, calls a spade a spade, language is pretty bad at times, but by golly, he knew what had to be done in 1967.''

Out at Downey, Bill Bergen went looking for John Healey, another of his old colleagues from Martin. Bergen had decided to put a senior manager in charge of each vehicle, ''one guy who was Mr. Spacecraft 101, one who was Mr. 102, and so forth.'' For the first manned vehicle, spacecraft 101—upon which not only North American's future but the future of the Apollo Program depended—Bergen chose Healey, a forty-five-year-old engineer. As Bergen reminisced, ''John became really quite a controversial person for a while. . . .''

There was, for example, Healey's first briefing by North American's Apollo management team in Downey on September 29, 1967, when ''to get their attention for the century,'' as Healey put it, ''I said, 'Hey, I think you guys are full of shit. You're in trouble, and you ought to stop acting like it's going to go away.' '' There was, for example, Healey's first meeting with NASA four days later, when he announced to Phillips, Mueller, Low, Rees, and Gilruth—most of Apollo's senior management—that NASA had an unworkably unwieldy system, with multiple directions and parallel activities. Effective that day, Healey said, NASA had a choice. Either they could let the subsystems people call him direct or they could assign one guy to work with him. But nobody who had anything they wanted done with spacecraft 101 could go through anyone but Healey. And by the way, Healey added, he was tired of all these modifications that NASA kept asking for; from now on, no more ''If you're willing to pay for it, okay, Mr. Customer.'' If NASA wanted the spacecraft built, Healey told them, they had to stop asking for so many changes.

''And then,'' Healey recounted, ''I turned around and looked at all the North American guys smiling, and said, 'Now let me clear some things up for you guys. You can't keep on freely reacting to the customer's directions just because they seem to agree with your own idea of what you want to do. You think they're a problem? Well, you're an equal problem.' '' And so on. Healey had a wonderful time that day— ''rammed all kinds of new assholes,'' he said.

Out on the assembly facility, Healey came to grips with a peculiar reality of assembling a spacecraft: ''The thousands of people involved in the program were jammed right up into a small opening in a funnel of five people in the command module''—for that was the most that could fit into the spacecraft. Healey figured that at any given moment, the time of those five was more precious than anyone else's, and he went down onto the shop floor and told that to the technicians. No longer would they have to get out of the command module to find a new drawing or tool or part. He, John Healey, would make sure that they got the support they needed so they wouldn't have to waste that kind of time. ''I told the working people that, and they kind of smiled. I said, 'No, I mean it. I'll do it.' '' And he did. If a welder working in the spacecraft had a question that needed to be answered by the chief engineer, then the chief engineer would break off whatever he was doing, get himself down to the assembly room, climb the steps up to the door of the command module, and deal with it. ''I had vice-presidents chasing plans so the workers could stay in the command module,'' Healey remembered.

The momentum picked up almost instantaneously. Just a week after Healey came on board, Low was writing to Gilruth regarding some of the problems that Grumman was having with the lunar module and saying that Grumman ''needs a Healey'' to get things back on track. On May 29, 1968, eight months after Healey took over spacecraft 101, it left Downey on a transport plane and arrived the next day at the Cape. A few days later, the receiving inspectors reported to Houston that they had found ''fewer discrepancies than any on any spacecraft previously delivered to Kennedy.''

Telling the story of North American's recovery from the fire from the vantage point of Bergen and Hello and O'Malley and Healey captures only one small fragment. There is Kenny Kleinknecht's story, as he became ASPO's point man for the C.S.M. Or the story of Max Faget's engineers, who worked through the redesign issues after the fire; or of Frank Borman's tiger team; or of Eberhard Rees's team from Marshall. The measure of all their successes is this:

Spacecraft 101 was the first of fifteen manned Apollo spacecraft launched from the Kennedy Space Center. Those fifteen spacecraft functioned in space for a total of 280 days—in earth orbit, lunar transit, lunar orbit, docked with Skylab, and in rendezvous with a Soviet spacecraft on the Apollo-Soyuz mission in 1975. One was hit by lightning during the launch phase, another (through no fault of North American) lost most of its electrical power halfway to the moon, had to be shut down for more than three days in the cold of space, and was powered up again just before entry. All the astronauts returned safely, riding a spacecraft they had come to love.

17

"And then on launch day it worked"

The stages for the first flight of the Saturn V began arriving at the Cape in August of 1966, even before the fire. The third stage, the S-IVB, arrived by air in a special bulbous plane called the "Super Guppy." The S-IVB was the only stage that would arrive by air; indeed, the other two stages couldn't arrive even by ground. They were too big to be transported by either truck or train, so they were barged in—the S-IC from the big plant built especially for Apollo at Michoud, Louisiana, and the S-II from North American's facility at Seal Beach, California.

On September 12, 1966, Ike Rigell, Petrone's right-hand man for launch vehicle operations, watched as the S-IC stage of the Saturn V arrived on the barge *Poseidon* at the Saturn Unloading Facilities on the Banana River. Teams of technicians swarmed over the S-IC, checking to make sure that none of its elaborate wrappings and protections had been damaged during transit. The sole, specialized task of the men Rigell watched was to meet Saturn and Apollo hardware when it arrived at the Cape and see that it was safely transported to the O&C Building (for the spacecraft) or the V.A.B. (for the Saturn).

Only about ten years earlier, Rigell reflected, he and a dozen or so others in Kurt Debus's launch team had done the whole thing themselves, from loading the rocket up in Huntsville to the launch itself. A couple of them would stay with the Redstone as it snaked its way down through the little country towns of Alabama and Georgia on a long Army flatbed truck. The rest of the group would stow the launch consoles in the trunks of their cars and drive down to the Cape, where they would meet the Redstone, check it out, launch it, and go home. But it had been a different kind of bird in those days. Four men could stand at the base of a Redstone and join hands around it. Now, Rigell stood beside the S-IC lying on its side, his eyes at the level of the exhaust nozzles of engines 1 and 2; engines 3 and 4 were two stories above his head.

NASA's official designation for George Mueller's audacious first-time all-up test was A.S.-501: "5" for the launch vehicle, "01" denoting the number of the flight. Within NASA, the flight was being called simply "five-oh-one." To the public, the forthcoming flight was known as Apollo 4.*

1

The S-IC was taken first to the Low Bay of the V.A.B., where the components that would become inaccessible when the Saturn was stacked were examined and tested. After the test conductors from Launch Vehicle Operations were satisfied that the first stage was ready, it was moved into the High Bay, where lines were attached from the crane far overhead and the S-IC was slowly lifted into place aboard the mobile launcher. Later the second stage, the S-II, went through the same preliminary checkout in the Low Bay.

On February 23, 1967, they stacked the two stages, an operation that epitomized the extremes of large and small in the Cape's work. First the S-II, 90 feet long, was lifted by crane 280 feet into the air; then it was settled down onto the top of the S-IC with a precision measured in millimeters and ounces—the crane operators were trained to lower the

---

* The Grissom crew's flight was A.S.-204 (fourth Apollo flight on a Saturn II). After the fire, at the widows' request, "Apollo 1" was reserved for the flight that never took place. Then Low suggested retroactively naming the three unmanned Apollo/Saturn II flights Apollo 1A, Apollo 2, and Apollo 3, respectively. While that was being considered, A.S.-501 was named Apollo 4; subsequently, NASA headquarters decided not to rename the earlier flights after all. That is why Apollo 4 was the first flight in the Apollo series.

hook with such delicacy that the hook could be stopped after it was touching the object below, but before it exerted even the pressure that would crack a raw egg.* Once the two stages were aligned, it took a team of engineers and technicians eight hours to join the S-IC and the S-II. The two stages were joined not with welds, but with pins: three 12-inch pins at 120-degree intervals around the periphery of the stage and 216 1½-inch, high-strength fasteners at 6-inch intervals.

This process was repeated for the third stage, the S-IVB. Then came the tests of the assembled launch vehicle's electrical networks, fire detection, telemetry, tracking, gyroscopes, onboard computers, pumps, engines, transducers, valves, cables, plugs, hydraulic lines. There were 456 such tests altogether, and they took weeks—just how many weeks depended on how things went. In the case of 501, a new vehicle with birthing pains, the tests took almost four months.

In the O&C Building, the command and service module for the first Saturn V flight, the spacecraft 017 that Joe Bobik had examined with such indignation after the fire, was undergoing the same kind of testing and preparation. On June 20, 1967, five months after the fire, spacecraft 017 was gingerly loaded into its own specially designed cradle—no more plywood and mattresses—and transported from the O&C Building to the V.A.B., where it was mated to the top of the stack. For the first time, there was now a complete Apollo/Saturn flight article.

The tests and the inspections in the V.A.B. went on for another two months. By now the test conductors were checking to see how the components and subsystems of the launch vehicle, spacecraft, and ground support equipment worked together. Suppose that at the moment of launch one of the umbilical tower's nine swing arms failed to disconnect. Could the vehicle shut down safely? Answering that question required ten separate tests. Suppose that the hold-down arms failed to release. Another set of tests. Were the hundreds of wires connecting the three stages correctly joined? Weeks of tests. Was each pin among the hundreds in the umbilicals' electrical interfaces connected to the correct socket within the vehicle? More tests. Slowly, laboriously, the verification seals and the sign-offs accumulated as each of the thousands of items was checked off the list.

---

* The people at the Cape love to tell that story, and it is technically correct regarding the precision of the crane operators and the cracking pressure of eggs. But it appears, sadly, that they never conducted the exercise with real eggs.

2

On August 26, the crawler carried the 501 stack to Pad 39A. Petrone announced that the Countdown Demonstration Test (C.D.D.T.) for flight 501 would begin on September 20. A week later, he rescheduled it to begin on the 25th. It actually began on the 27th.

For later flights, a C.D.D.T. for a Saturn V lasted four days—sixty hours of actual tests and thirty-six hours of planned holds. Expecting some first-time delays, Petrone planned for this first C.D.D.T. to take six days. It took seventeen.

As in the old days with the early Atlases and Redstones, nothing worked quite the way it was supposed to. The devices for regulating the flow of propellants and gases into the Saturn's tanks kept giving them problems: The regulators were designed for such heavy flows that they didn't know how to handle small ones. The piping had to be modified to trick the regulators into behaving properly. Computers were now in charge of monitoring the propellant loadings, and the software was full of bugs. The Instrument Unit, the brains of the Saturn V, wasn't keeping the black boxes of electronics as cool as planned. Cable connections on the S-II stage shorted out because of humidity and moisture around the pad. It wasn't any one thing, but an unending series of delays in almost everything they tried to do. "If I asked a guy how long something would take, he'd tell me ten minutes and it would come up maybe an hour," recalled Rocco Petrone, who was directing it all. "Everything about the Saturn V was bigger. Getting anywhere was bigger. If you had to pick up a valve, you couldn't pick it up by hand, you had to get a forklift truck! Everything was one or two dimensions bigger." Petrone began planning his schedules in terms of what he called the "Saturn V minute," which he calculated was about five times a normal minute.

For Ed Fannin, who was by then chief of Mechanical and Propulsion Systems for Launch Vehicle Operations, it began to seem as if the C.D.D.T. had been going on all year. The delays were especially tough on the propellants team. Loading propellants was an intricate operation, with dozens of procedures required to prepare the cryogen lines (for LOX and liquid hydrogen), pumps, regulators, and the tankage; all of the valves and settings at the pad had to be set so that the actual loading, the most dangerous part of pre-launch operations, could be conducted by remote control from the Firing Room. Once loading began, there were dozens more procedures, and the process took hours—after all, they were pumping the equivalent of 144 trailer-truck loads of kerosene, liquid oxygen, and liquid hydrogen into the equivalent of a thirty-six-story building. Then, every time the C.D.D.T. hit a snag and had to be stopped, and most of the others on the launch team went off to get some rest, the propellants people had to stay in the Firing Room and off-load, which was even more tedious than loading. For them, the low point of the C.D.D.T. came on October 4, the eighth day, when they got to T − 45 minutes, tantalizingly close to the end. Then a computer failed, and they had to off-load 502,000 gallons of kerosene and liquid oxygen from the first stage.

Exasperated, Grady Corn, chief of Fannin's Propellants Branch, wasn't even sure that the effort was going to accomplish anything. By his own admission, he was one of the Cape's "great disbelievers." Standing at the top of the tower, he would look right straight down along the great length of the rocket to the launcher 363 feet below and know there would be "no way in the world for that thing to lift off." He was not joking, and he was not the only person who worked on the Saturn V who felt that way. It just didn't seem possible that something that big, that complicated, with so many things that could go wrong, would really work when the time came to light it.

Day after day, the C.D.D.T. crept on. The launch team would gain a few hours, encounter a new problem, gather in the "woodshed" (a little conference room off the Firing Room, so called because of the nature of the encounters people tended to have with Petrone when they were called there), devise a fix, set the clock back, and return to the count.

On October 9, thirteen days after the C.D.D.T. had begun, the count was again getting close, to within T − 5 hours, but the launch team was nearing exhaustion. Petrone himself began to lose track of time. He looked down at the consoles in the Firing Room and saw Ernie Reyes, one of the senior engineers for Spacecraft Operations.* The spacecraft people were having a comparatively easy time of it—they had already struggled up their own learning curve on the three unmanned Apollo flights launched on Saturn IBs. Petrone marveled at Reyes—he looked so clean! Reyes told Petrone that he'd gone home, slept all night, shaved, and come back to work. Petrone hadn't noticed that by now he and many of the launch vehicle people were in the twentieth hour of one long shift. "We just can't go any further," Petrone told them, and sent the launch team home for a two-day recess. They came back to work on October 11, whereupon a battery heater in the S-II stage failed. It was about then,

* Petrone was director of Launch Operations. Launch Vehicle Operations and Spacecraft Operations, which prepared the Saturn and the Apollo, respectively, both reported to Petrone.

Rigell recalled, that the question had to run through your mind: "Can we *ever* get that baby off? Can we ever get all the green lights at one time?"

On the afternoon of October 13, the count kept getting closer to $T-0$. It passed the $T-45$-minute mark, their previous best. Then it passed the completion of the power transfer test at $T-26$. Then it began the chilldown process for the thrust chambers in the S-II and S-IVB engines. Then, miraculously, they were at the beginning of the automatic sequence at $T-3$ minutes 7 seconds. Petrone sat back in Management Row, the raised row of consoles at the back of the Firing Room, and watched the clock get closer to $T-0$. At this moment, the Saturn V on the pad was a fueled, checked-out, fully operational vehicle, doing everything it would do on launch day except light the igniters. As the clock counted down toward $T-14$ seconds, the point at which the countdown was to be halted, Petrone turned to Ike Rigell. It was an informal tradition between them at this moment in countdown demonstration tests. "Ike," Petrone said, "are you sure we got all the igniters out of there?" And Rigell assured Petrone, as he always did, that the igniters were out and this bird wasn't going to fly away just yet.

The long ordeal had finally come to a close. The men in the Firing Room, too weary to celebrate, drifted out to go home—except for the propellants team, who still had to unload all the propellants. Finally, twenty-eight hours after he had come on shift, Grady Corn announced to Ed Fannin that the propellants were off-loaded. "I think I've suffered permanent brain damage," he added.

"We got through it," Petrone said, describing the C.D.D.T. as one might recall a battle which one had unexpectedly survived, "and we learned a lot. I mean, we *learned*." For Petrone, it was a historic event, whether or not the rest of the world knew about it. At the Cape, those seventeen days were when "the program came to fruition." They were the days when Launch Operations came to terms with the Saturn V.

3

At 10:45 on the night of November 8, 1967, not yet ten months after the 204 fire, the first Saturn V stood on Pad 39A awaiting a launch time of seven the next morning. As every night, the tower and the vehicle were bathed in lights, set off by searchlights that intersected at the apex of the stack. To a *New York Times* reporter, the Saturn looked like a crystalline obelisk. To visiting Soviet poet Yevgeny Yevtushenko, the Saturn and the red umbilical tower with its swing arms were a white maiden clasped by a monstrous lobster. Rocco Petrone was reminded of a cathedral.

Petrone watched through binoculars from across the marshy lake that separated the Launch Control Center from Pad 39A. The wind was blowing hard out of the north; even from where Petrone was standing, he could see the wind meter spinning at the top of the tower and bits of debris swirling around. The winds were exceeding thirty-two knots, too high to launch, but Petrone's weather chief assured him that they would be within limits by launch time. Petrone ordered the propellants team to begin loading. He remained at the Launch Control Center throughout the night, watching the tower. An hour before dawn, the wind began to fade.

When they had launched the first Redstones in the early 1950s, Albert Zeiler had crouched beside the slit window in the little blockhouse by the pad and watched as the engine fired up in pre-stage (with too little thrust to move the vehicle). He was deciding from the color of the flame whether the mix of fuel and oxidizer looked right. If the color was wrong, Zeiler told Debus (who stood beside him—the whole blockhouse was only a dozen feet square), and they shut down the engine and tried to figure out what the problem was. For the first launch of a Saturn V, 450 engineers and controllers were assembled in the Firing Room at the K.S.C. Launch Control Center, working at eight rows of consoles in a room 150 feet long and 90 feet wide. There was no human within three and a half miles of the vehicle itself, nor had there been since Corn's team had begun to load the propellants eight hours earlier.

George Mueller waited in the large glassed-in viewing area set off at an angle toward the back of the Firing Room. Four years and one week had passed since he had sent his teletype to Marshall instructing them to fly the first Saturn V all-up. With Mueller sat the others from headquarters, Seamans and Phillips among them. Not Webb, however, who always considered it his job to remain in Washington to cope with the political heat if something went wrong.

In the large Marshall contingent, sitting near Wernher von Braun, was Arthur Rudolph. Rudolph had been with von Braun for almost thirty-five years. The two of them had shared bachelors' quarters in the early 1930s, and talked about going to Mars before they could even get their first liquid-fueled rocket, four and a half feet long, off the test stand. It was Rudolph's sixtieth birthday.

Like everyone else at Marshall, Rudolph had thought that the all-up decision was madness—it was Rudolph to whom Mueller had said, "So what?" But whether all-up was madness or not, Rudolph was the man who had to implement it, for he was the program manager for the Saturn, and therefore was to Marshall and the Saturn what Shea and now Low were to Houston and the spacecraft. To get to this moment, Rudolph had surmounted years of engineering crises—the combustion instability on the F-1, the immature technology of the hydrogen-fueled J-1 engines in the upper two stages, endless difficulties in the construction of the S-II stage. Ever since Mueller's decision, Rudolph's worries had been augmented by the knowledge that, when this day came, he would be flying two stages that had never flown even once and a third stage that had never flown in this configuration. He hoped that he had been wrong and Mueller right.

Petrone was watching the countdown from his command post on Management Row. After the trials of the C.D.D.T., this countdown was proving to be startlingly smooth. In fact, as they approached $T-3$ minutes 7 seconds, they were right on schedule.

Some of the press buses didn't leave the motels in Cocoa Beach until six that morning, but no one on board seemed worried about being late. Even if they got held up in traffic—not unlikely, considering the number of people trying to get onto Merritt Island to see the launch—it was inconceivable that A.S.-501 would launch on time, if it launched at all that day. During the last few months, everything involving the Saturn V had been late. As seven o'clock approached, the buses were still inching their way up Route 3.

At $T-3$ minutes 7 seconds, control of the launch process was turned over to the computers. For almost three more minutes, Petrone would be able to stop the launch manually if he had to, but now the Saturn V was busy preparing itself to fly, receiving through the umbilical hoses still connecting it with the ground the helium that created the pressures within the propellant tanks necessary to feed the propellants into the pumps. At $T-30$ seconds, the 55,000-horsepower turbine that drove the S-IC's five engines powered up. At $T-8.9$ seconds, an electrical signal was sent to the igniters, and four small, silent flames lit within the combustion chamber of each of the F-1s.

From that moment through liftoff, there was nothing Petrone or anyone else in the Firing Room could do. If something went wrong, the sensors would know it before the news could reach the Firing Room, and the Saturn V would shut down its engines without waiting for sluggish human beings to instruct them. So now the men in Management Row simultaneously swiveled around in their chairs—they got stuck if they didn't do it at the same time—and put binoculars to their eyes. Through the bank of windows at the back of the Launch Control Center, they watched Pad 39A. Petrone kept his hand near the button that would close protective louvers over the windows if the Saturn V blew up, though he always suspected that, if it happened, he would just keep watching instead.

As soon as the sensors within the combustion chambers of the F-1s determined that the igniters were lit, the main LOX valves opened, releasing liquid oxygen into each combustion chamber where it combined with a fuel-rich and comparatively cool combustion gas, an exhaust product from the turbine. The gas was comparatively cool—only 800 degrees Fahrenheit—and would help cool the nozzle during flight; now, it prepared the interior of the chamber for the thermal shock to come. This process took three seconds. The combustion of the exhaust gas produced a thick orange smoke.

At $T-5.3$ seconds, as sensors within each combustion chamber determined that the pressure at the face of the injector had reached 20 pounds per square inch (p.s.i.), the main fuel valves opened and a torrent of kerosene burst through the painstakingly sized and angled orifices of the injection plate, past and through the copper baffles that had been redesigned so often. The streams of kerosene (a ton per second per engine) and liquid oxygen (two tons per second per engine) then impinged, formed their fans, and, mingling, ignited.

The viewing area for the press and V.I.P.s was across the road from the Launch Control Center. Bleachers were set up, with a corrugated iron roof to ward off sun and rain. At the top of a slight rise beside the bleachers, looking like so many unpretentious beach cottages built for a view of the sea, stood the little wooden studios with picture windows in front that the television networks had built for their launch coverage.

At $T-8.9$ seconds, the people in the bleachers could see an eruption of orange smoke pushing down and bouncing off the flame deflector under the launcher, then bursting out at either side. Then, a few seconds later, the flame directly under the engines turned to an incandescent white as the orange smoke billowed outward and upward, beginning to envelop the rocket. Still 501 didn't move. Astronaut Mike Collins, who was hoping to ride a Saturn V some day, wondered momentarily whether this one was just going to sit there and be consumed in the holocaust.

The noise of the preparatory burn that had created the orange cloud was inaudible across the four miles separating the viewers from the launch site. Even as the engines went to mainstage and they saw the incandescent white flame, the sound had yet to reach them. For the people sitting in the viewing stand, the first seconds of the pyrotechnic display on Pad 39A remained eerily silent.

The main fuel valves in the S-IC's five engines opened at slightly staggered intervals, so that neither launcher nor vehicle would have to withstand the pressure of all five engines coming to full power at the same instant. Now, as the fuel-injection pressure on each engine passed 1,060 p.s.i., a pressure switch sent a signal to the Instrument Unit (I.U.) high in the S-IVB stage of the stack, announcing that the thrust for that engine had reached 1.5 million pounds. At 7:00:00, the I.U., having tallied five good signals, sent a command from the vehicle through the electrical cables still connecting it to the earth, asking to be released.

At the base of the Saturn V, four hold-down arms restrained the rocket as the engines came up to mainstage power. The arms were massive, not so much to restrain the Saturn V from lifting off (the loaded weight of the Saturn V, over 6 million pounds, helped considerably to hold down the vehicle's 7.5 million pounds of thrust) as to lessen the rebound load on the launcher if the engines were to shut down after reaching full power. In dealing with a normal liftoff, finesse was at least as important as strength, for all of the forces restrained by the hold-down arms were transmitted back into the structure of the Saturn V. To avoid putting stress on the body of the Saturn, the arms had been placed with great precision—Glover Robinson, the perfectionist engineer in charge of that operation, had used optical equipment to sight them in—and they had been designed so that there was absolutely no doubt whether they would simultaneously and instantaneously release the Saturn V upon command. Now, receiving the signal from the I.U., a helium-gas pneumatic device actuated the release, which occurred in all four arms within 50 milliseconds. If the pneumatic actuator had failed, an explosive bolt in each hold-down arm would have triggered the release.

Still the Saturn V was not entirely free of the earth. When a vehicle producing 7.5 million pounds of thrust is suddenly and completely let go, the release itself produces an abrupt shock load. Rather than shockproof the vehicle to sustain this brief, one-time jolt, its designers tethered the

Saturn V to the hold-down arms with soft steel bolts. Each bolt was an inch in diameter and protruded into a bell-shaped socket attached to the Saturn V. As the rocket began to lift, the soft steel was extruded through the sockets, giving the Saturn V a lingering release and attenuating the shock.

For the first milliseconds of its ascent, the Saturn V retained its umbilical plates. These plates, which held the fuel lines and electrical connections that would permit the Launch Control Team to regain control over the Saturn V if the engines were to shut down, remained connected until after liftoff—a lesson learned from M.R.-1 seven years before. After that instant of liftoff, however, the umbilicals could come out, for there was no possibility of the Saturn V rising an inch or two and then settling uneventfully back onto the pad as the Redstone had done. Once the Saturn V had moved even fractionally, the engines had to keep going or the Saturn would fall back, collapse, and explode. As the vehicle left the pad, it tripped two liftoff switches. Whereas until that moment it had been imperative that the umbilicals remain tightly connected to the rocket, it was now equally imperative that they disconnect.

Most of the connectors holding the umbilicals into the side of the Saturn V were of a ball-release type, meaning that when a pin within the umbilical was withdrawn, the balls which had been held in place by the pin collapsed, making the connection small enough to slip out.* When the liftoff switches were tripped, the rods were pulled, the balls collapsed, and the umbilicals came free.

Now the swing arms, which carried the umbilicals and had given the Cape's workers access to the Saturn V on the pad, had to get out of the way. For the Saturn V, there were nine arms on the umbilical tower, each weighing between ten and thirty tons and designed to be swung away—on a 73-degree arc for the bottom eight, a 135-degree arc in the opposite direction for the topmost arm containing the White Room. The arms had given the Cape more trouble than any other item in the ground support equipment except the crawler. At the moment of launch, four of them were safely out of the way, already retracted. The other five were called "in-flight arms," meaning that they remained in place until after the hold-down arms had released and the vehicle was already in motion. First

---

* All the ball-release devices used an even number of balls. No one was quite sure why. All Don Buchanan knew was that long ago, when building the Jupiter, he had built a three-ball device, watched it work twenty-seven times in a row, then called von Braun over to demonstrate it; whereupon, on the twenty-eighth try, it hung up. Subsequently, Buchanan and his engineers had determined to their satisfaction that an odd number of balls didn't always work whereas an even number did.

the outermost section of each arm retracted, and then the arms themselves began to swing, accelerating rapidly to get safely away from the vehicle. As the Saturn V slowly rose into the air beside them, the arms braked abruptly to avoid smashing into the umbilical tower.

For Don Buchanan, watching from the Launch Control Center, the hold-down period had seemed endless, until he had finally begun to think that his hold-down arms must somehow have failed after all. Now, as the Saturn lifted and all five in-flight swing arms moved smoothly away from the side of the vehicle, he began to breathe again.

As the Saturn V moved off the pad, the sound finally reached across the marsh and slammed into the viewing area. It came first through the ground, tremors that shook the viewing stand and rattled its corrugated iron roof. Then came the noise, 120 decibels of it, in staccato bursts. People who were there would recall it not as a sound, but as a physical force. In the C.B.S. broadcast booth, the plate-glass window began to shake so violently that Walter Cronkite had to hold it in place with his hands as he tried to continue his commentary.

One second after lifting off, only a few feet above the launch platform, 501 began to maneuver, yawing away from the umbilical tower. For the many viewers who didn't know this was supposed to happen, the Saturn seemed to be tilting as ominously as the Vanguards and Atlases of only a few years earlier. Even for the more knowledgeable viewers, it was a nervous moment. If everything was going nicely, why interfere by trying to steer the behemoth so soon? And yet that was what the I.U. was doing, sending a preprogrammed command to the engines of the F-1, which in response were gimbaling and guiding the Saturn V away from the umbilical tower.

At the beginning, it seemed more a levitation than a liftoff—the Saturn rose so ponderously that it took more than ten seconds for it to clear the top of the umbilical tower. Then, as the Saturn got farther from the ground, the scale of the F-1s' inferno became more fully apparent: The rocket climbed, but the trail of flames continued to billow all the way down to the base of the launcher. Not until A.S.-501 was several hundred feet off the ground did its plume of flame lift from the launch platform.

Grady Corn, sitting on the lower level of the Firing Room in Propellant Row, was down too far to see the Saturn on the pad or the actual liftoff. All he knew was that the big windows in the Firing Room were vibrating violently and plaster dust was falling loose from the ceiling of the Launch Control Center onto his console. Now, as Corn looked back up at the window, the Saturn V came into view, rising majestically against a blue sky, and Grady Corn was cheering along with the rest of the team, jubilant that he had been so wrong.

Up in the V.I.P. viewing area, von Braun yelled, "Go, baby, go!" Arthur Rudolph decided he had gotten the finest birthday present of his life. George Mueller looked pleased.

Now the I.U.'s guidance system was controlling the rocket. Massive as the Saturn V seemed to an onlooker, it would not naturally go in a straight line. On the contrary, lacking a guidance system it would have been as unpredictable as a child's skittering balloon. The job of the guidance system was to ensure that the line of thrust of the launch vehicle was aligned with the center of mass. To that end, the guidance system checked the vehicle's position, attitude, velocity, propellant levels, and a few dozen other variables every two seconds, and then sent messages to the four outboard F-1s (the center engine was fixed). If an onlooker could have gotten close enough to see, and if he could have ignored the scale of the machine, the behavior of the F-1s would have seemed almost delicate, as each of the four engines swiveled briefly in small, tightly controlled arcs, a few seconds here, a few seconds there, not just maintaining the Saturn V in a steady climb, but guiding it through a complex trajectory that involved programmed changes in attitude as well as constant, ad hoc adjustments to compensate for wind.

At 135 seconds into the flight, as planned, the center engine of the S-IC shut down. Fifteen seconds later, the outboard engines did the same. Then a signal from the I.U. exploded a cord of explosive primer attached around the base of the S-II, cutting away the S-IC. As the final act of its two-and-a-half-minute lifetime, the S-IC fired eight small solid-fuel retro-rockets, slowing the S-IC so that the S-II would be safely separated when its engines ignited. The great S-IC quickly lost the rest of its upward momentum and, still carrying its exquisitely crafted pumps and piping and engines, fell back to crash into the Atlantic Ocean.

High above, eight small "ullage" motors on the S-II fired for four seconds to give the S-II a burst of acceleration and settle the propellants in their tanks. Then the five hydrogen-powered J-2 engines of the second stage came to life, developing a total of a million pounds of thrust. As the S-II accelerated, a second length of primer exploded, separating the "interstage"—the sixteen-foot part of the wall of the rocket that had covered the J-2 engines and connected with the top of the first stage.

The first Saturn V continued to perform perfectly. The five J-2 engines fired for six minutes, constantly gimbaling in their delicate minuet. They too shut down precisely at the planned moment, the primer cord exploded, the retro-rockets pushed the S-II back and away, and the single J-2 engine on the S-IVB stage fired.

And still everything worked. The S-IVB fired for two minutes and twenty-five seconds, putting itself and the C.S.M. into a perfect orbit 118 miles high, with a speed of 17,400 m.p.h.

Eleven and a half minutes after it had lifted off the pad, A.S.-501 was over for the people at the Cape. It wasn't over for the flight controllers at Houston—they would relight the S-IVB a few hours later, bringing the spacecraft back into the earth's atmosphere at an entry speed of 25,000 m.p.h. But it was over at Cape Kennedy, where the 450 men crowded into the Firing Room had cheered again with each new report of success and were now a little groggy.

A.S.-501, which the newspapers called Apollo 4, made the headlines the next day. But there was no way that the papers could convey what a von Braun or Petrone or Mueller—or, for that matter, a Rigell or Fannin or Corn—knew. Only a few years earlier, many of them had been hesitantly trying, often failing, to launch rockets with a single, small engine in each stage. Today, in its first trial, they had launched a rocket the size and weight of a Navy destroyer, carrying eleven new engines, new fuels, new pumps, new technology of all kinds, and had done it perfectly. There was simply no way to explain it. They could recite how heavy it was and how powerful and how many parts it contained, but that didn't capture it. "We fought that thing for seventeen days," Ike Rigell said, remembering the tortuous C.D.D.T. "And then on launch day it worked. It worked beautiful." Thinking back on it, Rigell, a man not given to excited exaggerations, could only shake his head and say, "It was fantastic. Unbelievable." It was that, and more. A.S.-501 had opened the way to the moon.

BOOK III

FLYING

HERE MEN FROM THE PLANET EARTH
FIRST SET FOOT UPON THE MOON
JULY 1969, A.D.
WE CAME IN PEACE FOR ALL MANKIND

—Inscription on a plaque attached to
the leg of the lunar module *Eagle*

18

"We're going to put a guy in that thing and light it"

In the fall of 1962, when Rice University gave NASA 1,000 acres of salt-grass pasture south of Houston to build the Manned Spacecraft Center, some of the university trustees thought that manned space flight might not last. A clause was inserted in the deed specifying that if space flight fizzled, Rice could reclaim the facilities.

Because the Manned Spacecraft Center was designed with this contingency in mind, NASA built a facility that was neat, innocuous, and gave no hint as to its purpose. Langley had its wind tunnels, Marshall had its massive test stand, the Cape had the V.A.B. and the towering launch complexes. M.S.C. had three dozen buildings, squares and rectangles of glass and white textured concrete, scattered around the perimeter of a central green with three irregularly shaped duck ponds. At the time the manned Apollo flights began in 1968, the trees were still scrawny saplings and the grass was patchy, burned to a dust-brown during the long summer.

Day to day, most of the work of M.S.C. was as undramatic as its surroundings. In the Center's tallest structure, Building 2, at the end of the boulevard leading into M.S.C., Gilruth, Faget, Low, and the Center's other senior officials worked in their suites on the upper floors—holding meetings, leafing through reports from the contractors, examining their huge charts of task schedules, and dictating memoranda. Their staffs were scattered among the other buildings, at drafting tables and computer terminals, carrying out the prosaic tasks that go into running a space program.

But on launch day, everything changed. Once the Saturn V cleared the top of the umbilical tower, control of the mission shifted from the Cape to Houston. From that moment until splashdown, the sustained drama of an Apollo mission was played out at M.S.C., for it was at M.S.C. that Flight Operations made its home.

1

"Flight Operations was not born effective, it became effective," said one veteran. Back in 1959, when the Space Task Group was first struggling to put the program together, nothing was yet Standard Operating Procedure. "We accept this Control Center and operations mode that we have now," said Glynn Lunney, who was part of the Operations Division from the Space Task Group's first day, "but the truth is, it easily could have [evolved in] any number of other ways, and it could easily have been a failure."

In the beginning, preparing for Mercury, it wasn't clear how much Operations would have to do. In the flight-testing of aircraft, which was the closest analogue, the ground's role consisted of getting the airplane into the best possible mechanical condition, spelling out the day's test objectives for the pilot, and retrieving the data from the instrumentation after the plane landed. During the flight itself, the people on the ground talked to the pilot and kept track of where he was, but beyond that they had a limited role.

Thus when the Operations Division at the Space Task Group began thinking in early 1959 about what their job really was, the possibilities were vague and open-ended. Because none of Mercury's systems was actually operated from the ground, many from the N.A.C.A. envisioned a flight-test operation that would check out the capsule before launch and then let the astronaut do the rest. It didn't make sense to acquire a lot of real-time data on the ground if nobody was going to do anything with it.

And yet it didn't seem adequate to stand on the ground with just a voice link to see how things were going. The operations people at Langley began groping toward another, more ambitious understanding of their role. "I don't know how to describe it exactly," Lunney recalled, "but we began to realize that, 'Hey, we're going to fly this thing around the world!' and then a number of things began to emerge." There was already the matter of range safety, for example. If you were launching an unmanned rocket from the Cape, you had "range-safety limits." If the telemetry told you that the rocket was outside the safety limits, you blew it up before it descended on downtown Cocoa Beach. The Space Task Group people began to conceive of similar kinds of limits that protected not "the range" or the population of the Cape, but rather the man inside the spacecraft.

Thinking about them quickly led the Operations Division to realize that, whatever these limits were, they would keep changing during the mission. For the first few minutes after launch, the capsule would have its escape rocket; then the escape rocket would be jettisoned and a new set of procedures would come into play. A new concept which focused on the alternative ways of getting the astronaut back at different points in the mission began to emerge. This was the concept of the "abort mode."

"We began to realize there were some things we could make decisions about," Lunney continued. "Like: Was the capsule in orbit or not? If not, we had to decide when to fire the rockets so it would land in a safe place. How you knew it was in orbit was a big problem at the time. Now we just say, 'You're go for orbit,' but then we didn't know how the hell to do that, or didn't know if we could do it fast enough or with enough accuracy."

However they finally did it, they would have to have people on the ground processing the information, and so emerged the concept of a room on the ground with not just a man talking to the astronaut, but many people analyzing tracking data and telemetry data on the status of the launch vehicle and the spacecraft. The name they eventually gave to this place was the "Control Center."

To get real-time tracking data, Flight Operations had to have tracking stations. At first, the idea around Langley was to rent some truck-mounted radar sets from the Air Force and take them to sites around the world. But they began to realize that it wasn't good enough to have isolated radar sets; the people back at the Control Center needed a network of linked stations, capable of receiving, processing, and reacting to a variety of voice, radar, and telemetry data.

These people in the Control Center and at the remote sites were named "flight controllers." Because many different kinds of data would be coming in, flight controllers would have to specialize. One man would keep track of the launch trajectory, one would calculate the retrofire time, another would monitor the performance of the astronaut's life-support systems. Thus the specific positions in the Control Center began to take shape.

The flight controllers would have to learn to use this information quickly and harmoniously. To be trained, they needed some way to practice. Since they couldn't practice with real rockets, they would need a good make-believe alternative. Pilots had flight simulators; the astronauts were getting a capsule simulator. Why not have something like that for the team on the ground? So began simulations for the flight controllers.

It was a process of constant invention. Countdowns, for example. The rocket people had countdowns for checking out pumps and valves and pressures. But in 1959 no one had ever written a countdown for a spacecraft. Tec Roberts, the Welshman who was part of the AVRO contingent, recalled sitting down with Chris Kraft, then the assistant chief of the Operations Division for "Plans and Arrangements," and in three weeks creating the first countdown that had ever been written for the Control Center. They looked at how the launch people handled the launch vehicle count and applied the same philosophy to their newly emerging concept of a Control Center, relying mostly on "a good imagination," in Roberts's words, and making it up as they went along.

By the middle of 1959, as the Space Task Group struggled to launch Big Joe, train the astronauts, and decide how the flights were to be conducted, it thus became increasingly obvious that "operations" was different from the kind of thing the Langley people were used to. Operations wasn't a matter of good design and engineering in the Langley tradition. It wasn't even a matter of good management. What operations needed, in the words of a Space Task Group member, was "a very tough kind of a guy who's good at getting things flown, used to dealing with pilots, used to the rough and tumble that goes with making a flight program work, making it safe, and not taking any horseshit." The Space Task Group also needed someone who could contend with the Air Force. Given the slightest opening, the Air Force, which was making life miserable for Scott Simpkinson's team crammed into half of Hangar S down at the Cape, would be more than happy to take over manned space flight. Somehow,

Project Mercury needed to develop a close working relationship with the Air Force and at the same time avoid getting pushed around, or perhaps even out. Abe Silverstein decided to call in reinforcements.

2

In 1946, the N.A.C.A. had sent a young Langley engineer named Walt Williams out to the Mojave Desert to open a facility for testing the X-1, the plane in which Chuck Yeager would break the sound barrier. It was called the High-Speed Flight Station, using the Air Force facilities that would later become known as Edwards Air Force Base. On July 30, 1959, Williams's fortieth birthday, Abe Silverstein called Williams to Washington to try to persuade him to return to Langley to work for the Space Task Group. On September 15, Williams reported to Langley as Gilruth's associate director for Operations.

Williams was indeed the "very tough kind of guy" that the gentlemanly engineers of Langley needed to deal with the Air Force's brand of bureaucratic infighting. By 1959, Williams was already known in the flight-test business as a man who could work, carouse, cuss, or fight as prodigiously as any test pilot at Edwards. He was also tough in the other ways that Operations needed. "He had the ability to walk up to the problem of putting a man on top of one of these Atlas vehicles, which are really just big metal balloons, and not be cowed by it," said Lunney of Williams.* "Williams just walked up and said 'Goddammit this' and 'Goddammit that,' and got everybody saluting and doing what they should do." He was a genius of sorts, Lunney reflected, "though if you had to go up against him, he didn't seem like a genius, he seemed like a bull."

In those days, he even looked like a bull—over 200 pounds, a powerful man with a square head, dark, close-cropped hair, and heavy brows. Gene Kranz, who himself would scare a few people in his time, never forgot his first encounter with Williams. It happened in 1960, just a few weeks after Kranz had arrived at the Space Task Group. Kranz had been sent over to brief Williams on some work he'd been doing. Kranz, who knew Williams only by reputation, got there early and slipped into a seat in Williams's office while another briefing concluded. Williams, who

---

* The walls of the Atlas were so thin that they would collapse unless the vehicle was pressurized, whence Lunney's reference to balloons.

was slouched behind his desk chain-smoking Winstons, looked a little like Broderick Crawford in "Highway Patrol"—big and rumpled and knowing.

The men briefing Williams were not having a good day. Williams sat behind his desk, scowling at the hapless briefers, and "whipsawed them," Kranz remembered. "Just cut them up. Sliced them off at the ankles, midcalf, knees, midthigh. They went down the tubes and the thing was over." The objects of these attentions put away their papers and filed out of the office, leaving Kranz and Williams alone. Kranz began his report.

Walt Williams had a curious habit of appearing to fall asleep in the middle of meetings. Williams himself said that it was a device: "I listen to the guy's voice," he explained. "If the guy's trying to bullshit you, or is uncertain, you can hear it in his voice. I don't want to see his bright blue eyes or anything else." Did he ever really fall asleep? There were, after all, reports of the occasional snore. Well, Williams said ambiguously, when people had given him a briefing paper in advance, he didn't "bother even listening to the buildup," but waited until the guy got to the good part.

Kranz knew nothing of this. After a few minutes, he was getting into his material when he realized that Williams was starting to nod off. His head was slumped on his chest, his eyes were closed. Unbelievable as it seemed, Kranz decided, Williams had gone to sleep.

"In a one-on-one session with a legend, who you have just seen completely assassinate somebody, what the hell do you do?" Kranz would ask later. He decided that the safest course was to pretend nothing had happened and keep on talking. So he did, feeling more confident now that it seemed no one was listening to him. In fact, he was feeling confident enough to sidestep a small issue that he wasn't absolutely sure about. But what the hell, the man was asleep.

Williams always kept a roll of Necco mints nearby to soothe his smoker's throat. As Kranz breezed on, Williams's hand reached out, groping slowly for the roll of Neccos. Williams shook out two of them—eyes still closed—and chomped on them for a while. "And then he proceeded to ask exactly that question that I thought I had skated through very cleanly," Kranz recalled. "It was just absolutely intimidating."

Williams had that effect on the Operations Division as a whole. First, he got together with the astronauts, who had been struggling with the design engineers to be treated as pilots, not guinea pigs. Williams had spent his adult life around test pilots, and was sympathetic. (In July, before being offered the Space Task Group job, Williams read in the newspaper that the astronauts were complaining about not having any planes to fly. The Langley engineers should have known, he remarked to his wife, that "you can't have tigers around and not have raw meat for them.") Furthermore, it was Williams's view that the astronauts were the most carefully selected people and probably the most talented people in Project Mercury and it was foolish not to make use of them. From then on, the astronauts were deeply involved in all aspects of the hardware and flight planning. At the same time, Williams conveyed to the astronauts, and made it stick, that in manned space flight there would be a crucial distinction between "command" and "control." The astronaut had control of the spacecraft, but the ground had command of the mission. This would mean a degree of minute-to-minute direction from the ground that had no precedent in flight testing.

Williams radically altered the approach to preparing the capsule. No longer, he decided, would the engineers and technicians who made the spacecraft be the sole judges of their own work. An independent corps of inspectors would watch every step and must literally sign off on it before work could proceed. Williams went down to Hangar S at Cape Canaveral and pulled Joe Bobik off Simpkinson's crew to come up to Langley and inspect the work on the capsule being assembled in the Langley shops. "Hey, they'll throw me out of there!" protested Bobik, and he was almost right; but Williams's system stuck and eventually grew into the intricate spacecraft inspection system that Bobik would direct during Apollo.

When it came to flight control, Williams took a look at the paper-thin layer of experienced Canadians and the hodgepodge of young engineers beneath them, and promptly set up classroom instruction, using the senior engineers to instruct the neophytes on the details of the Mercury and Redstone systems—"how they worked," Williams said, "and more importantly how they didn't work." He began talking about mission rules and countdowns and operational requirements. Most of all, Walt Williams made everybody aware that this was not some abstract engineering exercise. He had at least one astronaut attend every Ops meeting. "Look, what you guys have to remember," Williams would say to them, "is that there's gonna be a bright, clear morning at the Cape, and that Atlas is gonna be sitting on the pad, there's gonna be a capsule on top of it, and we're going to put a guy in that thing and light it! This is not a hobby. This is real! This is what we're going to do!'' To do it, he taught them, would take not just knowledge and good faith, but also a hard-bitten tenacity and self-discipline.

"Walt Williams became like the father," said Lunney. "We were all his boys and he loved that. He came knowing that an operations organization had to be set up and having a lot of the instincts about what characteristics it should have. Walt encouraged it to happen." But Williams was too busy negotiating with the Air Force and taking care of the preparation of the Atlas launch vehicle to "make it happen" from day to day in the flight control part of the business, Lunney said. "And then this young Kraft guy came along with a lot of clever technical skills and the charisma to pull it off. Chris made it happen."

His unlikely name was Christopher Columbus Kraft, Jr. He was named for his father, who had been born in New York City the week that Columbus Circle was dedicated. Chris Junior was born in 1924 in Phoebus, Virginia, just up the road from Langley—his Uncle August had been one of the surveyors who had laid out the original runways at Langley Field back in 1915. A standout baseball player in school, Kraft toyed with the dream of playing pro ball. But he wasn't quite good enough for that, and he was rejected for military service during World War II because of a hand that had been badly burned when he was a child. So after returning home with his engineering degree from Virginia Polytechnic, Kraft drove down the road from Phoebus to Langley and applied for a job. He reported for work in January 1945 and was assigned to William Hewitt Phillips's Stability and Control Branch, where he worked for another young engineer named Chuck Mathews.

In later years, Kraft would reminisce about his years under Phillips with a craftsman's pride in his work, but eventually he began to scrape against the edges of the Langley mold. By 1958, Kraft was a quietly frustrated young man with an ulcer. Then that fall, Mathews recruited him into the Space Task Group as his assistant in the new Operations Division. One day, as Mathews remembered it later, Kraft came to him and said, "There needs to be someone in charge of the flights while they're actually going on, and I'd like to be that person." Mathews agreed, and that was how the position of flight director in manned space flight was born. After Williams came aboard and Mathews gradually moved into the background, the original Operations Division was split into three new divisions, for Preflight Operations, for Flight Crew

Operations, and for Flight Operations.* Kraft was made director of Flight Operations. Ultimately, he would become the embodiment of Mission Control.

Walt Williams would throw out an idea and Kraft would pick up on it. One of the earliest and most important of these was Williams's notion of "mission rules." "He said we had to have them," Lunney said. "We didn't know what 'mission rules' meant, but we went out and made them up." It was really just about that simple, according to Kranz. "Honest to God, I hadn't been on board more than a couple of weeks when they said, 'Hey, go down to the Cape and write some mission rules.' I said, 'What are mission rules?' 'Well, if something goes wrong, we want to have some thinking as to what we're going to do about it.' I said, 'Jesus Christ, I don't even know where the Cape is, much less how to go down there and write mission rules.' But it was unquestioned. We'd all muster out and go do it." Kranz and Tec Roberts and Paul Havenstein, a Navy officer seconded to the Space Task Group, flew down to the Cape on one of East Coast Airlines' creaky old planes. They sat out on the porch of a house borrowed from an Air Force officer stationed at Patrick Air Force Base, just south of the Cape, and there they made up mission rules.

Walt Williams had brought the notion of mission rules from Edwards, where they were used as a way of preparing the pilot to react correctly in situations where there was no time to think through a problem or ones in which he might intuitively do the wrong thing. Mission rules for space flight were developed to help the controllers decide ahead of time, calmly and deliberately, with plenty of information and time to think, what was to be done in a critical situation.

Williams remembered meeting considerable resistance to the notion of mission rules at first—"It took table pounding." People argued that they would never know exactly what to expect and that the rules were unlikely to work for a specific situation. That's not the point, replied Williams. "If you'd worked your way through problems, then when you saw a problem, whether you knew exactly what it was or not, you started

---

* Williams's arrival made a difficult situation for Mathews. Williams came in as the associate director (Operations) of the Space Task Group. Kraft had already been given charge of the functions that later became Flight Operations. Mathews remained, as before, chief of the Operations Division, but the positions overlapped so much that Mathews spent most of his time putting together the worldwide communications network for Mercury and stepped aside to let Williams and Kraft carry on with the development of flight operations procedures. Subsequently Mathews became director of the Spacecraft Technology Division, then of Project Gemini, then deputy director of O.M.S.F.

recognizing the characteristics of it—and understanding the system, you could work your way out of it.''

Throughout Mercury and Gemini, mission rules continued to develop, and by the time of Apollo they filled a number of thick books. But over time, the mission rules also took on the elements of a philosophy—or, as Lunney liked to think of it, a set of values. "It's like being raised by parents in some faith. You get a set of values, then when you're an adult, you probably live by them most of the time. And it became like that with us.''

The most generic of all principles was that the crew must be protected from dying. But the principle had to be more sophisticated than "protect the crew at all costs.'' Just as Owen Maynard and Caldwell Johnson constantly had to make trade-offs between different kinds of risks, so also the controllers found that it wasn't possible to make decisions simply on the basis of whether a procedure was safe or not. "When we first started,'' Lunney recalled, "people would say things like, well, the spacecraft's got to be 'good' [before deciding to continue a mission]. But what the hell does 'good' mean?'' Slowly, the Flight Operations Division wended its way toward a subtle and complex style of thinking about this crucial thing called safety. In this process, Kraft became what Kranz and some of the others called "The Teacher,'' developing precepts that became the catechism for later generations of flight controllers.

The first of Kraft's precepts was simplicity itself: "If you don't know what to do, don't do anything.'' But it was simple as a koan is simple, for flight controllers were trained and conditioned to solve problems; the temptation was for a controller to think that he knew what to do when he really didn't. True wisdom in flight control lay in being able to recognize one's own ignorance. Much of the training of the flight controllers consisted of showing them how they could be fooled.

Another of Kraft's precepts was not at all simple. According to Kranz, "Kraft always believed that once you accepted the risk of launch, once those engines ignited, you had bought a good portion of the risk associated with that mission. Once you got into orbit, what you wanted to do was exploit the environment you were in. It was really a philosophy of risk versus risk and risk versus gain which we debated many, many times.'' For example, during the Gemini V mission, there was a risk of losing the fuel cells if the spacecraft remained in orbit. If Kraft brought the crew home, another crew would have to repeat the mission and accept the comparatively higher risks of the launch phase. The conclusion, according to the Kraft philosophy, was that although it was risky to use the fuel cells, it was safer in the larger scheme of things to continue the mission than to bring it home with its work uncompleted.

If "safety before everything" was impossible, what then was the guiding principle for deciding to abort a mission? The canon was voluminous, but the unifying tenet was based on the ability to tolerate one more major failure. As Lunney explained it: "You will continue [flying] only if the next thing that happens to you—and it's the worst thing you can think of to couple with the problems you already have—is still survivable." If you thought in those terms, the flight controllers pointed out, a great many of the more specific rules fell into place. "Sounds obvious as hell," Lunney added, "but it took a little while to figure that out."

While these principles of flight control were evolving, two other developments profoundly affected the way that flight operations would be conducted during Apollo. One was a process lasting for years; the other was a single, terrifying event.

3

The process was the shaping of Flight Operations into a brotherhood.

For an extended period from 1959 until the mid-1960s, the new Flight Operations Division was continually hiring new people. But whereas other divisions within the Space Task Group hired a mixture of new college graduates and people out of industry, the Flight Operations Division tended to hire predominantly new college graduates. Furthermore, Kraft looked for a particular kind of person. He wasn't worried about grades—a B or sometimes even a C average could be good enough. In fact, the straight-A student wasn't likely to be right for Flight Operations. Kraft wanted people who weren't locked into a standard engineering career path. He wanted people who enjoyed nosing around in many different areas, and who knew that they were going to have to work hard to succeed. Most of all, he looked for applicants who were fascinated by space flight and who couldn't believe their good fortune when they were given the opportunity to work sixty-hour weeks at a Civil Service salary—as long as they were working for the space program.

When these youngsters, twenty-one and twenty-two years old, came to work, they found themselves in one of the most glamorous jobs in the program. They didn't sit at drafting tables drawing electrical circuits or designs for valves; they were the guardians of the astronauts. They went to exotic places—Hawaii and Bermuda; Carnarvon on the northwest coast of Australia; Kano in the interior of Nigeria; Tananarive in the Indian Ocean; ships and islands throughout the world—to man the remote sites. They had as their leaders the redoubtable Williams, possessing everything to inspire awe in twenty-one-year-old males but a silk scarf around his neck, and Kraft, who, as Williams left more and more of the daily operation to him, became the revered leader.

Kraft was the model for flight-controller cool, taking in bad news without changing expression or tone of voice, making decisions quickly, controlling people and events as the situation demanded. He had the indefinable quality called "presence" as well. "I've seen people argue and argue and argue with him, trying to get their way," one controller recalled. "He'd look at them very casually, and he'd say, 'I have your input.' And it would just terminate the conversation—just terminate the guy into a dummy load, is what we used to say."

"People idolized Kraft, and wanted to be like him," another controller said, speaking of the youngsters coming into Flight Operations fresh out of college. For a while back in the early 1960s, the young Glynn Lunney wore the same kind of loafers, smoked the same kind of cigars, held his cigar in the same way as Kraft. Gene Kranz, with his fighter-pilot experience, saw himself in his early days with Flight Operations as Kraft's wing man. He would arrive at Mercury Control long before Kraft and test everything on Kraft's console, check out every communications loop, then make sure his own console was prepared so that in the event of a failure on Kraft's console, Kraft could step over to exactly the same setup on Kranz's. "I was probably overkilling it a bit," Kranz said, "but I wasn't going to let my lead guy get shot down."

So there they were—young, male, in a high-pressure job, often the only one they'd ever known, many of them single, spending most of their waking hours together, often in remote overseas outposts, led by men they idolized. The result was more or less what one would expect of such a mix of circumstances—male bonding (a phrase no flight controller would be caught dead using) on a grand scale, and a kind of closeness that many of them would never know again. Along with the excitement went the hell-raising that groups of young men are prone to indulge themselves in. In Houston, where some had families and they were working all the time anyway, it was pretty tame—volleyball and beer—or reasonably discreet. At the remote sites and the Cape, the hell-raising got a little more exciting, much of it involving cars (driven into the surf off Cocoa

Beach, for example), women (Australia was an especially popular duty station), and more beer. The controllers had their own legends of outrageous flight-controller behavior—Ed Fendell, John Hatcher, and above all John Llewellyn. They had their own language. While they were working a mission in the Control Center, they communicated in short bursts of acronyms and abbreviations and code words that made it impossible for an outsider to have any idea what they were talking about. After work, drinking beer at the Flintlock or the Singing Wheel, they would add to the acronyms and the jargon their own figures of speech. A Saturn didn't fly to orbit; it "went up the hill." One didn't make inquiries of someone; one "pulsed" him. To fall asleep was to "go to poo," referring to program P00 which reset the onboard computers to zero. Then, for reasons that remained unexplained, the controllers mixed their technological slang with medievalisms—"yea verily," for example, and "it came to pass," and "thou shalt." Sometimes it seemed that the controllers lacked only decoder rings and a treehouse.

Kraft consciously reinforced the band-of-brothers atmosphere in Flight Operations. As long as you were part of the brotherhood, he would defend you against any charge to anyone outside. In one of the classic John Llewellyn stories, Llewellyn, who had been assigned to coordinate the recovery team for one of the early missions, decided unilaterally that the Navy ought to add an extra ship to the recovery task force. The Navy did, and sent a bill for something like a million dollars to Gilruth. As the story is told by the controllers, Gilruth called Kraft to his office and asked what this was all about. Kraft got Llewellyn on the phone, listened for a while, hung up, turned to Gilruth, and said simply, "Pay the million." That's the way Kraft was with his people.

"He's got to be the best motivator of people I ever had anything to do with," said Rod Loe of Kraft. For example, said Loe: During Gemini, Loe, a flight controller for the environmental control system, was worried about one of his subsystems. He couldn't get the contractor in charge of it to pay any attention to him—he was just a little guy. Kraft somehow heard about this, and when a few days later the contractor people were discussing another problem with Kraft, Kraft casually walked over to Loe and put his arm on his shoulder and asked, "What do you think, Rod?" That's all it took. " 'Rod? Who's Rod?' Those guys didn't know who 'Rod' was," Loe said. "But from then on, those guys started looking me up, to get my opinion, because they thought, 'God, Kraft listens to this guy.' " It didn't take many such episodes to build a lot of loyalty among Kraft's people.

If, on the other hand, you were part of Flight Operations and decided that it was time to move on to something else, you were likely to be cast into outer darkness. Who could want to leave the best place to work in the whole space program? Kraft would let you know that you were making a mistake. "Young man—" he would begin (to be called "young man" meant you had better listen carefully; to be called "sir" meant that you were already in deep trouble), and then he would launch into an explanation of all the reasons why you were better off in Flight Operations than in this other job, no matter how tempting it might seem from a distance. If you persisted, perhaps you could get him to agree, and in that case he would help you get the transfer and maybe a promotion to boot. But under no circumstances was it advisable to try to go around Kraft and get yourself transferred without his blessing. A few people did that, and soon thereafter they decided to leave NASA, or, at the least, to get out of Houston.

It was all part of being in an exclusive and absorbing club called Flight Operations. "We were on an absolute high," Lunney said of those days. "You can't imagine how excited we all were." Also, being so young helped keep the controllers from being awed by the fact that, after all, they were only kids and there would be live astronauts up there. Lunney again: "We had a few guys who came into Flight Operations who had been around awhile, and they had trouble believing that they were able to do what they were doing. I had a guy who was uncomfortable with it, and he used to ask me, 'How can you have the confidence to do this?' My reaction was, 'Well, how the hell do I know? I don't go around analyzing that, I just do it!' The people who'd been around for a while maybe had been blunted. They didn't do too well. The people who came in new, out of school, they didn't know any better. They didn't know they shouldn't do things or couldn't do things, or that life was going to beat them up sooner or later." The confidence and the closeness were to be indispensable to the way that Flight Operations functioned during crises.

4

The single event that decisively shaped Flight Operations happened on the morning of February 20, 1962, when John Glenn made the United States' first orbital flight.

For practical purposes, Glenn's flight was the first test of the Flight

Operations system. Shepard's and Grissom's suborbital flights, which preceded Glenn's, were lobs in a ballistic trajectory that lasted only about fifteen minutes each and involved little in-flight decision-making. For Glenn's flight, the flight control team would have to determine whether he was in a stable orbit and instruct him accordingly, monitor his systems over a period of five hours, and do its part in bringing him safely out of orbit and into the landing area where the recovery ships were waiting.

The flight controllers had never expected it to be an easy flight. Based on the Air Force's experience to date, they could expect to lose one out of four Atlases during the launch phase. Moreover, the Atlas had recently been displaying a tendency to explode within a few feet of the pad, the most difficult of all times to use the escape tower to lift the astronaut safely away. And some of the explosions had been extremely sudden, giving less warning than they would need to activate the escape tower. Everyone worried about the launch phase. Kraft himself kept an eight-by-ten photograph of an exploding Atlas under the glass of his desk.

But when Glenn actually launched on Tuesday morning at 9:47 A.M., the Atlas worked fine. The problem came instead as Glenn was heading east over the Atlantic on his second orbit. A technician named Bill Saunders, sitting in a room to the side of Mercury Control, was scanning the bank of meters in front of him when his eyes fell on meter number 51—"segment 51"—which registered the deployment of the heat shield.

In the Mercury capsule, the heat shield was designed to fall loose after entry into the earth's atmosphere so that a bag that would absorb the shock of landing could inflate between the deployed heat shield and the bottom of the capsule. Saunders's needle should have been pointing at +10, indicating that the heat shield was clamped to the bottom of Glenn's capsule. Instead, it was pointing at +80, indicating that the heat shield was unlatched. "I've got a valid signal on segment 51," Saunders said into his headset.

An observer sitting in the glassed-in viewing area behind the Mercury Control Center would have seen nothing unusual, just Kraft conferring with his controllers and Williams walking unhurriedly out the side door. Williams himself would recount the reality later, using the present tense: "It's almost impossible to describe the raw tension that this news introduces into Mercury Control and continues to build relentlessly during the rest of the Glenn flight. . . . We are all in a state of shock at the enormity of the situation." For if the heat shield had deployed, they believed at first, there was not a thing they could do about it. Soon, they would have to tell Glenn what had happened. And shortly after that,

Glenn would die. What happened thereafter was an object lesson in how different space flight was from ordinary flight-testing, and how much was yet to be done to make Flight Operations ready to do its job.

Their one hope during the first minutes of the crisis was that the signal was false. Instrument readings were as likely to malfunction as the systems which they were monitoring, and nothing had happened that might have triggered the heat shield to release. It was possible, even likely, that the signal was false—but they had no second data point in the telemetry that would enable them to assess that possibility. Perhaps the microswitch for segment 51 hadn't been properly set before launch—but they had no way of knowing that.

Then they realized there was a possible way out, even if segment 51 was telling the truth. In the Mercury capsule, the retro-rockets that slowed the capsule for entry were part of an assembly called the "retropack" that was held against the bottom of the heat shield by stainless-steel straps. The straps themselves were clamped onto the main body of the capsule. Normally, the retropack would be jettisoned after it had slowed the capsule. Suppose instead that they left the retropack in place during the entry? The straps holding the retropack to the spacecraft would burn through, but by the time they did, the forces from the entry might hold the loose heat shield in place. It was at least a hope.

Now they were faced with a classic dilemma: If the segment 51 signal was valid, they would have to leave the retropack strapped on, because it was the only hope for saving Glenn. But if the segment 51 signal was false and they left the retropack strapped on, Glenn could be killed. The shock waves of the burning retropack might damage the heat shield. The extra inertia created by the weight of the pack might change the capsule's attitude. In short, they had no safe choices.

To choose among the risks, Williams and Kraft needed to know what the odds were that leaving the retropack on would cause a problem. That kind of knowledge was not available in Mercury Control, so the calls went out. Over at Hangar S, the McDonnell engineers feverishly began laying out long strips of schematics on the hangar floor. Telephone circuits were patched up with the McDonnell plant in St. Louis. Williams called Houston, to confer with the man who had been thinking about the aerodynamic characteristics of the Mercury capsule longer than anybody, Max Faget.

No one had ever tested a spacecraft entry with the retropack attached; no one had ever even thought about it. Faget talked through the problem

over the phone with Mercury Control. Later, Joe Shea would be
fascinated when Faget recounted his thinking. Faget had a "first-order
feel," in Shea's words, that leaving the retropack on wasn't going to be
a problem. Faget understood the nature of the pressures and forces on the
spacecraft, not by calculation but by apprehending the engineering gestalt
of the situation.

Eventually, Williams decided that leaving the retropack on was the
lesser of the evils. They would find out later that segment 51 had been
misleading them after all and that they could have conducted a normal
entry, but by that time John Glenn was safely home, a national hero, and
Mission Control was being applauded for its cool professionalism under
pressure. Professional, yes: The flight-control team had stayed calm and
had done the best they could with the information they had available. But
it had been an ad hoc, jury-rigged, scrambling effort, and from that
experience came a structure for trying to ensure that they would never
have to make those kinds of guesses again.

19

"There will always be people who want to work in that room"

During the rest of Mercury and throughout Gemini, the flights became more complicated and so did the resources for supporting them. Beginning with Gemini IV on June 3, 1965, control of the flights shifted from the Cape to new facilities in Houston. Meters gave way to computer screens. Ad hoc conference calls gave way to a nationwide communications network. The modest goal of simply understanding what was happening to the spacecraft gave way to ambitions for solving hardware problems while the flight was still in progress. By the time the Apollo manned flights began, the Mercury Control Center that supported John Glenn's Mercury flight had been supplanted by a system that was to Mercury Control as the Saturn V was to the Redstone.

1

During an Apollo mission, the action at the Manned Spacecraft Center centered in Building 30 on the northwest side of the central green.

Building 30 had two wings connected by a large lobby. One wing, which overlooked the duck ponds, housed in its three stories the Mission Planning and Analysis Division (MPAD, pronounced "em-pad"), the Flight Control Division, and the Flight Support Division. The other wing, which would have looked out over a parking lot and empty fields if there had been anything to look out of, was a large, windowless concrete block three-quarters the size of a football field, also three stories high.

The working entrance to the windowless block was inside the lobby that connected the building's two wings. Equipped with the proper badges, a visitor to this wing entered a slow elevator which rose to the third floor and opened onto a short, nondescript hallway with an alcove filled with vending machines. The short hallway debouched onto an echoing corridor with mustard-colored walls and twelve-foot ceilings. The lighting was flat and shadowless, and that, combined with a knowledge of the business of this place, could create an uncanny feeling—for some, the feeling of being behind a grand stage set; for others, a sense of entering a forbidden sanctum. For still others, it was like returning to a battlefield.

The corridor formed a large rectangle. The outer perimeter of the square was lined with large rooms containing a number of electronic consoles with C.R.T. screens (cathode-ray tubes, the kind used in commercial television sets). These were the Staff Support Rooms, known as the S.S.R.s or more commonly as the "back rooms." On the interior of the rectangle were other rooms, some filled with electronic equipment for supporting the back rooms. The SPAN was in this block—of that special room, more later—as was the room where NASA and Navy officials coordinated the recovery operation after the Apollo spacecraft splashed down. But the dominating room in the interior of the rectangle was the large square chamber, sixty feet on a side, formally designated the Mission Operations Control Room, known to the public as Mission Control, and known to the people who worked there as the MOCR.*

For a place that came to stand for advanced technology and played such a large part in the public's perception of the space program during Gemini and Apollo, the MOCR was not a comfortable room to work in. The

---

* "MOCR" rhymes with "poker." Once again, the nomenclature can be confusing. Usually the people who worked there called it the MOCR. But sometimes they called it "M.C.C." (for Mission Control Center) or the "Control Center"—which, strictly speaking, is incorrect, since M.C.C. or Control Center officially designated the entire wing of Building 30 in which the MOCR was located. Actually, there were (and still are) two MOCRs. All of the Apollo missions except the first flight were run out of the MOCR on the third floor, but there was a duplicate MOCR on the second floor, right underneath it, where the Gemini flights had been flown. During Apollo, simulations would often be under way on the second floor even as an actual flight was being flown out of the third.

C.R.T. screens were black and white, had poor resolution, and were hard on the eyes. The decor, with its gray walls and the subdued lighting (to avoid interfering with the controllers' view of their screens), was not cheerful. The constant buzz of voices in the controllers' headsets often caused hearing loss. "Controller's elbow," a kind of bursitis, was a common complaint. But none of this really mattered. As one controller said, "There will always be people who want to work in that room. Regardless of how many headaches it produces, no matter how physically painful, there will always be people who want to work there. It's in the blood."

The room was laid out like a small auditorium. The displays were on the front wall, dominated by a display twenty feet long and ten feet high that usually showed a world map while the spacecraft was in earth orbit, the lunar trajectory during the coasts between earth and moon, or a map of the moon during orbit. To either side of the world map were four ten-foot-square display screens that could be changed according to the phase of the mission in progress. During a mission's launch phase, the big twenty-by-ten-foot screen itself was replaced by two more ten-foot-square displays showing trajectory plots.

Below these displays, mounted behind long benchlike working surfaces where the controllers could stack their flight plans, books of mission rules, and the logs they used to record events, were four rows of C.R.T. screens, approximately ten to a row. The front row was at ground level and the other three behind were arranged on risers so that each row was higher than the one in front of it.

In the back row of the MOCR were four consoles, none of which had anything to do with the minute-to-minute conduct of the mission. At the far left, sitting underneath the stationary television camera that relayed pictures of the MOCR to the outside world, was the console of NASA's public affairs officer, the P.A.O.* He was "the voice of Mission Control" on the television broadcasts. The Department of Defense, which was involved in the landing and recovery operation and some of the communications support, had a console at the other side of the back row.

The rest of the fourth row was reserved for senior management, people

---

* The word "console" refers to the station occupied by a particular flight control function. This might involve more than one "position," or controller. Furthermore, one position might use two or even three C.R.T. screens—all of which explains why there could be ten C.R.T. screens to a row which had only four consoles.

like the Mission Director from Washington, the head of M.S.C.'s Flight Operations Directorate, the head of ASPO, and the M.S.C. center director. These were the "hummers," according to one of the flight controllers—as information about what was happening was passed to them, they would nod their heads gravely and say, "Hmmm." One of them, the Mission Director, was intended to have an important role. "They originally thought he was really going to be the director of the mission," a flight director reminisced. "The controllers just said, 'Aaah, get out of the way,' you know. They just didn't have any patience for that." The Mission Director quickly became a liaison person for telling headquarters what was happening, mildly useful and no longer in the way.

The viewing room was immediately behind the fourth row, separated from the MOCR by a large window. This was where important visitors— it could seat seventy-four people—were permitted to watch the action. Sometimes the viewing room would be filled with senior NASA people, sometimes with senators and assorted celebrities. A hummer could be helpful in keeping them happy. Bill Tindall, a resident of the back row after he became head of the Flight Operations Directorate, recalled that he was useless in the MOCR from an operational point of view—the last thing the guys at the consoles needed was help from the boss—"but the glass would be right behind you, and you'd knock on the glass to whoever was sitting back there and say [mouthing the words], 'It's all right,' that kind of crap." The V.I.P.s ate it up.

The action in the MOCR began in the third row and, as a general rule, got progressively more intense as the rows descended.

At the far left in the third row was TelCom, the communications systems officer in charge of monitoring the instrumentation and communications systems on board the spacecraft.* For the lunar landing missions, TelCom also controlled the television camera on the lunar surface (and of course became known as Captain Video). To his right was the operations and procedures officer, who was constantly checking activities during the flight against the mission rules and mission techniques established beforehand. The O&P officer also had a variety of peripheral duties such as overseeing the projection of displays on the front wall.

Next to the O&P officer sat the assistant flight director, filling the least popular job in the MOCR. "I hated it," remembered one man who

---

* During the middle of Apollo the name of this position was changed to instrumentation and communications systems officer (INCO).

briefly filled the position. "You weren't really an assistant flight director, you were assistant to the flight director." It felt a lot like being a flunky, something that no self-respecting flight controller could stomach.

Next came the flight director's console. The flight director, called simply "Flight" when addressed by a controller on duty, ran the MOCR and the mission.

To the right of the flight director was the flight activities officer, a representative from Flight Crew Operations who made sure that the activities being initiated from the ground were consistent with the routines the astronauts were trained to follow. On the right-hand side of the third row sat the network controller, who coordinated the ground stations around the world that acquired the telemetry and tracking data for transmission to Houston.

In the second row, at the far left, was the Surgeon, the physician who monitored the health of the astronauts in flight and tended to the medical needs of the controllers as well.* Next to the Surgeon and just in front of the flight director was CapCom, the spacecraft communicator, still called by the name he had held during Mercury days, when CapCom was short for "capsule communicator." CapCom was always an astronaut, usually one who had been on the backup crew for the flight in progress. He was the link between the crew in space and the people on the ground. The flight director could speak to the crew if he chose, as could Chris Kraft, and occasionally it happened, but otherwise (save for the occasional presidential phone call) no one on the ground except CapCom spoke to the crew.

Sitting to the right across a little aisle from CapCom was the controller in charge of the C.S.M.'s electrical and environmental control systems. Known as EECOM (originally standing for electrical, environmental, and communications—another archaic acronym, since he no longer dealt with communications), this controller monitored the spacecraft's life-support, electrical, instrumentation, and mechanical systems. An Apollo EECOM, John Aaron, acknowledged that "in the context of the image of the

---

* During Gemini and the early Apollo missions, when the Flight Operations Directorate was still gaining experience in its handovers and procedures, and wanted instant access to its controllers, the F.O.D. built a bunk room over the lobby of Building 30. The room had about a dozen bunks, showers, and a little cafeteria (with food of legendary awfulness). The result was that the controllers haunted the Control Center—they just wouldn't go home, and they refused to keep regular hours. They came off shift still taut and keyed up; then when they finally got to sleep, they would constantly be awakened by people coming in or going out. To keep the controllers functioning, the Surgeon dispensed "whoa and go pills," downers and uppers. Some of the Surgeons, especially the older military ones, thought the whoa pills were a lot of nonsense and prescribed little airline-sized bottles of "mission whiskey" instead.

Control Center and where the glory was, the systems guys had a back seat. We were the guys who were supposed to keep everything going and keep everything on line, so the flight dynamics guys could go roaring off where they wanted to go." But it was no contemptible job, Aaron continued. "When you look at EECOM being all the electrical systems, all the power systems, all the instrumentation, which is your eyes and ears into that spacecraft, all the life-support systems, all the cryogenics, all the fluids, you basically have to know the whole spacecraft. So in order to be a good EECOM you had to know the whole spacecraft and then know a lot about what the other subsystems did that you weren't directly responsible for. . . . To be good, you had to know how you affected everybody else in case something in one of your systems failed. That builds a lot of capabilities." A good EECOM, said Aaron, must above all be curious—curious "not only about what you were responsible for, but how you affected everybody else. The more curious you tended to be, the better EECOM you were."

The guidance, navigation, and control officer, G.N.C., sat next to EECOM. G.N.C. didn't keep track of where the spacecraft was or where it was going, but rather watched over the guidance hardware and made sure that it was working properly. G.N.C. also had to worry about the hardware for the in-flight propulsion systems such as the reaction and control system jets and the engine in the service module.

Sitting to the right of the EECOM and G.N.C. were two parallel consoles for the lunar module. Their names were Control (the LEM's G.N.C.) and TELMU (the LEM's EECOM).

Together, EECOM, G.N.C., Control, and TELMU formed the core of what were known as the "systems guys," as opposed to the "trajectory guys." The trajectory guys sat in the front row of the MOCR, which was known throughout the Flight Operations Directorate as the "Trench."

Three positions at the left-hand side of the front row were for the men from Marshall assigned to Houston to monitor the performance of the Saturn V. The lead man, called Booster 1, sat in the right-hand seat of the three. The three Boosters disappeared after the burn for translunar injection (T.L.I.), when the last of the three stages on the Saturn V had been expended. These outlanders from Marshall were conceded by the Houston trajectory people to be good folks, but they were only honorarily part of the Trench. The real Trench consisted of the next three consoles, Retro, FIDO, and Guido, all of which were part of the Flight Dynamics Branch.

Retro, short for retrofire officer, sat next to Booster. The name Retro went back to Mercury days, when Retro's main function was to calculate the time of the burn of the retro-rockets that would bring the spacecraft back to earth. On the first two-thirds of a lunar mission, Retro had a psychologically strange job: He spent hours on end at the console making plans that were continually becoming outdated, minutes after he finished them. If an emergency occurred that called for the mission to be aborted, it was Retro's responsibility to tell the flight director what had to be done to return the spacecraft to earth. Retro had to know the attitude the spacecraft must assume, the burn that would be required, and how to work around the malfunctioning system. Moreover, all of these parameters changed continually as the spacecraft launched to earth orbit, left earth orbit and headed out toward the moon, passed the equi-gravisphere, and entered lunar orbit. For the first part of the translunar coast to the moon, the spacecraft could come back to earth on a direct abort, meaning that the command and service module would be turned around and the big S.P.S. (Service Propulsion System) engine would be used as a brake. As the spacecraft prepared for lunar-orbit insertion, L.O.I., Retro had to plan L.O.I. aborts—meaning, what do you do if, right in the middle of the burn to put the spacecraft into lunar orbit, something goes wrong?

Only when the spacecraft was safely in lunar orbit did Retro perform the first task that he could be sure would be used, the planning for trans-earth injection (T.E.I.) And then, when the spacecraft had successfully begun its trans-earth coast, Retro could finally begin overseeing the entry. In the case of a lunar mission, this meant working out the trans-earth corrections needed to bring the spacecraft safely into its narrow entry corridor at 25,000 m.p.h., getting the spacecraft into the proper attitude prior to entry, and working with the recovery team to select the best ocean recovery area.

Next to Retro sat the flight dynamics officer, FIDO, the lead man in the Trench during a shift.* Among the members of the Trench, it was assumed that FIDO was also the controller who had the most fun job in the Control Center except (a reluctant concession) the flight director's. During powered launch, FIDO constantly monitored the trajectory for deviations and planned the maneuvers to get to orbit in the case of a malfunction (while Retro was planning abort maneuvers in case getting to orbit was not possible). FIDO planned the translunar injection. During translunar coast, he analyzed the trajectory and planned the midcourse corrections and L.O.I. During lunar orbit, he planned the lunar module's

---

* The "I" in FIDO didn't stand for anything. For some reason, the same organization that could pronounce "DPS" as "dips" without feeling obliged to stick in an explanatory vowel usually wrote FDO as FIDO.

ascent and descent, and its rendezvous with the command module. The astronauts were in control of the spacecraft (another reluctant concession), but the FIDOs decided where it was going and when.

Over at the far right of the Trench sat the guidance officer, "Guido" informally, "Guidance" on the loops.* Guido was the ground navigator for the spacecraft. It was Guido who monitored the position of the spacecraft as determined by the ground stations and constantly compared it to the position shown by the onboard guidance system. Guido suggested stars for the crew to use in checking their attitude. Guido watched over the second-by-second performance of the LEM's computers and onboard guidance system during the descent and ascent phases of the lunar missions, and prepared the command loads for the onboard guidance computers. The distinction between Guido and G.N.C., as one controller put it, was that "Guido was looking at it from the standpoint of 'Is the guidance taking me where I want to go?' and G.N.C. was looking at it from the standpoint of 'Are those systems working properly and telling us the right information?' "

The provenance of the term "Trench" is shrouded in the mists of time. Some people think that it started in the old Mercury Control Center at the Cape, where the first FIDO, Tecwyn Roberts, and the first Retro, Carl Huss, sat off to one side of the room by themselves, facing their plotters (there was no Guido then). Another theory is that it started after they opened the MOCR at Houston, where these controllers first began to sit below everyone else.

The story with the most votes is that John Llewellyn named the Trench. Llewellyn, who in other incarnations operated a flame thrower in the Korean War and tried to raise cattle in Belize, was a capable and conscientious Retro within the walls of the Control Center and prone to the most outrageous adventures everywhere else—"Butch Cassidy born a hundred years too late," as another controller described him. Even within the walls of the MOCR, Llewellyn had his own way of doing things. For example, Retro was supposed to count down to retrofire in the usual "ten, nine, eight . . ." pattern, but with Llewellyn, you never knew. Once he started at fifteen. Another time he began "ten, eight," and, when the puzzled FIDO looked over at him, quickly added "nine, seven. . . ." Sometimes he got behind, and so the count would end up

* During Gemini, Guido was called Guido all the time. But during a simulation for one of the early Apollo flights, it was realized that "Guido" sounded too much like "FIDO" over the headsets, with potential for creating a mixup during a flight. Thereafter, the guidance officer was always called "Guidance" during simulations and flights.

". . . five, four, one, retrofire!" But he always got to "Retrofire!" at the right time, and was otherwise an exemplary Retro—inside the MOCR. Outside was another story, or dozens of stories. He is said to have broken a few bones falling from a collapsing drainpipe as he climbed back to his motel room after the front door had been locked on him at the Nigerian remote site—or was it in Australia? He is said to have found a man with a lady friend, thrown him out of a second-story window, and then, intent on inflicting further damage, jumped out after him. There are at least three different stories, involving three different bodies of water, in which Llewellyn submerged cars. Llewellyn himself was no help in sorting out truth from fiction in all this, smilingly denying everything, occasionally throwing in a correction that was more improbable than the original story.

In any case, it is said that Llewellyn used to get mad at the O&P guy sitting up in the third row of the MOCR. O&P would inquire of him whether his retrofire times were completed yet, and Llewellyn would tell him belligerently, "Y'all oughta get your ass down here in the trench workin' this instead of sittin' up there," and the name stuck. However the Trench came to be, its members were clannish, cockily self-confident, and on occasion well-nigh insufferable. They lit their cigarettes using elegant cream matchbooks embossed in gold with "The Trench." Their business cards had "The Trench" printed neatly on the line after their job title. They thought of the systems guys as "mechanics" and the computer guys as "electricians." If the flight controllers in general had the esprit of men in an Army combat unit, the members of the Trench were the Marines.

The Trench got its swagger from the fact that they came closer than anyone else on the ground to actually flying the spacecraft. "You know," said Retro Chuck Deiterich, "the flight dynamics job is the neatest job in the Control Center. It's even neater than being flight director. The systems guys, they watch the systems, and as long as the systems are working right, they don't really have much to do. ['BULLSHIT!' an old EECOM exploded upon hearing of that remark.] But trajectory guys, they get to say, 'Well, we're gonna do this kind of maneuver today, this kind tomorrow, do this, do that.' They're actually doing stuff all the time to control the trajectory." Indeed, another member of the Trench recalled, sometimes you got the spooky feeling that you were actually flying the spacecraft, down there in the fluorescent gloom of the MOCR, with the big ten-foot-square plot displays in front of you and all the other controllers out of sight behind you.

2

The function of the men in the MOCR was to run the mission. The difference between their role and that of the astronauts lay in M.S.C.'s distinction between "command" and "control," as originally formulated by Walt Williams: The astronauts controlled the spacecraft, the MOCR commanded the mission.

The ground had to be in charge. The volume of information required to make many of the crucial decisions was so great that only a fraction of it could even be displayed on the spacecraft's panels, let alone be absorbed and acted upon by the three-man crew. Only one astronaut ever tried to question the ground's primacy, and not only did he never fly again, neither of his crewmates ever flew again either.* But while the flight control team in the MOCR were in charge of the flight, they saw almost nothing of it. Few contrasts are as stark as that which separated the way an Apollo launch looked to the outside world and the way it looked from inside the MOCR.

The controller's whole world during a flight consisted of the white numbers on a black screen, the plots on the boards, and the low voices in his headset. At the moment of launch, there was for him no slowly rising Saturn V, no roar of engines, no Walter Cronkite. For the controller at liftoff, there was just an uninflected voice counting down and saying, "Liftoff," and then some of the numbers on his display beginning to change.†

A problem seldom announced itself. For a few of the most important potential failures, the controller had a specific warning light that would flash on his console. But hundreds of things might go wrong for which no light would flash, and the controllers didn't have the luxury of a computer smart enough to state clearly that a diode was overheating or a circuit had failed. Usually, the only sign given to the controller was that a number was not what it was supposed to be.

Discerning that the number was wrong was made difficult by the fact

* The astronaut was Wally Schirra, on Apollo 7, and his crewmates were Walt Cunningham and Donn Eisele. Most of the controllers liked Schirra as a person and admired him as an astronaut. But for whatever reasons, possibly aggravated by a miserable cold that lasted for most of the flight of Apollo 7, Schirra got into a running battle with the ground over a number of matters, topped off by one occasion when he referred to a flight director's instructions as idiotic.

† Television was routinely projected onto one of the large screens for lunar surface activities and on-orbit activities in which the visual image was an important source of data. The people in the back row had the leisure to watch a televised image of the launch itself, and it was sometimes piped into the flight director's console as well. But the early flight directors frowned on the use of television unless there was a compelling reason for it—they believed the televised image of a launch was just a distraction.

that the screen was filled with numbers. It was made more difficult still by the fact that many of these numbers were constantly changing. It was up to the controller to notice that one number among the dozens wasn't right.

Even when the controller had pinpointed an errant number, it seldom told an unambiguous story. The computers were often overloaded and slow to get the data up on the screens, and the signals from the tracking stations were sometimes erratic. One set of numbers on the controller's screen could be fifteen seconds old while all the other numbers were fresh, creating the illusion of a problem. Therefore, in the midst of everything else he was worrying about, the controller had to remember to check his data-source slots at the top right-hand corner of the screen to see if any of them had an asterisk under them, which would indicate that the numbers associated with that source were no longer updating. And then he should glance at the clocks on his screen—if a broader problem were affecting all of his data, then the clocks would have stopped counting.

If the controller was reasonably sure he had an anomalous number on his hands, its meaning had to be deciphered in the context of several other numbers that were changing—or failing to change. It was also possible that other numbers that might explain the conundrum would not be showing on the current screen, and he would have to call up one of the other displays that were available to his console. Perhaps they would not even be on displays that normally belonged on his console, so the controller must have prepared for that portion of the flight by specifying other displays, normally on other screens, that he could call up, or by knowing which other controller to ask whether one of the numbers on that controller's screen was changing, or failing to change, in a particular pattern.

The controller was also plugged into a set of communications loops. One loop was the flight director's, another was the air-to-ground loop. A third was always one of those rooms on the outer perimeter of the square corridor, the controller's back room where specialists were watching the data affecting the controller's job in much more detail than the controller in the front room could watch for himself. The number and choice of other loops that the controller would plug into depended on a variety of factors, all of which changed as the phases of the flight changed.

A controller might easily have a half dozen or more loops buzzing in his ear at one time. He would have turned the volume up about three decibels on his primary loops, so that words spoken on them stood out, but voices from all the rest of the loops would be constantly talking as well, in bursts of a few words at a time, usually in the form of abbreviations. Seldom did a controller hear a full sentence, for sentences take time, and to speak unnecessarily was a breach of loop discipline. The function of the flight control team was to provide essential information to people who had to have it and to react to situations that required a reaction, not to chat about how things were going.

So even though no one said more than a few words at a time, the combined volume of information constantly being fed into the controller's headset was formidable. Because the messages were so brief, the entire content of one could be buried under the sound of another. And because much of the information was not specifically intended for him, he could not expect another controller to make sure he had heard. He had to be able to pick up from the many voices on the loops that one murmured bit of information that was pertinent to what he was seeing. Naturally, the volume of information and the confusion of voices became greatest when something unusual was happening, which was at just that moment when the controller had the most urgent need to concentrate.

Once the controller thought he had identified a problem, he could not go to the flight director with it too quickly, for Flight's attention was a precious and limited resource. Even if the controller did have a problem that Flight needed to know about, he had to be aware of the other problems that were already on the flight director's plate, to understand (without asking) if his problem could wait longer than some of the other problems, and to remain quiet until the appropriate moment came. The flight controller also had to be aware that if, during a time-critical phase such as launch or lunar descent, he spoke too quickly and was wrong, he could set off a chain of reactions that could ultimately lead to the needless abort of a mission. Nor, of course, could he afford to be too slow in reporting a problem to Flight, for there were times when the gap between the onset of a disastrous problem and the time when it was too late to do anything about it was measured in seconds.

For the flight controller, that was about all there was to it. All he had to do was recognize a problem hidden in his displays, assess it correctly in the context of a mass of additional information, be neither too precipitous nor too hesitant in reporting it to the flight director, and be prepared to do all this instantly for hours on end.

20

"The Flight Director may take any necessary action"

At the vortex of the MOCR, third row center, sat the flight director—"Flight."

It is difficult to think of another role in the modern era that is the counterpart of flight director for an Apollo lunar mission. Many other jobs carry with them some measure of the same unrivaled power—the captain of a navy ship, for example. But even that comparison is inapt, for the captain of a modern naval vessel exercises his power under the orders of others who are in constant radio communication with him. The Apollo flight director experienced no such direction during the course of a flight. By explicit statement in the mission rules, the flight director's authority was sweeping: "The Flight Director may, after analysis of the flight, choose to take any necessary action required for the successful completion of the mission." Theoretically, either Flight's immediate boss, the director of Flight Operations, or the Mission Director could intervene. And when there was time, a flight director would as a matter of courtesy tell the senior executives in the back row what was happening. But none of them, not even Chris Kraft, ever changed a flight director's decision, for all of them knew that they could not possibly read the technical situation during the course of a flight as well as he.

Along with the power of command came breathtaking exposure in the event of error. The lunar missions were, after all, landmarks not just in the history of the United States, but—without hyperbole—in the history of humankind. A flight director prepared for the manned Apollo flights in the knowledge that his decisions during the course of his shifts in the MOCR could put him at the center of the world's attention. And yet the rewards for success and the penalties for failure were not symmetrical. With success, the flight director remained comparatively anonymous. On the other hand, if a crew were ever lost that could have been saved by the flight control team, a flight director had to be aware that he would be testifying before congressional committees and explaining his actions to boards of investigation, in addition to trying to live with himself. Formally or informally, he would be marked forever as the man who had failed to save the astronauts. "It is the finest job in the space-flight business," said Gene Kranz of the flight director's post. But it was not one for just anybody.

1

Even before Mercury had ended, Chris Kraft had become the model of the flight director that all the others would try to copy throughout Gemini and Apollo. He underplayed his work, advertising his theory that the flight director was like the conductor of an orchestra. "The conductor can't play all the instruments—he may not even be able to play any one of them," he would say. "But he knows when the first violin should be playing, and he knows when the trumpets should be loud or soft, and when the drummer should be drumming. He mixes all this up and out comes music. That's what we do here." Around the MOCR, it was known as the piccolo theory ("The conductor doesn't have to know how to play the piccolo . . .").

All this was nonsense, according to his controllers. When Kraft was talking to one of his controllers and began the sentence by saying, "I don't understand your discipline that well, but . . . ," the controller would cringe, waiting for the zinger that was about to come. The fact was, in their opinion, that Kraft knew exactly how to play the piccolo; he could play every instrument in the orchestra if he had to. One FIDO remembered that Kraft would come down to the Trench during his off-shift and pull up a chair and sit there and chat about something that had come up on the preceding shift or about something he knew was coming up on the next—revealing in the process an intimidating technical knowledge of what the FIDO had done or was about to do. Or when he was at Flight's console up on the third row and not much was happening, he'd sit there and bring up the displays that were showing on your console (he could do that with every position), just to keep his hand in. "Kraft is incredible," said another controller. "Mind-boggling. Here was a man running a whole directorate. Every mission was different, had different ground rules, different flight rules. Each one had different things that were set up that way because of different constraints that went with them. And we'd be in a meeting discussing them before we deployed out to a site, and one of the crewmen would say something, and Kraft would say, 'No, we're not going to do that. Here's what we're going to do, and here's why we're going to do it.' I'd sit there and say to myself, Where the hell did he ever find time to be able to do that? Where did he find time to understand all that?" The other flight directors followed suit, making self-deprecating jokes about their own ignorance of the controllers' specialties even as they unobtrusively burrowed into the nuances of every console in the room.

Kraft also set the tone for one of the most striking features of Flight Operations, unquestioning trust—not of superiors by subordinates, but the other way around. In the flight control business, where the consequences of mistakes could be irretrievable, this level of trust was sometimes an awesome thing to receive. Kranz remembered his first shift as Flight, on Gemini IV. "I was looking for a handover, some general sense of direction, because I was picking up the shift from Kraft. And he left no handover. He didn't brief me. He just said, 'You're in charge,' and walked out." During Apollo 5, which was an unmanned mission testing out the lunar module's descent engine, an early engine cutoff triggered a set of convoluted contingency plans. Kranz, the flight director, was coping, but he was being stretched to the limit. Kraft walked over and stood beside Kranz's console for a few moments. Finally he said to Kranz, "I don't know what in the hell you're doing. I don't know why in the hell you're doing it. But I hope to hell you pull it off." And he strolled away, while Kranz did.

The trust that Kraft confided in his flight directors was to characterize the relationship of flight directors to controllers. Thus during the launch of Apollo 12, after lightning had struck the spacecraft, flight director Gerry Griffin, a former G.N.C., was absolutely convinced that his G.N.C. ought to call for a certain circuit breaker to be pulled. It was a crucial situation, and Griffin wanted the G.N.C. to make the call. He hinted. He nudged. He prodded. But he would not—as he could have—unilaterally order the circuit breaker to be pulled. Finally his G.N.C. decided that was the thing to do, and things worked out fine.

Give a lot, expect a lot: That was the credo Kraft left for the other flight directors. "Chris Kraft was the kind of guy who would leave you alone, and let you do your job," FIDO Jerry Bostick said. "But without him ever saying anything, you knew you'd better not screw up. You'd better get it right. Don't try to fake it. Because he didn't give people a second chance." The key was not so much being perfect—the nature of the controller's job meant that sometimes he was going to make a mistake. The key was being smart enough to recognize the mistake, correct it, and then never repeat it. ("To err is human, but to do so more than once is contrary to F.O.D. policy," was another of Kraft's sayings.) And above all else, when you found out you had made a mistake you had to admit it immediately. "The flight director's looking at a lot of data, and sometimes the information [to be inferred from the data] is not clear," an EECOM once explained. "Your data are one thing, information is another. So if you were trying to tell Kraft something that he might use to make a critical decision, like reenter the spacecraft, you had to give him very tailored, specific information. One day I saw a guy actually give him some bad data. He just tried to bullshit his way around a problem. And Kraft knew. Kraft went down and put his hand on the back of this guy's neck and told him to leave the Control Center. That was it, for that guy."

2

Kraft was the first flight director in manned space flight and the only one until the last flight in Project Mercury, Gordon Cooper's. Cooper was scheduled to be in orbit for thirty-four hours, which was longer than even Kraft could handle on his own, so he appointed his deputy, Englishman John Hodge (part of the AVRO group), to be the second flight director. They adopted colors to identify themselves. Kraft was Red Flight, so the men working his shifts would be the Red Team; and Hodge was Blue

Flight. When Gemini began in 1965, Kraft added three more flight directors. The first two were Gene Kranz, White Flight, who had been acting as Kraft's assistant flight director during Mercury, and Glynn Lunney, Black Flight, the youngest of the original members of the Space Task Group. After Gemini VII in December 1965, Kraft stepped down as a flight director to leave more time for his responsibilities as head of the Flight Operations Directorate. Cliff Charlesworth became the next flight director, choosing green as his color.

In early 1968, John Hodge left to work on post-Apollo programs and three more flight directors were added: Gerry Griffin (Gold), Milt Windler (Maroon), and Pete Frank (Orange). There were other additions and subtractions before Apollo ended. But just as Kraft provided the master model of the flight director, Charlesworth, Lunney, and Kranz defined the basic subtypes.

Of the three, Charlesworth seemed the closest to normal. A Mississippian, Charlesworth was a little older than the others (he had reached the advanced age of thirty-five when he was promoted), and in many ways acted older as well, with a lazy voice and a deliberate manner. Charlesworth didn't give much away, watching and reacting rather than trying to get out front and pull people along. "Cliff is a very wise man, in a student-of-human-behavior kind of way," one of his colleagues observed. "He could have been a pool shark or a Mississippi riverboat card player." As a flight director, Charlesworth was what another colleague called "the laissez-faire flight director," leaving his controllers alone until a problem arose that required him to act.

In contrast, Glynn Lunney's problem, if it was a problem, was making himself hang back long enough to let his controllers figure things out for themselves. By common agreement among controllers and the other flight directors, Glynn Lunney had the quickest mind in a business where quickness was a supreme virtue. "Lunney was always so quick that many times the controllers would get behind Glynn," said another flight director. "Any time we'd get into a simulation, whether it be in the systems area, or trajectory, or whatever, Glynn would start pulsing the controller with questions and leading him into paths to look at." Since a main point of the simulation was to see whether the controller would figure it out for himself, this got in the way of the training. But it was hard for Lunney to rein himself in. "Lunney could whip everybody up into following more lines of corrective action than you really want to get into at any given time," said one FIDO. "There are always several solutions to a given problem, and when Glynn was on, the controllers would all be furiously attacking every solution, at the same time."

Lunney, only twenty-eight years old when he became a flight director, had a sunny personality and an open exuberance about his work that made him a favorite among the controllers. If sometimes Lunney got ahead of them, they didn't get upset—"People who worked with him over a period of time became aware that he was just doing what's normal for him," a colleague said. "He couldn't help it if it's not normal for everybody else."

The third member of the triumvirate, Gene Kranz, was similar to Lunney only in his open relish of the job. Lunney looked as if he belonged in a Campbell's Soup ad, whereas Gene Kranz looked like a drill sergeant in some especially bloodthirsty branch of the armed forces—hair cropped to a regulation military brush, with a wedge of a face cut into rough planes.

Kranz was as relentless as he looked. "If nothing was going on, he would invent something," said one of his controllers. "If there's five minutes of spare time, then he'll think up something else, another contingency. There was almost a constant chatter on the flight director's loop, talking to somebody about something." During Gemini, Kranz acquired the nickname "General Savage," after the hard-driving hero of a contemporary television series (earlier a movie), "Twelve O'Clock High."* When an anonymous controller made a stencil of the name and hung it outside his office door, Kranz left it there as long as he was in the office. He loved it.

Yet despite the hard, unyielding aspect of his reputation, Kranz wore his heart on his sleeve more than any of the other flight directors. A devout Catholic, Kranz didn't hesitate to pray for help before going on shift. An unabashed patriot, he played tape recordings of "The Star-Spangled Banner" and Sousa marches every morning in his office. They got him pumped up, ready to give his all, which was the only throttle setting on Gene Kranz's personality.

More than the other flight directors, Kranz had a fierce loyalty to his team, the White Team, even though its members changed from flight to

---

* In the movie *Twelve O'Clock High,* the unrelenting General Savage replaces a sensitive commander who was loved by his pilots even though under his leadership they couldn't hit their targets and got shot down in large numbers. General Savage turns the squadron around and is finally recognized to be a terrific guy underneath his tough exterior. Kranz had in fact flown F-86s and F-100s for the Air Force during the mid-1950s.

flight.* Early in his career as flight director, he began wearing a different white vest for each flight, sometimes new vests for important shifts within a flight. After a while, this put a strain on his wife, as she combed the Houston fabric stores trying to find a white brocade or silk or twill or something that she hadn't already used to make one of Gene's vests. But it was all part of Kranz's abiding attachments—to God, country, family, his band of brothers in the Flight Operations Directorate, and the MOCR itself.

In an odd sort of way, Kranz was also a democrat. "Gene assumes that everybody knows as much as he does," said one controller. "He doesn't think that anybody's dumber than he is; he doesn't think anybody is not as involved in their job as he is. That's the way he looks at people. So he assumed you were doing everything he was doing." But asking of others only what he demanded of himself could in his case be scary. Kranz's credo, what Kranz labeled "premise number one," was stark: "Any error that a flight director or a team makes is unforgivable."

It was an intimate business, this directing of space flights—or, as Kranz liked to say, "You've gotta be knowledgeable of the human." When John Glenn was waiting to launch in 1962, Walt Williams told Al Shepard, the CapCom, to chat Glenn up a little—Williams wanted to listen to Glenn's voice, calibrate his state of mind, as later flight directors would listen to the voices of astronauts and flight controllers alike. Each crew member or controller was a little different in personality, in the systems he knew best, in the way he put things, and the flight director had to take that into account. "In some cases, you might know you were getting a little out on the fringe of what the guy really might be able to put together and give good recommendations, so you took a little more time with it," Lunney recalled, "to be sure that the facts were there and the guy had thought it through. There were other guys who were just like this [Lunney snapped his fingers] and I'd pass it through almost as they said it. It was a calibration process that everybody engaged in. The controllers did the same thing with flight directors."

"I think the biggest role of the flight director was asking the right questions, which is an art more than a science," said Gerry Griffin. "It's anticipating what questions need to be answered." The objective was to stay as far ahead of the spacecraft as possible, to have already thought

---

* Team members would change from mission to mission (for particularly specialized functions, a controller might even work on different teams during the same mission). But the intense training that preceded each flight caused its different teams to meld and cohere into distinct working entities.

about a problem before the problem demanded an action. And then, having asked the right questions, Flight had to listen carefully to every word and nuance of the answers. "The most likely error we can make in the business is not listening," said Kranz. "We've got very smart people [in the controllers]. We breed them to be very smart, we breed them to give opinions. We breed them to work in an arena of conflict. Sometimes they may disagree with you. Sometimes they may not be too smooth in words. Sometimes they may come in on an untimely basis and disrupt your train of thought." But none of that can interfere with your listening and understanding precisely what it is that you are hearing. If you can't listen to each person and understand exactly what that person is saying, said Kranz, "we're going to screw up. As you get down to having to work a problem in twenty seconds, you've got to have that relationship with the people."

To coordinate the work of the entire MOCR, a flight director had to keep in his memory a prodigious amount of material. By the time of the Apollo flights, for example, the mission rules books ran to hundreds of pages. Neither Kranz nor Lunney nor Charlesworth can remember ever having to refer to those pages during the course of a flight. This does not mean that the flight directors had literally memorized them.* Actually, different flight directors had different attitudes toward the written-down rules. Tec Roberts, the first FIDO and a mentor of both Kranz and Lunney, put it this way: "Glynn is a freewheeling kind of individual. 'Rules?' he'll say. 'What rules?' If they make sense or suit what he's doing at that moment, they're tolerable. If they don't, they're obviously meant to be ignored. For Gene, everything is by the book." Of course, Roberts reflected, when it comes to Kranz you have to remember also that he wrote the book.

But at least the books were always there if they had to be used. The more critical demand on the flight director's memory had to do with short-term memory. All of the complexities and ambiguities of the controllers' understanding of what was happening were being funneled into the flight director's ear. "All sorts of different considerations apply," said Lunney. "You have a problem with the life-support systems and you have a problem with the landing weather, so when are you going to come down? And don't forget you had this problem with a jet earlier." All these things had to be cranked into the decision. "The flight director's

* The memories of the best flight directors were extraordinary, however. Years later, Kranz insisted to a visitor that he had a poor memory. Lunney had an amazing memory, he said, but not he. This was said a few minutes after Kranz had referred with total recall to a minor point that the visitor had raised in a conversation eight months earlier.

role is to integrate all that stuff and make it come out right and then order it so that what goes up and down to the crews is also ordered." And always, under the pressure of time. "You've got a limited window, you've got to get it all done. You've got to really think through doing things in a priority order. They might not even be the most important, but they're the things you have to get done now."

And once Flight had remembered, calculated, ordered his priorities, and made the decisions, he had to communicate them. "It's very important to be able to communicate what you want to get done in as few words as you can," Cliff Charlesworth observed. "I used to spend time on my own thinking about how to do that—'How can I say what I want to say so that this guy will understand in the fewest words?' Because you didn't always have a lot of time to sit there and laboriously go through it." Thinking about what to say "slows you down," Charlesworth continued. "I used to sit [beside the flight director's console] when, say, Glynn was on the console, and even as quick-witted as he is, I'd get ahead of him sometimes, when the problems started coming in. I'd scratch my head and say, 'Why is that?' It's because I wasn't having to talk, and he was."

So it wasn't just the authority and the visibility that set the flight director apart, but the job itself. It was the flight controller's job writ large: Know in technical detail one of the most complex machines ever made. Master a complex flight plan and a huge body of mission rules. Piece together tiny and often unconnected bits of information from multiple sources coming to you at the same time. Do all this under the gaze of the world in situations that might give you only seconds to make life-and-death decisions.

If it was not a job for just anybody, it was also a job that had no equal. Reminded of all the ways in which working in the MOCR was tiring and even physically painful, Lunney once replied, "I can only say that I loved it. I thought it was wonderful. Remember the Patton movie? He's standing by the tank with the battlefield full of dead soldiers and he says something like, 'May God help me but I love it.' And when he said that I said to myself, I can relate to that."

CHAPTER

21

"There was no mercy in those days"

By the fall of 1967, the apparatus for controlling manned space flight—the MOCR, its support network, the mission rules, the skills—had been evolving for eight years. Mercury had been elementary school, teaching the neophyte flight control team the rudiments, and occasionally showing by harrowing example how much remained to be learned. Gemini, consisting of ten manned flights with two-man crews during the period from March 1965 through November 1966, had given the controllers a chance to become proficient in advanced concepts such as rendezvous, extended flight, and extra-vehicular activity.*

Despite all that had been learned, however, and despite the sophisticated apparatus that was in place by 1967, it remained for the Flight Operations Directorate and ASPO to learn how to fly the Apollo missions. North American Rockwell (as it was known by the end of

* Gemini was conceived as a preparatory program for Apollo. Many of the design and ASPO engineers thought it a waste of time: Focus on getting the Apollo spacecraft ready first, they argued, and practice rendezvous and E.V.A. with the hardware you're going to use for the lunar flights. The Flight Operations people vehemently disagreed, contending that the ten Gemini flights were an indispensable training ground for them.

291

1967) had produced a new, much more complicated spacecraft: an Apollo command module with myriad capabilities—switches could be thrown, dials turned, data computed, thrusters burned, engines lit, in numberless permutations. Grumman was producing a lunar module that was even more novel and equally complex. These machines were going to be the first in history to carry men out of the earth's environment and to an alien place. The question to be answered was: Now what?

1

First, all the maneuvers and activities had to be planned down to the last switch setting and data entry. Then everything had to be orchestrated between the crew and ground control, because, in effect, each spacecraft had two cockpits: the one in space and the one in Building 30. "It got to be a big flap," said Jerry Bostick, who as a FIDO was in the middle of it, "a really bad situation between NASA and the contractors and between the flight controllers and the astronauts about who does what and who gives who what data before the flight and during the flight and who does what on the ground and who does what on board and how much onboard data do you need to put on telemetry to give to the flight controllers, how much do they need displayed. . . . The whole thing was a big mess."

So George Low created a new job. On August 3, 1967, he posted a memo on the bulletin boards at the Manned Spacecraft Center announcing that Howard W. (Bill) Tindall, Jr., had been appointed to a position called "Chief of Apollo Data Priority Coordination." It was a typical NASA title, one that could have meant just about anything. "Specifically," said the announcement, "his job is to determine the operational rules and procedures for properly utilizing the Apollo systems, including primary and backup systems; to investigate the system capabilities and constraints and to evaluate their accuracies; to establish the criteria for system selection during various phases of the mission; and to establish the proper spacecraft and ground displays and use of these displays." It went on like that for a few more lines. In short, Bill Tindall was supposed to figure out how to fly the missions. His function cut across ASPO functions and Flight Crew Operations and Flight Control Operations. How he was to do it wasn't quite clear to anyone. Just do it, Tindall was told. The name that was eventually attached to what he did was "Mission Techniques."*

Bill Tindall had been in the space program since Mercury days, when he had helped to set up the communications network. In the Gemini program, Tindall had been the man who had figured out how to do a rendezvous in orbit. The mathematics of orbital mechanics and of rendezvous itself were well known (Buzz Aldrin, the Apollo 11 lunar module pilot, had written his M.I.T. dissertation on the mechanics of rendezvous). But no one had applied these theoretical findings to the world of hardware and tracking stations, where it was essential that the ground be able to monitor and control a rendezvous. Bill Tindall was responsible for solving what proved to be an extraordinarily complex task. "It was a big deal," said an official close to the process, "in my mind, one of the major accomplishments of the space program."

During Apollo, Bill Tindall had continued to be a troubleshooter, officially attached to the Mission Planning and Analysis Division but in reality a free-floating resource. At the time that the Mission Techniques task came up, he had been babysitting the arduous and often tangled development of the spacecraft's computer software.

Tindall had the exuberance of a seven-year-old on his way to a circus. His friends were "super" and "gangbusters" and "giants" and "really neat" and "just absolutely outstanding," and he didn't seem to have any adversaries worth mentioning. He saw the Apollo Program as one great long stretch of fun that had by some miracle been given to him instead of work ("It's just incredible that we'd get paid to do what we were doing!"), and all the stories about the brutal hours and travel schedules as just a cover for men on a lark ("We weren't working overtime, we were playing!").

In the process of working with M.I.T. on the guidance software, Tindall had arbitrated among the FIDOs and Guidos and the software people about what displays had to be available to what places. This experience turned out to be ideal for planning mission techniques, partly because computer capacity was at the heart of deciding how to run a mission, and partly because the process of designing the displays had been so contentious. Apollo-era computers had severely limited storage— the computer capacity of the mainframes in the Control Center was smaller than that of some of the desktop systems of the 1980s, and the

---

* Tindall was working for George Low and ASPO, but he drew Flight Operations into the process as well. Over in F.O.D., Rod Rose was performing a complementary role for Chris Kraft, as head of a smaller group that met monthly and was called Flight Operations Panel, or F.O.P. Rose's group outlined the mission for F.O.D. with a much broader brush than Tindall's.

onboard computers in the command and lunar modules had less capacity than some modern pocket calculators. So the competition for room was fierce. "Every time we'd get a new capability in the computer systems, the flight controllers would start laying their requirements on it," Cliff Charlesworth said. "In short order, we'd overflow the boxes.* And Kraft would get mad. He'd say, 'Goddammit, get it back there where it'll fit, we can't get any more computers!' So he'd go get Tindall and he'd say, 'You go fix that.' "

Tindall had started holding what were known as "Black Friday" meetings. All the people competing for computer capabilities would crowd into one of the conference rooms over in Building 30 and explain in fervent detail just why their particular needs were crucial to the success of the mission and the safety of the astronauts. "We'd all get in there and defend our requirements, and then Tindall would cut them," Charlesworth continued. "And then we'd cuss him." And Tindall would grin, and cuss back, and laugh his loud, infectious laugh, and keep right on going.

Tindall took from that experience a conviction that the only way to resolve so many competing interests was to get everybody into the same room and let them fight it out. There was a method to the madness he created. For John Aaron, then a young EECOM, Tindall's technique was a model for his own use in later years when he became a senior NASA official and had to do the same thing. In Aaron's view, Tindall recognized that what matters most to people is not that they get their way, but that they feel they have had a chance to make their case. Somebody would voice an idea, and Tindall would field it, treating it seriously and yet at the same time managing to dispose of bad ideas quickly without putting anybody down. Ideas that weren't so bad were improved on around the table without bruising egos. Tindall was able to get some very proud engineers to say "Well, I'll be damned" when it turned out that their pet ideas weren't as good as they had thought they were, Aaron remembered. They might walk into Tindall's meetings "polarized and cultures apart," but Tindall built them into teams.

Tindall himself confessed to no such cerebral theories. As far as he was concerned, the real secret to his meetings was that people weren't just sitting around making recommendations to forward to the higher-ups. "Everyone knew we were making decisions right there," he said, decisions that would govern the way the missions were run, and if people

---

* "Overflow the boxes": Put in more requirements than the computer memory and computing capacity could handle.

wanted to put in their two cents' worth, they had better be there. It led to a madhouse, sometimes, but Tindall loved playing lion tamer to a room packed with arguing engineers.

For one of the big meetings, there might be as many as a hundred design engineers, astronauts, controllers, and mathematicians from MPAD crowded into the conference room in Building 5, thirty or forty of them crowded around the long table and another fifty or sixty sitting or standing around the perimeter of the room, sometimes spilling into the projection room as well, their faces peering out through the little projector windows. They'd hammer away at each other for hours, filling the blackboards with numbers. "It became a battle for the blackboard," Tindall recalled. "Whoever would get the blackboard and the chalk, that's the guy who could win. I was really good at that."

In all, there were twelve distinct phases to each lunar-landing mission: launch phase, earth orbit (while they checked out the C.S.M.'s systems), translunar injection (T.L.I.), midcourse on the way out to the moon, lunar-orbit insertion (L.O.I.), lunar orbit (while they checked out the LEM), lunar descent, lunar ascent, lunar rendezvous, trans-earth injection (T.E.I.), another midcourse phase, this time on the way back to earth, and entry.* Each phase had dozens of specific elements to be worked out.

By Tindall's estimate, only about 10 to 20 percent of the work involved specifying the techniques to be employed under nominal circumstances ("nominal," in NASA, means "without abnormalities"). Most of the time they were worrying about mission techniques under abnormal circumstances—which in turn drove most of the specific mission rules for the Apollo flights. When is a failure a failure? Tindall's meetings had to ask. If the guidance platform is drifting, but not completely wacko, how much drift is acceptable? How much of the backup capacity has to be available to proceed to the next step?

Tindall raced from one question to another. "He'd take people like me, from Operations, and the theoreticians, and the contractors, and they'd sit in his meetings for hour after hour after hour," recalled Steve Bales, one of the Guidos who attended. "He'd say, 'Okay, look. We've just got to the moon.' And he'd draw a big circle on the blackboard. 'Okay, here's Rev 1 [the first revolution around the moon]—what are you guys going to do?' " And they would begin to work out in exacting detail everything that needed to be done during the first revolution around the moon—which was one small part of one portion of the lunar-orbit phase of the

---

* Tindall did not have to deal with lunar-surface activities, which were planned primarily out of the Flight Crew Operations Directorate.

mission, which was in turn one small portion of the mission as a whole. "Then," Bales continued, "he'd have another meeting once a week that said, 'Here's what we're going to do in the ascent phase, from LEM takeoff to insertion.' Another meeting two times a month on 'Here's how we're going to do the rendezvous.' One on 'Here's how we're going to do guidance procedures on the surface.' The guy was incredible. Had a thousand-ring circus going all the time." Not all the meetings had a hundred people at them. He would have smaller ones, almost daily, of half a dozen people. For critical phases like the final descent from 47,000 feet to touchdown, he might hold special meetings twice a week until they had nailed the techniques down the way he wanted.

Out of these meetings came Tindallgrams, so called because they read like no other memoranda in the Apollo Program. They owed their origin to a secretary, Patsy Saur, assigned to Tindall soon after he came to Houston. "You do dictate, don't you?" she said to Tindall the first time she met him. "No, I don't," said Tindall. "Well, you'd better learn, because I'm not going to lose my shorthand proficiency," she announced peremptorily. "I'm a meek guy and she was tough," Tindall reminisced, and for a few days he wrote out his memos the night before and then held them under the desk the next morning, pretending to dictate. But finally he got used to the idea of dictating, and then he found that the memos were sounding the way he talked. They became a sensation around M.S.C.

Who else but Tindall would, in an official NASA communication, describe the magnitude of a required change in spacecraft velocity as "teensy weensy"? Who else would entitle a memo "Vent bent descent, lament!" or "Let's move the recovery force a little," or, simply, "Some things about ascent from the moon"? What other engineer at Tindall's level would, in an official NASA communication, call a top NASA official's proposal "unbelievable" and proceed to treat it as if it were the work of a crackpot?

The main thing about Tindallgrams was that they said what they meant. "They were wonderful," Bostick said of the Tindallgrams. "They didn't have any bullshit, this bureaucratic stuff." Since everyone loved them (including George Low, who refused to let his secretaries give him a summarized version of anything Tindall wrote), one might have thought that everyone would have started to imitate them. Some tried, but they quit after a while; it sounded as if they were trying to be cute, while a Tindallgram just sounded like Tindall. About a fuel warning light on the lunar module display, Tindall wrote:

The present LM* weight and descent trajectory is such that this light will always come on prior to touchdown. This signal, it turns out, is connected to the master alarm—how about that! In other words, just at the most critical time in the most critical operation of a perfectly nominal lunar landing mission, the master alarm with all its lights, bells, and whistles will go off. This sounds right lousy to me. In fact, [astronaut] Pete Conrad tells me he labeled it completely unacceptable four or five years ago, but he was probably just an Ensign at the time and apparently no one paid any attention. If this is not fixed, I predict the first words uttered by the first astronaut to land on the moon will be "Gee whiz, that master alarm certainly startled me."

The humor and the breeziness of the Tindallgrams were window dressing for a serious endeavor, however. From the first Mission Techniques meeting in October 1967 through the first lunar landing in July 1969, Tindall's collected works comprised eight thick loose-leaf volumes of highly technical memoranda. These were backed by stacks of compilations of specific techniques broken down by mission phase. During Apollo 11, this is what Guido Gran Paules carried with him to his console in just his "Descent" binder alone:

"Apollo Mission Techniques, Mission G, Lunar Orbit Activities"
"Apollo Mission Techniques, Mission G, Lunar Descent, Revision A"
"Mission G, Abort from Lunar Powered Descent and Subsequent Rendezvous, Techniques Description"
"A User's Guide to the LUMINARY Lunar Landing Programs"
"Ground Monitoring of Guidance Velocity Residuals During Powered Descent"
"Topography Profiles of Potential Apollo Lunar Landing Sites"

Inserted into these volumes were another hundred pages of dense technical memoranda updating and revising the material in the books.

2

It was impossible that anyone should sit down, absorb this Torah of flight control, then go out and be a controller on a mission. Nor could anyone

* By this time, the written designation for the lunar module was "LM," still pronounced "lem."

expect to learn it all through actual mission experience. As Apollo was gearing up for its first flight in 1968, there had been no manned flights since November 1966. What the flight control team needed above all else was practice, and it was to this end that the simulation people plied their trade.

The Simulation Control Area (S.C.A.) was a room to the right of the MOCR proper, down on ground level with the Trench. A long picture window opened onto the MOCR to let the simulation controllers in the S.C.A. watch what was happening to their victims. Inside the room were two U-shaped work areas of consoles that looked just like those in the control room. This was where the simulations—sims—were played out, making the mission rules and the mission techniques come to life. The simulation supervisor, called SimSup ("Sup" rhymes with "loop"), was for the controllers what the flight instructor is for the novice pilot.

The simulations had four distinct functions. First, they were a way of keeping in shape, the equivalent of five-finger exercises for a pianist. Second, they gave everyone a chance to see whether the mission rules and techniques that worked on paper also worked when put to a test. Third, they gave the controllers experience in how to deal with the particular set of emergencies portrayed in a particular sim. And fourth, they made the controllers familiar with fear, putting them through such hair-raising, gut-wrenching adventures that nothing they encountered in an actual flight could seem more terrifying or hopeless.

The one point emphasized by everyone who was in the Apollo MOCR is that simulations never felt like games, or even practice runs. Their verisimilitude, and the seriousness with which they were taken, reproduced even the emotions of the real thing—controllers swore that, with momentary exceptions, they felt no more anxiety during missions than they had during the sims. To Gene Kranz, the sims were the reason the missions succeeded. "I could say it's hard work, perseverance, all of those things. But it was the training process that did several things to you. First of all, it humbled you. And you'd have your successes too, and you felt good about them. You learned to pay exquisite attention to detail. You learned the nuances of the voices that are talking to you on these loops. By the time you're finished, they have worked you over so much and so well and so thoroughly that you never considered failure."

In Mercury days, the sims had been primitive. Harold Miller, a young Langley engineer from Tennessee, had been put in charge of developing Project Mercury's simulation capability. He didn't know anything about simulations—nobody did—but somebody had to do it, and, as he later recalled, he didn't know enough to say no. He began to put together a team—Dick Koos was one of his early additions—and they tentatively began to build digital tapes that would, for example, simulate a trajectory that was not quite up to orbit. Then they would see if the tape produced realistic effects on the flight controllers' meters (in those days, everything was on meters instead of C.R.T.s, which were not yet available). Sometimes it worked and sometimes it didn't. Lightning from the afternoon thundershowers at the Cape, attracted by all the antennae on the roof of the Mercury Control Center, would produce power surges that set the meters back to their zero state. The computers broke down. And sometimes the men in the sim room broke down: Because they didn't yet have programs that could automatically adjust the scenario to respond to the actions of the flight control team, the simulation controllers had to do it by hand, frantically turning knobs on a panel, trying to calibrate their responses so that the telemetry streams looked reasonably plausible to the people in the control room. They got more adroit as time went on, but it remained pretty rough.

The remote sites gave the Mercury and Gemini sims a special flavor as well. Though Mercury Control at the Cape had voice links with the remote sites, they couldn't transfer flight data back and forth in real time. Each remote site therefore had to be a miniature Mercury Control, with its local director and FIDO and CapCom, passing on instructions from the Cape but also prepared in an emergency to take action on its own.

For the sim guys, the absence of a data link meant that they had to prepare separate data tapes for each of the remote sites. They would have a worldwide countdown, so that the tapes would all begin at the same moment and everyone around the world would have a view of the same problem at the same time. But this also put a considerable burden on the SimSup to anticipate events during the course of the sim. "The SimSup had to guess what you were going to do and when you would do it," one of the controllers recalled, "which was the tough thing to do. You'd make a decision to abort now, but the tape didn't say abort now, so the damned thing would keep flying."

In those days, simulations were also a prime place for practical jokes. As the space program grew, the jokes tapered off, but in the early years nobody was sacrosanct, not even Chris Kraft. Things happened in the Mercury and Gemini sims that would have been unthinkable five years later. For example: Chris Kraft's console in the Mercury Control Center included a screen that showed a live television shot of the launch pad.

During simulations, the Redstone or Atlas of course just sat there, but the camera was nonetheless always turned on. One day a controller named John Hatcher substituted a tape of a launch for the live picture. Hatcher synchronized the tape with the simulated countdown and waited for the moment of launch. As always during Mercury, Chris Kraft was the flight director. At $T-0$, as Kraft pushed the little gear lever that started the clocks, the Redstone on the television screen belched smoke and fire and lifted off the pad.

The gear lever wasn't hooked up to anything that could conceivably have launched a rocket, but the sight was too compellingly realistic to be discounted. "Look at that!" Kraft yelled in dismay to Kranz, who was sitting beside him at the assistant flight director's console. Kranz, in on the joke, sat expressionless. "Did you see that?!" Kraft cried out again, pointing insistently at the screen, and a story was born that would be told and embroidered upon for years to come.

The sobering presence of Chris Kraft normally kept such behavior within bounds. Out at the remote sites, life was more unbuttoned. Ed Fendell was one of the many free spirits sent out to run a remote site. As local director at the Hawaii site during sims for Gemini VII, he was dissatisfied with his CapCom, an Air Force officer in training for what was expected to be the Manned Orbiting Laboratory (later canceled). The guy thought that all he had to do was to pass on what Fendell told him; he couldn't understand that he had to dig in and learn for himself how the systems worked. So Fendell took it into his head to show the Air Force CapCom that he had to learn how to take charge. Fendell got on the phone with Carl Shelly, a SimSup back in Houston (where the Control Center had moved after Gemini III), and arranged that during the next sim Fendell would fake a heart attack. Shelly gave his approval. Fendell told his Surgeon what he was up to. At the appropriate point during the simulation, Fendell clutched his chest, groaned loudly, and toppled out of his chair. The Surgeon rushed over, announced that Fendell was *hors de combat*, and motioned for CapCom to take over.

The indolent CapCom was shaken. "Honolulu, this is Gemini Seven; Honolulu, this is Gemini Seven," he called in confusion, and began to find out why he was supposed to be on top of things. Fendell was pleased. In fact, he was loving it, until a second physician waiting to come on shift, an elderly Navy doctor whose presence Fendell had not anticipated, rushed over to examine him. Fendell held his breath. "I think he's dead," the old gentleman announced, and began to administer cardiopulmonary resuscitation.

It still would have been okay, except that, back in Houston, Shelly had forgotten to tell Chris Kraft that it was all a charade. For the next few hours there were messages coming in from Houston inquiring solicitously about the status of Fendell. "Later on, of course, it came out that it was all part of the sim," Fendell said, "and Kraft just went berserk. He like to killed Shelly."* But it worked, Fendell added. "The CapCom took over, ran a series of passes, and we had a good time."

For all of the jokes, the sims were deadly serious business, no more so than in the elemental, life-and-death decision named "abort." An abort during Mercury and Gemini usually meant de-orbiting and coming home early. An abort during Apollo could mean coming home early or it could refer to halting an attempt to land on the moon and returning the LEM to the command module in lunar orbit.

The worst abort situation would be during powered flight, with the astronauts strapped to a Saturn rocket that could kill them within seconds. "I approached each launch and rode through each five minutes of powered flight in a state of petrified terror," recalled tough old Walt Williams of the Mercury launches, and it didn't get any easier for the people who controlled the Gemini and Apollo flights. Of all the possibilities of catastrophe during powered flight, the worst was what the simulation people called "the pad fall-back case," in which the booster lost power within seconds after liftoff.

If that were to happen, things would go bad quickly. If the booster was only a few feet off the pad, it would explode as soon as it hit the ground again. In that case, the decision to fire the escape rocket would probably have been made by sensors within the booster that were supposed to detect the incipient catastrophic failure, even before the Control Center or the astronauts had time to react.†

If the launch vehicle was a few hundred feet above the pad, then Booster had about four seconds to look at his data, decide something was wrong, and call an abort. Booster therefore had an independent abort capability on his console, consisting of a toggle switch with a red cover to light the onboard abort light and a voice link with the spacecraft (all

* Kraft, who was sometimes said to lack a sense of humor, played his own role of exasperated leader so perfectly for so many of these shenanigans that one wonders whether he wasn't engaging in a little simulation of his own.

† Could the astronauts on a Saturn V have survived an explosion of the fully loaded three stages lined up beneath them? Difficult as it is to believe, the answer is probably yes. The escape tower on the Apollo generated 134,000 pounds of thrust on an 11,000-pound spacecraft for three seconds—an experience that would have left the astronauts battered by the g forces, but would have gotten the spacecraft away from the Saturn before the fireball engulfed them. Or so the designers calculated.

aborts required both cues). To Kranz, "that was the worst, worst case," and the sims included one every so often, to keep Booster mentally prepared to give up so abruptly on the mission.

After the vehicle had gained some altitude, the abort decision became more leisurely. Two people in addition to Booster had abort switches, and now they played decisive roles. One was of course Flight himself. The other was FIDO. Neither one of them actually caused an abort to happen by flipping his switch; instead, they lit an alarm light informing the crew that an abort situation had arisen, leaving it to the crew to take the action to initiate the abort. The abort itself could take one of two forms, depending on the situation: separating the spacecraft from the rest of the stack and firing the escape rocket, or casting off a malfunctioning first or second stage and continuing to ascend on the S-IVB. Unlike Booster during the first moments of launch, FIDO did not have the authority to call an abort unilaterally. His switch was to be reserved for situations in which Flight and the flight crew had already decided to abort but were waiting to reach a particular range and velocity before they did so. In such a case, FIDO would use his switch to tell them when that moment had arrived.

But these were only a limited subset of the many situations in which an abort might become necessary. During powered flight, the rule of thumb was that the flight controllers had from about fifteen to twenty seconds to make an abort decision. For anything more exigent than that, the crew would take action based on their onboard displays. But for a problem that involved control of the vehicle—multiple engine failures, a hard-over engine (meaning an engine that had gone out of control and gimbaled to its limit), or trajectory deviations—the astronauts would have, in Kranz's words, "no clue in the world what the hell had gone on," and the controllers would have to take the lead. If the controllers took longer than twenty seconds, a spacecraft caught in a turning motion would probably break the tension ties that held the C.S.M. to the rest of the vehicle, and then they would have a whole new set of problems to worry about. "So what we had to do was get our timing down for that case where we had lost control just due to engine hard-overs," said Kranz. "And we had to get the crew off before we had structural failure." The SimSups "kept working, working, working that case."

The difference between the earlier manned programs and Apollo was that all of the old problems had multiplied. In Mercury and Gemini, there had been one period of powered flight, the launch phase, and three abort modes: return to earth using the escape tower to separate; at higher altitudes, separation without the escape tower; and, in the latter stages of powered flight, "abort to orbit." In Apollo, there were six separate instances of powered flight: the launch phase, translunar injection, lunar-orbit insertion, lunar descent, lunar ascent, and trans-earth injection. During all of those phases except lunar ascent and trans-earth injection, there were multiple abort modes to worry about. In Apollo, the flight control team had to be prepared to fly two completely separate spacecraft, each with its own maze of systems.

Against this backdrop, the SimSups began to prepare the flight control team and the astronauts to land on the moon. Since Mercury, all the sims had become "closed-loop," meaning that the computers driving the sims could react to the controllers' actions, changing the scenario as the sim played out. Occasionally, they ran long-duration sims, where the goal was to exercise the flight controllers in a routine, systematic fashion— give them problems that were not imminently mission threatening or life threatening, and make sure that everyone was fully on top of his system.

But the computers' increased sophistication was more useful for the other kind of sim, the ones that focused on the four most difficult of the mission phases—launch and entry, ascent and descent—and left controllers damp with sweat.* Now the SimSups could double and triple and quadruple the load on the controllers. A failure in the EECOM system could be followed by an independent failure in the reaction and control system, and that one could be followed by static in the primary voice link. The point was to load the flight control team with several problems at the same time, to approach the controllers' absolute capacity to absorb and react to a constantly deteriorating situation. "They always tried to run those in double-time, or close to it," said one controller, throwing perhaps two dozen problems at a single controller in the fifteen minutes prior to landing. You had to field the problem, snap out a call, and move on to the next problem almost instantaneously, or you were going to be working on an old problem while a new one had caused the LEM to crash.† The logic behind this brutal routine was that just such a combination of failures would create a disaster in a real flight.

---

* In flight operations for Apollo, the words "descent" and "ascent" always meant specifically descent to and ascent from the lunar surface, never the earth's surface. Several similar potential confusions of wording had to be resolved during the Apollo missions. The old familiar "go" and "no-go" convention, for example. What would the message "go" mean to a crew that had just landed on the moon? That all systems were "go," meaning okay? Or that they should "go," meaning leave? Tindall decreed that "stay" and "no-stay" be substituted while the LEM was on the surface.

† A "call" refers to the information that a controller gives to Flight about a specific event. It might be a diagnosis of the problem or a recommendation for action.

The simulation controllers liked to think of themselves as being imaginative and creative. To the controllers, "devious" and "sneaky" seemed more apt. But the men who devised the sims had to be that way. They could assume that any controller who was promoted into the MOCR was cool and quick or else he wouldn't have gotten to the front room. There was no point in simulating the obvious malfunctions, because he would swat those down instinctively. A good sim was one in which the nature of the problem was masked, in which there was an obvious answer (wrong), a more subtle answer (still wrong), and a right answer which only the most suspicious and persistent mind would see as a possibility.

The one constraint on the SimSups was that, by unwritten rule, they could not give the flight control team a problem for which there was no answer. No matter how bad things got, the controllers could always say to themselves that somehow, somewhere, it was in their power to save the lives of the crew. Within that single constraint, the SimSups could subject the controllers to any combination of horrors.

"We would dig into the system and find little things," recalled Dick Koos, who ran sims from the first Mercury days on into Apollo and was nominated by one veteran controller as "the most devious SimSup." Koos would flip through schematics of the systems, looking for some obscure little diode, and then work out what would happen if that diode should fail. The more utterly unconnected the diode and the resulting failure, the better. The more work it made for the controllers, the better. "Okay, that'll keep 'em real busy for a while," Koos would say when he found a good one, and watch with anticipation from behind the window in the Sim Control Area to see how the controllers reacted.

Just because a failure seemed impossible didn't mean that the SimSups couldn't go ahead and make it happen anyway. Thus one ebullient and notoriously inventive SimSup, Bob Holkan, was poring through the procedures books looking for arcane rules, and found one for console failure in the MOCR. The procedure said that if a console failed, the controller at that console was to move over to the next console, and so on down the line, the theory being that the last controller, Retro, would move onto Booster's console, which was empty after the launch phase. Holkan found that the way to create the "maximum motion," which was his objective, would be to cause a failure in Guido's console at the far right-hand side of the Trench. But Holkan couldn't figure out how to make it fail; not surprisingly, the designers of the consoles had gone to great lengths to prevent such an event from occurring.

Holkan found that the only way to shut down a console was to throw

its circuit breaker. So Holkan got hold of one of the MOCR's main-
tenance men and arranged with him to come in early one morning.
Together, they pulled up some of the floor panels, ran a nylon cord from
the S.C.A. up into the back of the Guido console, and wrapped the end
of the cord around the circuit breaker. They replaced the floor panels,
hiding the cord, and Holkan retired to the S.C.A. to wait.

Steve Bales was Guido that morning, and they were running a
simulation of the ascent from the moon. At the right moment—which is
to say, at the most delicate part of the ascent phase—Holkan pulled the
string. "And goddang, if it didn't work."

"I heard this big pop," Bales recalled, "like somebody shot a gun in
front of me, and the whole console went blank." He had to do
something—he was sitting in a critical chair for that part of the
mission—"but I couldn't. I froze." He looked at Jay Greene, the FIDO
who was sitting next to him. Greene shrugged; he was busy with his own
problems. "I've lost my console," Bales told Flight forlornly, and then
just sat there for about thirty seconds, frantically casting about in his mind
for something to do. He had never encountered the obscure musical-
chairs rule that Holkan had found, but the SimSup always left you a way
out, he told himself. He finally thought to ask the O&P controller to put
the Guido screen up on one of the ten-foot-square displays at the front of
the MOCR—not the prescribed solution, but a fine demonstration of
controller ingenuity. Holkan smiled at the memory. "We went to great
lengths to put malfunctions in."

Usually these were "integrated sims," in which the astronauts would
work from their full-scale simulators in Building 5 or at the Cape, and the
flight control team would interact with the crew just as they would during
a real flight. By the time of the Apollo flights, the simulators for the flight
hardware—both the command module and the LEM—had become highly
realistic. The astronauts not only had working versions of all the
spacecraft's switches and levers, they had a view out the window of the
LEM simulator showing the lunar surface as it would appear during the
descent phase. In another section of Building 5, M.S.C. technicians had
built a plaster-of-Paris model of the lunar landing site. As the LEM
simulator descended, a camera mounted above the model of the landscape
would mimic its movements. Occasionally, this became gulpingly
realistic. There was, for example, the time when the sim scenario called
for a radar altimeter failure: The astronauts descending in the LEM were
led to believe that they were at a ridiculously high altitude for that point
in the descent. Failing to recognize the hardware breakdown, the crew

accelerated their descent so much that they were unable to recover when, in the last few seconds, watching the lunar landscape come rushing up at them, they realized their mistake. The camera came down so hard that it smashed the lens against the plaster of Paris. To the astronauts, the impending crash looked all too real.

Sometimes the sims proceeded without the astronauts in the trainer, and one of the people in the S.C.A. had to serve as an AstroSim. The AstroSim was, by the rules of the game, supposed to be able to do anything a real astronaut could do. His job was not a sought-after one. "An astronaut's got literally hundreds of switches," Bob Holkan observed, "and he's got valves he throws, and he's got sixty pounds of flight-data file that he responds to. This guy [the AstroSim] is sitting there, he's got a C.R.T., and the ground controllers are asking him, 'Would you reconfigure the switch on panel A4? And could you tell me the status of the little talk-back over on the left side, and while you're there, throw the valve.' " That's the way they tried to do it during a flight—wait until they had several items to package, so they didn't talk to the crew too much. "Well, that's nice if you're in the spacecraft and you're floating around doing it," Holkan continued. "But the AstroSim's got this one C.R.T., and he doesn't know where in the vehicle this stuff really is. And he's asking all these other guys in the sim room, 'What do I tell them? What do I tell them?' "

In these situations, the people behind the picture window were not above lying—"Well, just tell them the switch is already in the right position," they'd advise the AstroSim. What could the poor flight controller know? "There's always the possibility that it was a transducer problem, or telemetry problem," a FIDO pointed out. "The system had enough nodes in it that there was always a possibility that the crew could be telling you one thing and your data could be saying something else and there was a logical explanation."

Therein lay the continuing, inevitable advantage of the simulation people over the controllers. Under the tough flight directors—Kranz was especially strict on this point—the controllers had to assume that the problem was real. Once in a while, however, an AstroSim would carry things too far. There was, for example, the AstroSim who, during a Gemini sim, was asked by Guido to look out his window and report the position of a particular star. The AstroSim, with no idea where that star was supposed to be, claimed that he was unable to comply with Guido's request because of cloud cover. That one didn't wash. But short of that, the AstroSims could get away with a lot. "Play it as it happens," Kranz preached. "Never accept the fact that it's a sim screwup." And so the flight controller would call down to the back rooms and get readouts on obscure possibilities. Finally, when the controllers had thrashed around long enough and had proved to their satisfaction and Flight's that the situation in the spacecraft couldn't possibly look the way the AstroSim had told them it did, the AstroSim would say, "Oh yeah, I just found out I hit this other switch with my elbow when I was over here getting out my lunch." The sims had to be survivable, but they didn't have to be fair.

Each simulation was followed by a debriefing, also known as the "wake." The simulation team remained behind their picture window while Flight went to each of the controllers and asked him to describe what had led him to make the calls he had made, and whether in retrospect his calls were the best ones. It was a group confessional, with one controller telling his story, another chiming in to explain how his remarks had influenced the situation, the people in the back rooms describing through the intercom what had been coming up on their screens—"cleansing their souls," they called it. The debriefings could be as harrowing as the flight, especially when they still had the remote sites. "If you were sitting down in Australia and you screwed up," Ed Fendell recalled, "Mr. Kraft, Mr. Kranz, or Mr. Hodge would get on the line and commence to tell you how stupid you were, and you knew that every switching center, every office at Goddard, ships at sea, everybody and his mother, everybody in the world was listening. And you sat there and took it. There was no mercy in those days."

Day after day they went on, one after another, eight or ten launches in a day, eight or ten lunar descents, each with hideous, multiple failures, each with its debriefing and one's shortcomings laid out for inspection. Bill Tindall was not a flight controller, but he couldn't stay away. "I used to plug in and watch them go," he recalled. "It was like going to an exciting movie, watching those people work. Talk about exhausting! Handling of emergencies in the missions was an absolute piece of cake by comparison. The real missions were easy."

CHAPTER

22

"You've lost the engines?"

The morning of April 4, 1968, was one more day in a year that was turning the United States inside out. The Tet Offensive, which had so intensified the debate over the Vietnam War, had occurred two months before. Six weeks before, Minnesota senator Eugene McCarthy, running for President on an antiwar platform, had shown surprising strength against Lyndon Johnson in the New Hampshire primary. Just the preceding Sunday, on March 31, President Johnson had announced that he would not seek reelection, and in the days since then the newspapers had been filled with speculation about whether New York senator Robert Kennedy would enter the race. Also in the news, though less prominently, was Dr. Martin Luther King, Jr.'s current efforts on behalf of the striking garbage workers in Memphis, Tennessee. This Thursday morning, King was in Memphis, awakening to the new day in his room at the Lorraine Motel.

Within the world of Apollo, the outside world looked completely different. "I missed the entire Vietnam War," said one, typical of many. "I watched no television, read no newspapers, came to work at six in the morning and worked until nightfall, six or seven days a week for years." The people of Apollo were barely aware that the Vietnam War was going on, barely aware that this was a presidential election year, barely aware that there was such a thing as a War on Poverty or L.S.D. or *Sgt. Pepper*

308

or race riots. Had one of Chris Kraft's flight controllers been asked about the most important event so far of 1968, he would probably have said that it had occurred on January 22, when the unmanned Apollo 5 had carried a lunar module on its first test flight. Today, April 4, was going to be the next important date of 1968: the unmanned flight of Apollo 6, the second flight of a fully operational Saturn V.

That morning, Bob Wolf was Booster 1. The Boosters, whose job lasted only the eleven and a half minutes of powered flight it took to reach orbit on a Saturn V (plus another six minutes for translunar injection, for the lunar missions), were always in an odd position in the MOCR. Just when the rest of the flight control team was getting down to work, the Boosters went home.

Bob Wolf had been pressed into service for this flight only a few weeks before. But he had been working with the lead Booster for several months and felt no special apprehension as the countdown approached zero. Jay Greene was FIDO, Neil Hutchinson was Guido, and Cliff Charlesworth was Flight.

At one second past seven o'clock in the morning, Cape time, the five F-1 engines on the Saturn S-IC lifted the spacecraft off Pad 39A. All five engines operated nominally at first.

At the 125th second of powered flight on the first stage, the accelerometers in the S-IC began recording lengthwise oscillations in the launch vehicle, producing a chugging motion like a car with bad spark plugs, decelerating and accelerating. This type of vibration was known to rocket scientists as "pogo," after the motion of a pogo stick.

Pogo had been encountered in earlier space flights. The Titan that flew the Gemini spacecraft had experienced pogo. The Saturn V on Apollo 4 had shown a mild pogo effect. But never had the pogo been so severe. The vibrations on Apollo 6 were vicious, alternating backward and forward at the rate of five or six cycles per second, producing stresses on the order of plus or minus 10 g's. The pogo continued for ten seconds, then stopped. The only people who were aware of the pogo in real time were the Marshall engineers in Huntsville, who were watching the detailed data coming in to their own version of the MOCR, the HOSC (Huntsville Operations Support Center). The rest of the S-IC burn was uneventful. In the MOCR, everything looked normal.

The Saturn V staged and the five J-2 engines on the S-II ignited and ran up to full power. Into the fourth minute of the S-II burn everything still looked normal. Then the screens in the Trench began acting up, showing erratic data. This was not unusual. The Trench got its tracking data from several sources, and one of FIDO's main jobs during a normal launch was to work with a controller who sat downstairs in Building 30's Communications Center, selecting the clearest channel. As FIDO Jay Greene shifted to a new data source at four and a half minutes into the burn of the S-II, Wolf's console began showing data that might either reflect a data problem or indicate that engine 2 was faltering. He couldn't tell which. Seconds later, the console showed that the engine had shut down altogether.

This was bad enough, but then came the shocker: 1.3 seconds later, without any warning, engine 3 shut down. "That was beyond anyone's imagination," Wolf said later. The chance of one of the five engines on the S-II stage shutting down were small. The chance that two of these independent systems would shut down simultaneously was infinitesimal.

Jay Greene got his first word "over the airwaves," as the controllers referred to speech not spoken into headsets. "We got two engines out!" exclaimed an off-duty Booster standing behind Wolf. Wolf was staring at the screen, thinking he had to be getting bad data. Then he looked up at the trajectory plot board. Until then, the course of the Saturn V snaking its way up and across the plot board had overlain the planned trajectory almost exactly. Now, it was visibly deviating, falling under the expected altitude.

Because the likelihood of a dual failure was so remote, the Marshall engineers had not spent much time analyzing how the Saturn V would behave with two engines out. But even the preliminary analyses seemed to indicate conclusively that the vehicle would tumble out of control. They had a euphemism in the space program for such unlikely failures: "a bad day." You didn't bother to plan for bad days. By definition, they were the kind of thing that the gods did to you sometimes, and the best thing to do was go home and have a drink and come back the next day to pick up the pieces.

Back in the third row, Charlesworth had not heard the exclamation from the off-duty Booster. Wolf's voice finally came over the flight director's loop.

"Flight, Booster," That was the protocol: Say the name of the person you were calling on the loop, then identify yourself.

"Go."

"We've lost, uh, engine two and engine three." Wolf's voice was expressionless.

"You've lost the engines?" Charlesworth said, with a perceptible "oh shit" tone in his voice.

"That's affirmative," said Wolf.

"Roger. It's your action."

In the unmanned Apollo flights, pushing Booster 1's abort switch would actually cause the Saturn V to abort. Booster 1 had the authority to push it on his own initiative. Given the point of the burn at which the shutdown occurred, the mission rules told Wolf to do so. But as he watched the trajectory tracing on the big screen, the deviation wasn't as bad as the analyses had predicted. There wasn't any tumbling.

Three seconds of silence passed, as Wolf tried to figure out what was going on.

"Are those adjacent or opposite?" Charlesworth finally asked. The five engines of the S-II were hung from cross beams in the same arrangement as the F-1s in the first stage, with the fifth engine in the middle of the cross beam. If the two engines which had shut down were opposite each other, control would be easier to maintain than if the engines were adjacent. They were adjacent, unfortunately, but Wolf was too busy to answer. Another three seconds of silence. Finally Booster 2 came on the line.

"Flight, Booster 2, we seem to have good control at this time."

"Roger."

"Guidance system performing nominally, Flight." It was Neil Hutchinson chiming in from the Guido console.

"Roger," said Charlesworth, who by now hadn't heard from Booster 1 for an eternity. Charlesworth prodded him:

"You sure, Booster?"

The language of the MOCR was one of the most economical in the world, for encapsulated in those three words was a long list of questions. "Are you really sure that the Saturn V is under control, Booster?" Charlesworth was asking. "Are you sure that you shouldn't hit that abort switch, Booster?" "Are you sure you know what you're doing, Booster?" All of that and more was in those three words. But from his tone, Charlesworth might have been asking Booster whether he really wanted another cup of coffee.

Down on the front row, Wolf had been going through quiet agonies. The philosophy of the German team at Marshall, instilled in each of its Americans as well, was all in one direction: Be disciplined, be professional, and don't improvise, especially during an actual flight. The heritage of Marshall was unmanned, automated vehicles, and "the Marshall guys were very nervous about these flight controllers," as one Marshall engineer recalled. "The idea of somebody being able to send a command to the [rocket] was a little bit strange to them." On the first all-up flight of the Saturn V the preceding November, Booster had sent just such a command, not planned in advance, when there had been a little glitch in the propulsion system of the S-IVB. "He did the right thing," the Marshall engineer said, "but it was a little bit questionable at the time and some people questioned it." They questioned it so much, in fact, that Booster 1 for that flight had had a nervous breakdown, and that's why Bob Wolf had been pressed into service so abruptly.

In this case, the pre-planned, disciplined, by-the-book action was clear-cut. Two adjacent engines out is an abort situation. But, contrary to all the predictions, the thing seemed to be flying. So, in response to Charlesworth's question—"You sure, Booster?"—Wolf just said: "Data indicates it, Flight."

And they rode it out. The guidance system on the Saturn worked better than anyone had figured it could under such circumstances. They extended the burn on the remaining three engines for a minute, using the propellants that hadn't been used by the two silent engines. Then they staged, and that's when FIDO Jay Greene began to earn his money.

After the two engines had gone out, the vehicle had maintained a pitched-up attitude known as "chi-freeze" for far longer than it would have under ordinary circumstances. "Well, the S-IVB lit up," Greene recalled, "and the first thing it said was, 'Omigod, I've got too much altitude.' And so it pointed its nose straight at the center of the earth." This battle between the guidance system and the gimbal limits on the engine continued for about eighty seconds, with Greene getting closer and closer to an abort call of his own. When the S-IVB finally gave up trying to get to the altitude it wanted, it had a flight-path angle that was unacceptably low. "So then the little devil said, 'Well, this is bad, I've got to pick up the flight-path angle,' so it started pitching up, and as it started pitching up it said, 'Now I'm overspeed,' so it actually went into orbit thrusting backward."

The plot boards showing where the Saturn had wandered looked as if a drunk had been drawing the trajectory. It was without question the most exciting powered launch anybody in the MOCR had ever witnessed. "A fascinating flight," Greene said tersely—his very first shift in the MOCR. What was he doing all this time? "Puckering."

There was more to come. After waiting for two orbits, it came time to relight the S-IVB for a simulation of translunar injection. And it didn't relight. The S-IVB had now failed, and the controllers used the smaller Service Propulsion System (S.P.S.) engine on the service module as a substitute. They managed to complete most of the maneuvers in the flight plan, and the spacecraft was returned safely to earth, but there remained two sobering, unassailable facts: The pogo in the first stage was so severe that a crew might have been injured, had to abort, or both; and three separate engines had failed.*

Later that day, in the early afternoon, Martin Luther King was shot and killed as he stood on the balcony of the Lorraine Motel, and the United States moved from a time of troubles to one of domestic crisis. The rest of the world paid little attention to the Apollo 6 flight one way or the other. Insofar as anyone noticed, it looked like a success—NASA announced, accurately, that all of the mission's major flight-test objectives had been achieved. Within NASA, Apollo 6 was deeply disquieting.

1

Slowly, each of Apollo 6's malfunctions yielded an explanation. The pogo effect had occurred because the natural vibration of the thrust chambers on the F-1 engines, approximately 5.5 hertz, was too close to the structural vibration of the vehicle as a whole, which peaked at about 5.25 hertz. A system of "shock absorbers" was installed to de-tune the engine frequencies, in effect, so that they would never again get into synch with the structural frequencies.

The pogo did not cause the engine failures, however. The S-IVB engine failure and the first of the two failures on the S-II stage had both occurred for the identical reason, ruptures in a six-foot length of stainless-steel fuel line, five-eighths of an inch in diameter, that in each stage carried liquid hydrogen to the starter cup of a J-2 engine. The line had two short vibration-absorbing "bellows" sections in it, and the bellows sections had a wire braid shielding on the outside. When the line was tested on the ground, it worked perfectly, but only because— unbeknownst to the engineers—the liquid hydrogen running through the line caused frost to form on the braided shield, helping to dampen the

---

* Still another problem occurred up in the Spacecraft Lunar Adapter, the SLA, the section which during a lunar mission contained the LEM—a section of it ripped away shortly after the pogo problem. For a time, it was thought that the cause of this problem might have been a cause of one or more of the launch vehicle's problems as well, but it turned out to be an independent event, difficult to diagnose but comparatively easy to fix once the nature of the structural failure in the honeycomb material of the SLA had been found.

vibration in the bellows. In the vacuum of space, where there was no moisture to form a protective frost, "those little old bellows just sang like a rattlesnake," as Jerry Thomson described it (the veteran of the F-1 fix was called in on this one too). They quickly fatigued and failed. The bellows section was eliminated in favor of a new, stronger design.

The third of the engines that failed hadn't really failed at all. Through a manufacturing error at the Michoud Assembly Facility for the S-II, two wires connecting engines 2 and 3 in the S-II stage had been crossed. The shutdown of the sick engine 2 had triggered the shutdown of a healthy engine 3.

Each of these diagnoses and fixes took many weeks—NASA later calculated that 125 engineers and 400 technicians spent 31,000 man-hours on the pogo problem alone. And no matter how exhaustive the tests, the fact remained that while the flight of the first Saturn V had been splendid, the second one had been a cripple that by all rights should have tumbled into the ocean. Had there been men on board Apollo 6, the crew probably would have aborted the mission during the pogo, when they would have been so violently banged around that they couldn't have operated the spacecraft. If the pogo hadn't happened, Charlesworth probably would have had the crew abort when the two engines on the second stage shut down. It was all right for Wolf to wait and see what happened with an unmanned vehicle, but it would have been dangerous to do so with a crew on board.

It was in the context of all this that, shortly thereafter, George Low decided that the next launch of the Saturn V should carry men—that it should carry them, in fact, all the way to the moon.

2

Between flights, the center of gravity at Houston was in Building 2, the nine-story headquarters building at the south end of the complex. The higher the floor, the higher the position. Thus when George Low became manager of ASPO, taking a demotion from his old position as deputy director of the center, his office shifted from the top floor, the ninth, down to the seventh.

Like other senior staff at M.S.C., Low had a suite. It included a large, paneled office with space for a conference table and windows looking out over Clear Lake to the east of M.S.C. In the reception area were his three secretaries, headed by Marilyn Bockting.*

During the week, the three secretaries worked in overlapping shifts, with Judy Wyatt arriving first at about 7:15, a little after Low, and the last of them leaving at eight in the evening, usually before Low himself went home. On Saturdays, Low worked a shorter day, only until five or six. The secretaries traded off Saturdays.

On one spring Saturday in 1968—probably shortly after Apollo 6, though she couldn't recover the exact date—it was Judy Wyatt's turn. Frank Borman, the astronaut who was in charge of the "tiger team" that ASPO had sent out to Downey after the fire, was in Houston that weekend and wanted to report to Low on the status of the redefined Block II command modules that were being produced by North American. Wyatt arranged a thirty-minute appointment for nine o'clock. Borman was in there an unusually long time. When he finally emerged, Low called her into the office to take some dictation.

Wyatt usually enjoyed taking dictation from Low—he was the only man she had ever worked for whose subjects always agreed with his verbs—but this day she was fretting over the mess that Borman had made of the Saturday-morning schedule. It wasn't until Low was finished that she realized what she had been writing down. The memo was to Chris Kraft. Recent progress at Downey had been faster than they had hoped, Low was saying. Borman thought that the command module was about ready to go to the moon. Would Chris Kraft please get together a small group of people and find out very discreetly where they stood on this possibility from a Flight Operations standpoint—lunar trajectories, mission techniques, and so forth? Low told Wyatt not to make a copy even for his own files, and to tell Betsy Bednarcyk, Kraft's secretary, that this was an "007." Low was a James Bond fan, and "007" meant that after Kraft read the memo, his secretary was to destroy it.

3

Owen Maynard had devised the flight schedule for Apollo through the first landing, assigning a letter to each type of mission. "A" designated the unmanned Saturn V test flights (Apollo 4 and 6 were both "A"

---

* Marilyn Bockting was a secretary with a reputation so awesome that when she died an annual "Marilyn Bockting Award" for the top secretary at M.S.C. was instituted in her honor.

missions) and "B" was the unmanned test of the LEM, Apollo 5. "C," scheduled for the autumn of 1968, was the first manned mission, a repeat of the one Gus Grissom was to have flown. "D" was the first manned mission using both the command module and the lunar module, in low earth orbit. "E" was another mission using the command module and LEM in combined operations, but in a high earth orbit that would take the spacecraft as far as 4,000 miles away from earth. "F" was the first mission in which a spacecraft would go to the moon; it would enter lunar orbit and exercise the LEM but not land. "G" was the first lunar landing. Each mission had its own reason for existence. None could safely be skipped.

In the spring of 1968 when Low sent his 007 memo to Kraft, he had been thinking of making E Mission into a lunar-orbit flight instead of a high earth-orbit flight, something that he and Kraft had been discussing off the record for some time. By May 24, Low felt ready to tell Gilruth in his daily "Apollo Notes for Dr. Gilruth" that "Chris Kraft and I agreed . . . that we would pursue an E Prime Mission which would be a lunar orbiting mission." But as early as June, another more radical possibility had occurred to Low, which he still kept to himself. At the end of July, he left for a vacation in the Caribbean. On Monday, August 5, he returned to the office.

"He came back on Monday morning, and things just started popping," Judy Wyatt recalled. "He said, 'I can't tell you what it is right now, but I want you to keep a log of people I talk to. And I want it kept under a secret cover sheet.' " Over the next few days, as she kept the log, she began to see a pattern. One morning she was in the washroom, brushing her hair, idly thinking over the people her boss had been seeing, when it hit her. Wyatt dashed back into the suite. "Marilyn, we're going to the moon!" she said. Bockting, one of a handful that Low had let in on the secret, laughed and said nothing.

To George Low, the reasoning was simple. Wally Schirra's crew was going to fly the first manned mission, C, known publicly as Apollo 7, in late September or early October. Low assumed it would be a success, that the C.S.M. would come out of Apollo 7 certified flight-worthy for a lunar mission. To get to the moon during 1969 and meet the Kennedy deadline, the D mission certifying the flight-worthiness of the LEM had to fly before the end of 1968. But to fly D, they had to have an operational lunar module, and the LEM was running behind schedule; D couldn't possibly fly before the end of the year. None of the subsequent missions could be skipped. Therefore, why not switch the order of the D and E missions?

They wouldn't have a LEM by December, but they would have a good C.S.M. and a Saturn V to lift it. They didn't need the lunar module for that first deep-space mission. They needed to obtain deep-space experience in translunar navigation, lunar orbit, communications, and thermal conditions, and all of that could be gained with a crew in the command module. So: They could fly the E Mission, but without the LEM, in 1968, let the D Mission slip until early 1969 when the LEM would be ready, and then proceed directly to the F Mission. That was the logical half of Low's plan. The audacious half was that George Low was proposing to fly to the moon on only the second manned Apollo spacecraft, the first manned Saturn V, and the first Saturn V to fly after the failure-ridden Apollo 6.

During the first half of the week after he got back from the Caribbean, Low called Scott Simpkinson, now promoted to ASPO's assistant program manager for Flight Safety. Could Apollo 8 go to the moon without compromising Simpkinson's safety rules? Looking at the hardware, Simpkinson couldn't see any reason why not. It looked to him as if the real problem was navigation, which wasn't his specialty. Low called Bill Tindall, whose specialty it was. Tindall thought the idea was gangbusters. As for navigation, they were in great shape. His Mission Techniques meetings had already worked through the procedures for launch, earth orbit, translunar injection, and entry. All they had left were the techniques for midcourse corrections on the way to the moon, lunar-orbit insertion, and trans-earth injection to come home again, and Tindall told Low those would be a piece of cake.

Low talked it over with Chris Kraft. Kraft, who had always been Low's ally in trying to shift E Mission from high earth orbit to a lunar mission, was receptive, but what Low was asking now—for Kraft to get his flight controllers ready for a lunar mission in only four months— presented a new magnitude of difficulty. Pledging them to secrecy, Kraft met with Jim Stokes and Lyn Dunseith, his two leading experts on the status of the computer software for a lunar mission. If the software couldn't be ready by December, there was no point in proceeding. Stokes and Dunseith went off to conduct a private inquiry.

On Thursday, August 8, Low flew to Kennedy Space Center for a meeting on work schedules, still saying nothing about his idea. He let Rocco Petrone do his work for him.

Petrone was getting frustrated. At Mueller's last program review in Houston, it had sounded to Petrone as if Mueller still wanted Petrone to push for a December launch of a spacecraft with a LEM, and Petrone knew it couldn't be done—the LEM just wasn't going to be ready. But so far he hadn't gotten the message across. Mueller's Apollo Program manager, Sam Phillips, was giving him the meeting at the Cape to convince him—and through him, Mueller—that December was impossible. Petrone called together his staff and told them to prepare a straight-out briefing. "I don't want any emotion in this thing," Petrone told them, "and I don't want you picking on anybody, because then they always feel you're trying to shove a hot poker up their ass. Just lay it all out for them." So they did, and Phillips listened. "Sam," Petrone concluded, "there's no way that thing is going to launch before the first of February; I don't care if you give it to God." Low flew back to Houston reconfirmed in his determination to send a LEM-less Apollo 8 to the moon.

Early Friday morning, Kraft gave Low another green light. Dunseith and Stokes had determined that the schedule for the Control Center software was tight, but it could be met. At 8:45 that morning, August 9, Low walked from his office up the two flights of stairs to Gilruth's office on the ninth floor and for the first time revealed his plan to the center director. Kraft and Slayton came in and added their endorsements. By temperament a cautious man, Gilruth nonetheless thought this was a great idea—"It took me ten seconds to respond," he said.

Low called Phillips, who was still at the Cape, while Gilruth called von Braun. Phillips was intrigued, but he reserved judgment. He suggested that they convene a meeting of the program's key people to discuss it. In a remarkable demonstration of how rapidly NASA could react in those days, a meeting was arranged for 2:30 that same afternoon in Huntsville. Only five hours after anyone outside Houston had heard of the idea, most of NASA's senior officials were gathered in von Braun's office—Sam Phillips and George Hage from O.M.S.F.; Kurt Debus and Rocco Petrone from K.S.C.; Eberhard Rees, Lee James, Ludie Richard, and von Braun from Marshall. Representing Houston were Gilruth, Low, Kraft, and Deke Slayton.

Low's plan was an easy sell—the initial reaction, Phillips recalled, was "guardedly positive." Von Braun promised that the Saturn could be retargeted ("It doesn't matter to the launch vehicle how far we go"). Debus promised that the Cape could be ready for a lunar launch by December 20. The mission would be designated C′, spoken as "C-prime.'' For the time being, the code phrase for it would be "Sam's budget exercise.''

The only two NASA officials who didn't express their enthusiasm were Jim Webb and George Mueller, and they were 4,000 miles away attending a conference in Vienna, Austria, on the peaceful uses of outer space. In fact, nobody told them what was happening. The enthusiastic cabal in von Braun's office decided that there wasn't any point in bothering Webb and Mueller with this new idea quite yet, not until they had all had a chance to take a closer look at it.

23

"It was darn scary"

By August of 1968, Jerry Bostick was holding Glynn Lunney's old job as head of the Flight Dynamics Branch, making him the leader of the Trench. If Kraft sent a spacecraft to the moon in four months, it would be the Trench that would be responsible for seeing that all the burns happened at the right times, for the right durations, with the spacecraft in the right attitudes. It would be the Trench that would have to oversee navigation to and from the moon and prepare the dozens of abort contingency plans. So Bostick was among a very small number of people who were let in on the secret. Cliff Charlesworth, in whom Kraft had already confided, called Bostick on Friday evening after the Huntsville meeting and told him to go to Kraft's office the next morning. "It's about the C Mission," was all that Charlesworth would say. "I think you'll find it interesting."

It was a small meeting that Saturday—Kraft, Charlesworth, Kranz, Arnie Aldrich (chief of the C.S.M. Systems Branch for the Flight Control Division), and Bostick. Kraft told them that George Low wanted the next flight to go circumlunar. Aldrich thought it sounded like a "clever strategy." Bostick was shocked. "How do you think we're gonna do that?" Bostick protested. "We're not ready to do that!"

"No, don't take that attitude," Kraft replied. He, too, had thought it was crazy when Low first suggested it. He just wanted Bostick to think about it until Monday morning.

"I left there still thinking that, geez, I've heard of some stupid things, but that's crazy," said Bostick. He spent Sunday with a few people in MPAD (the Mission Planning and Analysis Division) who had been alerted. They decided they wouldn't have time to get everything into the Mission Control Center computers by December, but that was okay. MPAD could run the calculations off line, in their own computers, and then feed the data into the back rooms. It wouldn't be perfect, the Trench wouldn't have all the displays in front of them that they might like, but the MPAD people could walk the hard copy from the office wing of Building 30 over to the MOCR, and they could read it off the printouts. "By about noon Monday," said Bostick, "I had concluded that, yeah, we can do that."

Slowly, still limited to a few key people, the word spread, and the effect was electrifying. Years later, Aaron Cohen remembered as the highlight of his illustrious career the day when Low called him and said he wanted to "pull off a coup" on Apollo 8. Cohen, then working as Owen Maynard's top engineer for the C.S.M.'s systems, was directed by Low to see whether the C.S.M. was ready to take on a lunar mission. Rod Loe, running the Communications and Life Support Systems Section in Flight Operations, was getting too far up the hierarchy to act as a flight controller any more. His boss, Arnie Aldrich, kept threatening to move him out of the MOCR and into the Spacecraft Analysis Room, the SPAN, a more senior position but less fun. Loe had been resisting. After a meeting in Kraft's office where Loe first heard that they might be going circumlunar on the December mission, Loe and Aldrich walked back to their offices in Building 30. On the way, Loe made up his mind. Just let me be your lead EECOM on this flight, he said to Aldrich, and he would never argue with him again about going into the SPAN. Loe would go there for the rest of his career, he said, if only he could be in the MOCR for this one.

1

Sam Phillips had gone back to Washington after the Huntsville meeting to tell Tom Paine, the acting administrator in Webb's absence, about the idea. "Sam," Paine said, "we've been arguing whether we could put a man on the Saturn V. The last time, it had pogo troubles, the SLA came apart, and three engines shut down. . . . Now you want to up the ante. Do you really want to do this, Sam?'' "Yes, sir!" Phillips said. Paine instinctively liked the idea, but it seemed too good to be true. Yet Phillips's judgment was always solid, and if he was this enthusiastic then C′ must be feasible. "We'll have a hell of a time selling it to Mueller and Webb," Paine said.

When Sam Phillips finally got hold of him by phone on the Tuesday, August 13, George Mueller was "skeptical and cool."* The risks were obvious. The technological gains of a lunar flight over an earth-orbit flight were not immediately apparent. Mueller would think about it.

It wasn't until Thursday, August 15, that they reached Webb on a conference call to the U.S. embassy in Vienna. Mueller had not yet mentioned C′ to him. "Sam and I sat down in two offices and managed to put over the idea to Jim," Paine recalled. "He was horrified." "He was pretty crisp and clear in his disbelief," Phillips agreed. "If a person's shock could be transmitted over the telephone, I'd probably have been shot in the head."

For Jim Webb, Low's brainchild was more than he wanted to face. By the summer of 1968, Webb had been administrator of NASA for more than seven years, negotiating a path for NASA in the midst of Johnson's growing preoccupation with Vietnam and increasing congressional hostility toward Apollo. Webb had borne months of criticism after the fire. Now, he was being asked to approve a mission that jumped the schedule everyone had agreed to and that, if it were to fail, could fail in a particularly ghastly way, with astronauts marooned in lunar orbit at Christmastime. Furthermore, Webb knew he was going to be out of his job on January 20, 1969—neither Humphrey nor Nixon would let him stay on through the first moon landing.

All this led Webb to two conclusions, mixing loyalty and political canniness. If something did go wrong, he wanted to be able to defend NASA and his engineers aggressively. If he were administrator, anything he said would sound self-serving. Also, there was nothing to gain by staying. If C′ were a success, it would not add much to his experience at

---

* It originally had been planned that Phillips would fly to Vienna to talk to Mueller, to avoid Soviet interception of transatlantic telephone conversations. One purpose (though not publicly emphasized) of going circumlunar so quickly was to beat the Soviets to the punch. By 1968, NASA was no longer worried about the Soviets landing on the moon before the Americans, but a Soviet attempt to fly around the moon before the Americans landed seemed a very real possibility. Paine finally decided that for Phillips to show up in Vienna would set off more warning signals than the telephone calls.

NASA; if it were a failure, he would not only be the obvious target of recriminations, he would have to respond without the resources of the administrator's office to support him. Staying with NASA as of the fall of 1968 was for Jim Webb a high-risk, low-gain proposition.

On September 16, Webb went over to the White House and renewed a suggestion he had made before, that perhaps it was time for him to retire. L.B.J. took him up on it, suggesting that Webb resign immediately. Webb was surprised that Johnson wanted to do it that very day, but he was glad to be released. James E. Webb, NASA's second administrator, its longest-tenured, and in the minds of many its finest, left NASA on October 7, 1968, four days before the first men flew in an Apollo spacecraft. He was replaced by Tom Paine.

Rocco Petrone was sitting with Sam Phillips and George Low in the Firing Room, watching a dry run with the Apollo 7 astronauts, when the news of Webb's resignation came over the wire. It had been a rough day, Petrone remembered—the dry run had encountered the usual series of problems, unpleasantly reminiscent of another dry run with the crew of A.S.-204, nineteen months earlier. Everybody was edgy. Then Jack King, the Cape's public information officer, came over to Petrone's console and told them that Jim Webb had just announced his retirement. It came as a surprise to all of them; this was, they recognized, the passing of an era. Still, there was another side to it. George Low turned to Sam Phillips and said, "This makes C-prime possible."

It had already been announced that a new mission was going to be inserted between Schirra's flight and Jim McDivitt's test of the LEM. Phillips had given a press conference in late August, revealing that the new Apollo 8 might be more than just a low-earth-orbit flight. NASA was preparing a "flexible mission," he said. Perhaps the spacecraft would have a high earth orbit, with an apogee several thousand miles out. Conceivably, Apollo 8 could involve a circumlunar flight, but Phillips buried that possibility under so much talk about the "flexible mission" that only Tom O'Toole of *The Washington Post* put the circumlunar option in his lead. Most of the others came away from Phillips's briefing thinking that a lunar mission was such a remote possibility that they didn't even mention it.

By the time that Apollo 7 was launched on a Saturn IB with Wally Schirra, Walt Cunningham, and Donn Eisele on October 11, 1968, the press had discovered that the possibilities of a circumlunar flight were greater than they had realized. Then Apollo 7 turned out to be close to a perfect mission, discounting the head colds that plagued all three astronauts and some wrangling between the crew and ground. The eleven-day flight gave the controllers a chance to wring out all of the Apollo spacecraft's environmental and control systems. The S-IVB that would have to push Apollo 8's spacecraft into a translunar orbit worked perfectly. So did the S.P.S. engine on the service module that would slow the spacecraft into lunar orbit and later speed it out of lunar orbit and back toward the earth.

After Schirra's crew landed on October 22, the speculation about a lunar flight for Apollo 8 was fully in the open and sometimes being treated as certainty. Within NASA, however, a final decision had yet to be made. Mueller had continued to play a skeptic's role throughout the fall, unconvinced that the gains were worth the risk. But the people at Houston kept coming up with counters to his objections. On October 28 and 29, a critical review was held. After two long days of briefings, Mueller said tersely that he had concluded that "there is no technical reason not to fly Apollo 8 as a lunar-orbit flight."

But still he was dissatisfied. On November 4, Mueller sent a letter to Bob Gilruth urging one long last look. "There are grave risks to the program as a whole, not just to the Apollo 8 mission," he wrote. He was satisfied that the risks "from a purely technical aspect are probably reasonable and acceptable," and he recognized that "the greatest single advantage" of flying Apollo 8 to the moon was the way it had galvanized people. "Yet," Mueller pointed out, "you and I know that if failure comes, the reaction will be that anyone should have known better than to undertake such a trip at this point in time."

Mueller got the additional long last look he wanted, but by this time the psychological momentum was nearly palpable. On November 7, the C' Crew Safety Review Board came in with its favorable recommendation. On November 10, the Apollo executives were briefed by Phillips, Low, and other senior NASA managers on the benefits and risks of Apollo 8, and unanimously went on record in favor of a lunar flight (with the McDonnell-Douglas representative recommending that it be circumlunar instead of lunar orbit). On November 11, there was a NASA management meeting to go over everything one more time, then a smaller meeting of Paine, Mueller, Phillips, and the center directors, then a third meeting of just Paine, Homer Newell (the associate administrator), and Mueller. The next morning, Tuesday, November 12, 1968, acting administrator Paine appeared at a press conference and announced that Apollo 8 would be a lunar-orbit mission.

2

Reflecting on it years later, Mike Collins wondered whether the most historic moment in the Apollo Program might have occurred not on July 20, 1969, when Neil Armstrong and Buzz Aldrin became the first men on the moon, but at 9:41 A.M. Central Standard Time, December 21, 1968. On that morning, Collins was CapCom. If it hadn't been for a bone spur requiring surgery the preceding July, Collins would have been up there himself—he had been a crew member on Apollo 8 until the surgery had made him lose too much training time. Collins had been reassigned to a later mission, Apollo 11.

Now, they were two hours and fifty minutes into the flight of Apollo 8, carrying Frank Borman, the commander; Jim Lovell; and Bill Anders. So far, George Low's imaginative leap had been vindicated. There had been no pogo during launch, no premature engine shutdowns, no problems with the SLA. The spacecraft had been checked out in orbit and its attitude had been meticulously aligned and double-checked for the next step, the procedure known as translunar injection. As CapCom, it fell to Collins to pass up the word. "Apollo 8," Collins said into his headset. "You are go for T.L.I." From the spacecraft, Jim Lovell answered, "Roger, understand. We are go for T.L.I."

Collins, a man with a sense of both poetry and history, felt even as he spoke that the words weren't enough. Here is one of the most historic things we've ever done, may ever do, he thought to himself, and there ought to be some recognition of it. And what do I say? "Apollo 8, you're go for T.L.I." But in the MOCR, that's just the way "Mankind, the time has come to leave your first home" was said. The S-IVB relit as programmed, firing for five minutes. It increased the spacecraft's speed from 25,000 to 33,500 feet per second, sufficient to take the spacecraft out of earth's gravitational field.

The flight to the moon was outwardly serene. During the third night of the translunar coast, the boys in the Trench entertained themselves by calculating the precise moment when the spacecraft would reach the "equi-gravisphere," the balance point between the gravitational fields of earth and moon. When the spacecraft reached that invisible point in space at 2:29 P.M., C.S.T., December 23, they called up hard copies of their displays for mementos. But other than that, all seemed quiet. The course was good and the television newscasts told that all was well until the next big event, insertion into a lunar orbit (L.O.I.).

Elsewhere in the Control Center, however, there was a degree of concern. The first night out, CapCom had received a cryptic message from the crew of Apollo 8 to listen to the dump of the onboard crew conversations (periodically, the data recorded on board, including the crew's conversation, were transmitted—"dumped"—back to earth). Charlesworth didn't get to it right away. A while later, the crew repeated its suggestion, and so Charlesworth went to a back room and listened to the playback. The message the crew had recorded on the onboard tapes, so that it couldn't be monitored by the world at large, was that the commander of the mission had a medical problem—vomiting and diarrhea. Because Borman didn't seem a likely candidate for space sickness—he had never been ill on his fourteen-day Gemini mission— one of the first explanations that occurred to Charlesworth and medical director Charles Berry was that Borman's illness had something to do with the radiation belts Apollo 8 had traversed as it headed outward toward the moon. Much of the next day was spent playing through the implications if either of the other crew members were to come down with the same symptoms, and rethinking the abort rules for medical problems. But Borman recovered, and the crisis passed.

By that time, however, the design engineers were getting fidgety about the status of the Service Propulsion System, the S.P.S., the engine that had to work to take the spacecraft into and out of lunar orbit. Just two days before launch, engineers doing ground tests on the S.P.S. engine had brought Aaron Cohen a disturbing piece of news. The S.P.S. engine had two legs, or routes, for feeding propellant into the combustion chamber. Either alone would be adequate to do the job, but mission rules demanded that both be functional before the flight director committed the crew to lunar orbit. The problem was that tests showed a high spike—a potential detonation—if they initiated both legs without first wetting the combustion chamber. "Wetting," which wasn't difficult, consisted of firing the engine for a brief period, but no one had been asked to incorporate the kind of wetting they needed—using each leg separately— into the flight plan, nor had any provision been made for firing the S.P.S. en route to the moon if the firing wasn't needed for a course correction. They would have to be ready to fire the engine (a "burn") twice, using each leg separately, and to arrange the two burns so that they canceled each other out. Such a requirement should have been planned and incorporated into the flight plan and the mission software weeks earlier, and Aaron Cohen had a few tense minutes when he told George Low about it. In fact, recalled Cohen, the usually calm and measured Low "chewed me out something royal." Still, there was nothing else to do but incorporate the two corrections into the flight plan, and so at the last minute they did.

When they did perform the two burns, they got a jolt when the data revealed that the engine had not performed up to expectations: The degree of acceleration, known in the Control Center as delta-V, was a little short.* Teams at Rockwell and Houston spent "a pretty active three days," as Charlesworth put it, while Apollo 8 was on its way to the moon, assuring themselves that it was a minor problem. Helium bubbles that had been loaded with the propellants were causing the engine to miss during very short burns.

Ed Pavelka was sitting at the FIDO console when the time came to pass up the data for the burn that would put Apollo 8 into lunar orbit. This was done by voice. Pavelka provided the numbers to CapCom, CapCom read them to the crew, and then the crew read them back to the ground to verify. Pavelka had done it many times before during simulations, but he had a funny feeling about giving out the numbers for real. It was, he decided, a sobering feeling.

It was a time for getting nervous about things they knew they shouldn't be nervous about. Hitting the moon, for example. The people in MPAD had calculated the trajectories to fractions of degrees and independent computer programs were constantly checking their answers. The algorithms themselves had been tested and retested. "I know we have figured this right," Charlesworth kept telling himself. "And I know all our guidance systems are accurate, and we tracked it properly, and all the mathematicians in the world have looked at this thing . . ." Still, sixty miles was, in Charlesworth's mind, pretty close when you scaled sixty miles to the diameter of the moon, and Charlesworth, who was off duty for that shift, continued to pace nervously in the back row of the MOCR.

John Mayer, director of MPAD, was in a back room that night— MPAD staff supported the consoles in the Trench—and he found himself the center of attention. "Suddenly Mueller and Gilruth and Kraft came walking into the room, saying, 'How sure are you we're going to miss the moon?'" Mayer remembered. "I'm real sure," he answered, laughing.

Emil Schiesser, one of Mayer's young navigational whizzes, couldn't figure out why people were so edgy. "Shoot, we had been to the moon

---

* The Greek letter delta, Δ, is a mathematical convention signifying a change in the value of a variable; V is the standard representation for velocity.

with unmanned spacecraft on a number of occasions, had flown the lunar orbiter, and they were worried about us hitting the moon!'' Schiesser said. "They never once asked us how we were gonna not burn them up when we came back, and nobody had ever come back [from the moon] before. I thought that was funny. Why were they so concerned about us smattering them all over the moon and not worried about us burning them up?'' Schiesser was referring to one of the most delicate phases of the mission, the entry back into the earth's atmosphere, which would occur at an unprecedented speed of 25,000 m.p.h., and which required the navigational system to bring the spacecraft into a narrow, ten-mile-wide corridor. But somehow the entry was familiar, at least psychologically. The flight controllers had brought lots of astronauts back to earth before, ever since they had sweated out John Glenn's landing from orbit during Mercury. Entry was known terrain, whatever the new complexities, whereas consigning a crew to lunar orbit was not.

Mayer could understand why it seemed tricky to a layman. It was comparable to taking a rifle outside his office in building 30, he would acknowledge, and aiming it at a basketball in downtown Houston, some twenty-six miles away. Actually, it was even tougher than that, more like aiming at a point that was one-sixteenth of an inch to the side of the basketball. So, yes, Mayer could see why a layman might think it was tricky. Pavelka, who wasn't a layman, decided it was legitimate to be nervous. "We had simulated the hell out of the stuff, but there was never proof that it all really fit together, you know?''

It was not quite four o'clock in the morning on December 24, 1968. In the viewing room, a hundred people were packed into a space meant for seventy-four. One of them was Robert Sherrod, a well-known journalist who had been covering the space program after a colorful career as a war and foreign correspondent that had begun before World War II. "I looked up at the big center screen beyond the banks of flight controllers' consoles," he wrote later.

> Suddenly the familiar map of the earth vanished from the big plastic screen, and in its stead a map of the moon appeared. The effect was overwhelming. I can recall a few similar heart-stoppers in a lifetime of looking and listening—the view of Angkor Wat's timeless, brooding temples by moon- light, hearing Giulietta Simionato leading the Anvil Chorus at the Met, seeing for the first time the coast of Asia from a bomber after more than three years and many blood-drenched battles across the wide Pacific. None of these equaled the shock of seeing that illuminated map on the screen.

A lighted sign above the viewing room's glass partition began flashing "Quiet Please." Through the intercom, they could hear CapCom Gerry Carr say, "Apollo 8, you are riding the best bird we can find." Jim Lovell replied, "Thanks a lot, troops. We'll see you on the other side." Tom Paine, another of the watchers in the viewing room, muttered, "Jim, I hope so. I hope so."

Dick Koos, watching with a crowd of other SimSups through the window of the sim room, was as nervous as an expectant father. Up until then it had been fun. He had been brought in on the secret early so that he could plan the sims. If anybody had known they were preparing sims of lunar-orbit insertion, the word about C' would have gotten out, so they had worked without telling even their own supervisors. Then after Schirra's flight, when the secret had come out into the open, there had been the big push. Again and again they had run the T.L.I. sims, the L.O.I. sims, the T.E.I. sims. The fall had turned to winter and the days had gotten shorter. Once an entire week had gone by when Koos, driving to work before dawn, running sims all day, never so much as stepping outside the windowless Control Center, driving home after nightfall, had not seen the sun. Now, as Apollo 8 disappeared behind the moon and its radio signal fell silent, Dick Koos felt lightheaded. He wondered vaguely whether he was going to faint.

Jerry Bostick was Retro. As director of the Flight Dynamics Branch, Bostick wouldn't ordinarily have been sitting at that position, since Retro wasn't the senior console in the Trench. But Chris Kraft had asked Bostick to do it. Retro handled abort modes, and if the spacecraft reappeared on the other side of the moon after anything except a nominal burn, Kraft wanted Bostick sitting at the Retro console. For Bostick, the moment when the spacecraft went behind the moon was depressing. Controllers had something of the mother hen in them, and now they were helpless. There weren't any data to watch, nothing at all they could do to help the astronauts if something went wrong. Bostick didn't reason all this through. He just felt a sense of loss.

Bostick was still brooding when he heard flight director Glynn Lunney say, "Okay everybody, this is a good time to take a break. Everybody back by 69:20." Bostick's first reaction was that, on the contrary, this was a hell of a strange time to be taking a break. But by that time Lunney already had his headset off and was walking down the middle aisle in the MOCR. He stopped and put his hand on Bostick's shoulder. "How's it going, Retro?" he asked, smiling at the notion of Bostick working the

Retro console again, just like the old days. It finally came to Bostick that it was no good being a mother hen just then, that there wasn't a damned thing to be done anyway. So he followed Lunney out the door and got in line for the restroom. There was no shortage of people to cover the consoles while the Black Team was gone—it seemed to Bostick as if every flight controller in F.O.D. had congregated in the MOCR for this moment.

No one lingered. Within a few minutes all the controllers were back, watching the digital clocks on the front wall to time the exact moment of the spacecraft's reappearance around the edge of the moon. Bostick had set up two clocks: One counted down to the time when radio contact would resume if there had been no engine burn behind the moon and the spacecraft was on its way back to the earth on a free-return trajectory. The other was based on time-to-acquisition for a nominal burn that would put Eight into the planned lunar orbit.

The moment when the spacecraft would have reemerged if there had been no burn came and went, and then it became a waiting game. Some sort of burn had occurred, but what kind remained to be seen. Bostick had never seen the room so still. It was jammed with controllers, and yet there wasn't a sound. Lunney stood behind his console, apparently unperturbed.

The numbers on TelCom's console revealed that they had acquired the radio signal before the crew actually said anything. The timing was to within a second of the predicted time for acquisition after a nominal burn. To Bostick, the exactitude of it was "like a religious experience"—God, Bostick decided, had a lot on the ball. The silence in the MOCR was broken as the controllers resumed murmuring into their headsets and CapCom talked to the crew.

On Christmas Eve, in lunar orbit, Frank Borman read a prayer for the congregation at St. Christopher's Episcopal Church back in Houston. Later, on their television broadcast, the three crew members took turns reading from Genesis. It came as a surprise to the controllers in the MOCR, as it did to the millions watching on television, and it was just as overpowering to the controllers as to the rest of the world, this magnificent poetry about the creation of the earth, read by the first men to see the earth whole. Frank Borman, the commander of Apollo 8, had the last verses:

" 'And God said, Let the waters under the Heaven be gathered together unto one place, and let the dry land appear, and it was so. And

God called the dry land Earth, and the gathering together of the waters He called Seas, and God saw that it was good.' '' Borman paused, then concluded: ''And from the crew of Apollo 8, we close with goodnight, good luck, a Merry Christmas, and God bless all of you—all of you on the good earth.''

Rod Loe, sitting at his EECOM console, working this last special flight as a controller just as Arnie Aldrich had promised he could, found his eyes welling over with tears. He bent over his flight log, embarrassed, hoping that no one would notice.

Religion and the Christmas season were on many people's minds during the flight. In its preflight coverage, *Time* magazine drew a comparison between the journey of Apollo 8 and the stations of the cross. There were five key phases for the flight: launch to orbit, translunar injection, lunar-orbit insertion, trans-earth injection, and entry. During the first three, there had been a way to abort and go home. It was the fourth, T.E.I., trans-earth injection, that had worried people since August. This was the moment that some said had caused Webb to leave NASA. It was quite simple: The S.P.S. engine had to work, bringing the crew out of lunar orbit, or else the crew would circle the moon, with a nice, clear radio link back to earth, for about nine days until their oxygen was exhausted and they died. For people watching around the world, this was the moment of the most awful fascination with the flight of Apollo 8.*

Accordingly, the S.P.S. engine had been the object of morbid scrutiny by the media. Diagrams of the engine were shown in popular magazines. Reporters breathlessly described the meticulous checkout that the engine was receiving. This, they emphasized, was the one system that had no redundancy. There was only one S.P.S. engine, no backup. If the S.P.S. failed when a spacecraft was in earth orbit, the crew could use the small jets in the reaction and control system to slow the spacecraft enough to de-orbit. In lunar orbit, the same small jets did not have nearly enough power to speed the spacecraft out of lunar gravity into an earth-bound trajectory.

Within NASA, confidence in the S.P.S. was high. ''This is not to say that we didn't recognize there was a degree of risk in doing C-prime,'' said Rod Rose, one of Kraft's deputies. ''In fact, I guess one of our basic

* The possibility of the astronauts being marooned preyed on everyone's mind. A few days before the flight, a senior official in O.M.S.F., still uneasy about the decision to go into lunar orbit, was heard to ask, ''Just how do we tell Susan Borman that Frank is stranded in orbit around the moon?'' Tom Paine, a no-nonsense type, heard of this remark and responded, ''I guess I would have sat down, held her hand, and said, 'Susan, Frank is stranded in orbit around the moon.' ''

homilies that we drilled into everybody at every conceivable opportunity was, 'Look, fellows, when you light up the S-IVB for the second time [for T.L.I.], you're going to be three days away from home, and don't anybody ever forget it.' You've got to have that much more confidence in your systems." Especially, they had to have a lot of faith that the S.P.S. would fire when the time came to head back. But they had reason for faith: "We had three eminently successful flights as far as the S.P.S. was concerned. We had burned the thing under all manner of conditions, some of which it was never really designed to do."

Marty Cioffoletti, the Rockwell engineer who had helped disassemble spacecraft 012 after the fire, was by this time conducting the checkout of the S.P.S. at the Cape. He thought that the obsession with the S.P.S. as a one-shot, nonredundant system was a lot of hype. "When they say 'no redundancy,' that's a misnomer," he said later. "There was only one engine bell, of course, and only one combustion chamber, but all the avionics that fed the signals to that engine and all the mechanical components that had to work, like the little valves that had to be pressurized to open the ball valves, and so forth, were at least single-fault tolerant and usually two-fault tolerant. . . . There were a heck of a lot of ways to start that engine." And of course they had indeed checked it out carefully before the flight, but nothing they didn't do for any other mission.

All this was still correct as of Christmas Eve, 1968. And yet it ultimately didn't make any difference to the way many of the people in Apollo felt. Caldwell Johnson, speaking as a designer of the spacecraft, explained it. He knew about the checks and balances and all the other people working on the design who were bound to catch a major error. But still, he said, "after a while, you really become appalled that you've gotten yourself involved in the thing. At first, it's an academic exercise. And then the first thing you know, there's people building these things, and they are really getting ready to do it, and you start thinking: Have I made a real bad judgment somewhere, and the damn thing is just not going to work at all?" Marty Cioffoletti, despite knowing all the ways that the S.P.S. was supposed to be foolproof, concurred. For him as for the most unsophisticated citizen watching on TV, "it was darn scary."

As the spacecraft disappeared behind the moon on its last revolution, the one when it would fire the S.P.S. to free itself of lunar gravity, the MOCR once again filled with off-shift controllers. Once again the room fell silent. Chris Kraft later reported that waiting to reacquire the signal was his most apprehensive moment since he had joined the space program, worse even than Glenn's entry after the heat shield problem.

FIDO Jay Greene reported the news: "Flight, we have U.S.B. [upper side band] data. Initial residuals look good."

Then came the first words from the crew. "Please be informed," astronaut Jim Lovell intoned across the 240,000 miles, "there is a Santa Claus."

In the FIDO flight log, Greene wrote in small, neat writing, "Burn status: Burn time $3+23$. Ignition on time. Attitude nominal. V_x $-.5$. V_y $+.4$. V_z 0." And below that, in big, sloppy, joyous script, "WE IS COMING HOME."

For many of the people in the Apollo Program, Apollo 8 was the most magical flight of all, surpassing even the first landing of Apollo 11. For some, like Mike Collins, Eight's momentous historic significance was foremost. For John Aaron, an EECOM, it was simpler than that: "When you're twenty-five and caught up in the thing, and the MOCR's the only environment you know, you don't tend to view things that way." For Aaron, it was the sheer excitement of going to the moon for the first time. Or as FIDO Jay Greene put it, Apollo 8 was the time that they stopped "just running around in circles. Apollo 8 *went* someplace."

Around the world, people looked at the moon with a special wonder. So did the controllers. In Jay Greene's case, it occupied most of the rest of the night. He went home and took a bottle of scotch and some ice and went out beside the pool in his apartment complex and lay back on a lounge chair for hours, all by himself, getting happily drunk and watching the moon.

Dick Koos wanted to do something like that, but he couldn't. On Christmas Eve after the reading from Genesis, Dick Koos didn't get back home from the Control Center until three o'clock in the morning. When he did, he had to hurry to put together his five-year-old daughter's Christmas present, a bicycle. Just as he finished, he heard a little girl's voice say, "Santa Claus has been here!" He never did get any sleep that Christmas.

The people of M.S.C. always waited until after splashdown and recovery were safely over to hold the parties, and so it wasn't until the afternoon of December 27 that NASA 1 Boulevard exploded in a no-holds-barred moveable feast that went on through the night. John Aaron and Rod Loe went to the Flintlock. They were taking a break, standing in the downstairs entrance as people streamed up the stairs to join the mob on the second floor, when a friend passed by. What was Rod doing down here? Why wasn't he upstairs partying?

There were only four days left in that sad and chaotic year of 1968. Things hadn't gotten any better after April 4. Since then, Robert Kennedy had been shot and killed, the nation's cities had been torn by riots and burnings, and unprecedented bitterness over the Vietnam War divided Americans from one another. Still, at that moment all Loe could think of to say was, "I'm just standing here being very proud to be an American."

CHAPTER

24

"We . . . we're go on that, Flight"

Dramatic as Apollo 8 had been, and as novel as the lunar journey had made it seem, the crew had stayed in the command module and performed the kind of maneuvers astronauts had been performing since the first orbital flight in 1962. As 1969 began and preparations for Apollo 9 drew to completion, that was about to change. Beginning with Nine, a brand-new spacecraft was going to be flown, one that functioned very differently from the command module—and, for that matter, differently from any machine that man had ever flown before.

1

The lunar module was a spidery, flimsy contraption. Since the days when Caldwell Johnson and Owen Maynard had drawn designs for the landing gear based on the assumption that the moon would be just like Arizona, the Grumman engineers working on the lunar module had taken the craft through a series of incarnations.

It had been hard for the Grumman engineers to get used to the idea that,

335

because the lunar module would function exclusively in the vacuum of space, it could "be any shape it wanted to be," as one of the early designers put it. "The realization of this was slow to come for some of us, slow to come that it didn't have to be round-edged and smooth, that it could be square. It could have corners on it. Things could stick out at odd angles, it didn't make any difference." It was disorienting. "Suddenly we were in a very, very free-form world of engineering. There wasn't any precedent. And we developed a shape that at first looked ridiculous, and looked more and more ridiculous as we worked on it."

It wasn't just the shape that was different; the materials were also different. The moon's gravity is only one-sixth of earth's, and the stresses that the LEM would have to endure were proportionately smaller. The result was that the LEM's materials were so light they felt like paper to the workers at the Grumman plant. Because they were so light and flimsy, fittings often couldn't be stamped out without creating stress lines, so much of the LEM was made by hand, with the Grumman technicians taking a block of metal and milling it until it fit the blueprint.

To complicate matters, the propellants for the LEM were so incredibly volatile, one technician remembered, that he could dip a stick into the oxidizer and flick a few drops onto snow on the ground—and the snow would catch fire. The internal pressure in some of the tanks was in the neighborhood of 6,000 pounds per square inch. This combination of lightweight materials, caustic gases, and high pressures was particularly dangerous. One day, a technician at White Sands working with a fully fueled test article was filling out a report and absentmindedly clicked the end of his ballpoint pen on a fuel tank. The pen exerted just enough extra pressure on the tank to open a pinprick leak. Eventually, or so the story goes, they found the pen embedded in a fence post, along with a nub of finger bone. The wire-thin stream of propellant had sliced the finger off as neatly as a scalpel.

Because it was called the "bug" and because the command module was thought of as the mother craft, people tended to think of the LEM as being much smaller than the command module. In reality, it was twenty-three feet high compared to less than eleven for the command module, and with a larger internal volume as well.* It consisted of two separable parts. In the

* All spacecraft until the shuttle (discounting Skylab, which was not a spacecraft) were claustrophobically small. The Apollo command module, spacious by comparison with Mercury and Gemini, had three men lying shoulder to shoulder, with the control panels only a few feet from their faces, for up to fourteen days. The much-talked-about "lower-bay area," which the press sometimes described as if the astronauts had the equivalent of a little room to go to, provided enough extra space for the astronauts to move about—not much, but a little—during the course of the flight. In the Mercury and Gemini capsules, they could hardly move at all.

upper portion, along with the ascent engine and its fuel tanks, was the cockpit where the astronauts stood while they descended to and ascended from the lunar surface. They also used this area to eat, sleep, and change into their lunar-surface suits and backpacks. The descent engine, its tanks, the landing gear, and the storage areas for the experimental apparatus (and in later missions the Lunar Rover used to explore the lunar surface) were all in the lower portion. On liftoff from the moon, pyrotechnics and a guillotine apparatus separated the two portions of the LEM. The ascent stage returned to the command module; the lower portion remained on the lunar surface. After the two spacecraft had rendezvoused and all the lunar samples had been moved from the lunar module to the command module, the ascent stage was set adrift, and only the C.S.M. left lunar orbit for the trip home.

The lunar module was thus in and of itself a formidably complex vehicle, a self-contained system designed to perform multiple functions under unique, never-before-experienced circumstances. Apollo 9 was the first of the missions to attempt to operate it.

Apollo 9 was commanded by Jim McDivitt, the man many within the Apollo Program thought should have commanded the first lunar landing. Jim McDivitt was that good at everything. The luck of the rotation had instead left him with Apollo 9.* This was too bad for Jim McDivitt, if he harbored dreams of becoming famous, but it was not such a bad thing to have one of the best astronauts commanding Nine. In the judgment of people around MSC, Apollo 9 was, with the exception of the first landing itself, the most difficult of the Apollo flights. It involved not only the first test of the LEM, with all the hazards and difficulties associated with a maiden flight and a rendezvous, but also an extra-vehicular activity (E.V.A.) by astronaut Rusty Schweikert to test the backpack that would be used for exploration of the lunar surface. Such E.V.A.s, known as "space walks" to the media, were one of the few activities in manned space flight that were even more dangerous than they looked. On top of all this, Nine would be flown in a low earth orbit, which, for

---

* There was a long-standing disagreement between ASPO and the astronauts' office on crew assignments. Should the crews be handpicked for specific missions? Or should they be assigned by rotation? Deke Slayton, who ran Flight Crew Operations, felt strongly that anything except a rotation system was an insult to the astronauts and destructive of morale. Others in ASPO and the Flight Operations Directorate felt just as strongly that some astronauts were more equal than others, and it was foolish not to pick the very best of the best for the lunar landing. McDivitt and Frank Borman were thought to be especially strong candidates for the first lunar landing. Early in the flight planning for Apollo, a roster of missions had been laid out so that the rotation was likely to turn up one of the two of them for that mission, but first the fire and then the insertion of C′ into the schedule had thrown off those calculations, and Slayton held out strongly against suggestions that the roster be revised. It is a measure of McDivitt's stature within NASA that he was chosen to succeed George Low as head of ASPO when Low moved to the deputy administrator's slot after Apollo 11.

technical reasons involving both communications and navigation, was more difficult than trying to carry out the same activities on a real lunar mission. Apollo 9, the D Mission in Owen Maynard's alphabet schedule, was known within the program as the "connoisseur's mission."

Demanding, dangerous, crucial to the success of the program, and also perhaps the most anonymous of the Apollo missions, Apollo 9 was launched on March 3, 1969, carrying McDivitt, Schweikert, and David Scott. The mission lasted ten days, during which the LEM separated from the command module, fired both its descent and its ascent engines in a variety of modes, and performed without a hitch. Technically, Apollo 9 was a resounding success. Outside NASA, the media and the public paid little attention. Within NASA, Apollo management penciled in the G Mission for a July launch.

2

The F Mission, a lunar flight like Apollo 8, but with a LEM that would be manned and flown to within 47,000 feet of the lunar surface, came next.* It was a controversial mission. Many in the space program—George Mueller among them—thought it foolish to go all the way out to the moon, take all the risks associated with the journey, and then stop nine miles short of the surface. If everything looked good, why not be prepared to take advantage of success and go all the way down?

Owen Maynard hadn't included an F Mission in his original schedule. As far as the hardware was concerned, there was no need for it. D and E—or, as things worked out, C' and D—had exercised all the systems under all the conditions they would have to face for a landing. But Maynard was part of ASPO. Over in the Flight Operations Directorate, Rod Rose and Carl Huss had been discussing the same problem and were insistent on having an F Mission. "We said that operationally we'd like to have everything else S.O.P. from beginning to end so that [the astronauts] had a storehouse of experience and knowledge," Rose remembered, and there was a good reason for it. Learning to take the LEM from the command module down to 47,000 feet was a big job in itself, and Rose and Huss felt that the less that was new when the time

* The cutoff point was a natural one. This was the altitude at which the LEM would fire its descent engine for the final descent, the altitude at which the LEM's landing radar became effective, and the minimum altitude at which it was feasible for the C.S.M. to come to the assistance of the LEM if it were to become disabled.

finally came for the first landing, the better. So despite some spirited arguments within NASA itself, Apollo 10 with a crew of Tom Stafford, Gene Cernan, and John Young blasted off on May 18, 1969. Three days later, Stafford and Cernan undocked the LEM they had named Snoopy from the command module and descended toward the lunar surface.

One hair-raising moment in Apollo 10 occurred then, with the LEM far from the command module, circling low over the moon, as the crew tried to separate from the LEM's descent stage. One of the two astronauts had accidentally mis-set a switch, causing the abort guidance system to begin searching for the command module, which threw the LEM into wild gyrations. "Sonofabitch!" Cernan yelled, disrupting NASA's unremitting effort to make the astronauts come across as squeaky clean in every way.* Stafford took over manual control of the LEM before the guidance system locked altogether, and managed to bring it under control. The test of the LEM in lunar orbit ended uneventfully, and Ten was another in NASA's growing string of successes.

All of the intermediate steps had been taken. More than half of 1969 remained. To fulfill the commitment that John Kennedy had made eight years earlier, the one remaining task left to NASA was an actual lunar landing.

3

With the manned missions coming at the rate of one every two or three months, it was not feasible for a flight controller to work every mission. Each mission had three flight directors, with a fourth available to fill in occasional gaps, and each Flight had his own team for each mission. One flight director was designated the "lead" flight director for each mission, responsible for its overall supervision. The mission everyone wanted to work was, of course, G. And within the G Mission, the shift everyone wanted to work was the lunar descent.

Cliff Charlesworth was the lead flight director for the G Mission. He

---

* Cernan caused NASA's public-relations officials to blanch only momentarily. Other astronauts caused months of fretting. One, a cheerfully uninhibited fellow who used four-letter words as grace notes to his everyday conversation, was a special project. Look, they finally told him, hum instead of talking. (Another version of the story is that NASA actually sent him to a psychologist, who hypnotized him and left him with the posthypnotic suggestion: *Hum.*) That's why, in the tapes of his mission, one Apollo astronaut can be heard going "dum-te-dum-te-dum-te-dum" as he bounces across the lunar surface.

didn't hesitate in picking the flight director for the lunar descent. It had to be Kranz. Kranz had been Flight on Apollo 5, the first unmanned test of the LEM. He had worked the LEM maneuvers on Apollo 9, the first manned test of the LEM. No one else had nearly the experience with the lunar module that Kranz had, plus something else. "Gene was the guy you wanted on the headset when there was trouble," Charlesworth said later. "When the term 'flight controller' is used, the first person I think of is Kranz."

That decision made, Charlesworth asked himself who had the most experience with rendezvous. The answer was Lunney, so Lunney got the lunar ascent. That left the launch phase and the lunar-surface activities for Charlesworth. Milt Windler, one of the three newest flight directors, would round out the team.

The experience with the LEM that Kranz's White Team brought to the G Mission was of course limited to the LEM systems and how the machine flew. But accomplishing the actual landing would be different. The flight profile after leaving the 47,000-foot level that Apollo 10 had reached was tricky even if nothing went wrong.

The first key point was P.D.I., powered descent initiation, when the crew had to fire the lunar module's descent engine (DPS, "dips," the Descent Propulsion System). After that moment, the command module was useless as a rescue vehicle. If something did go wrong, no matter what the failure might be, the astronauts would have to use the LEM to get back up again for a rendezvous.

During the first part of the burn, the descent engine pushed forward against the LEM's momentum, to slow the orbiting spacecraft, keeping the LEM in a leg-forward position almost parallel to the moon's surface. During this phase, the crew would be able to look down at the lunar surface, but they would not be able to see where they were going. At about four minutes into the burn, the LEM would rotate so that the critical radar systems would now be looking down, acquiring the data that would guide the spacecraft to its planned landing site. At approximately six and a half minutes into the burn, the crew would throttle DPS back to about 55 percent power. At eight minutes, now only 7,500 feet from the surface, the LEM would pitch forward so that it came nearly to the vertical, enabling the crew to see out their windows toward the landing site. Then it would be time for the landing itself.

The LEM was designed to fly like a helicopter with a rocket engine instead of a rotor. And like a helicopter, the LEM was not easy to fly. On the ground, the crew had three different ways to practice. One was in the simulators at M.S.C. and the Cape, which didn't fly but could rotate and tilt and show lunar views out their windows. The second was an ingeniously constructed tower facility at Langley, where the crew could practice the final seventy feet of the descent in a simulator suspended from the top of the tower in such a way as to duplicate one-sixth earth gravity.* Finally, there was the "flying bedstead," a free-flying machine that had given everyone scares for the last year. First Armstrong and then another test pilot had been forced to bail out when the machine crashed. It had been grounded for many months, and only because Armstrong insisted was he permitted to resume using it for practice in the spring of 1969. To some degree, design faults made the flying bedstead liable to crash, but there was more to it than that: A machine that truly mimicked the task of flying a lunar module was going to be tricky to fly.

Lunar descent was a complicated procedure, and there wasn't much time to practice. From the first Apollo flight through Apollo 11, the people building the simulators were caught behind the curve. The flight hardware was always at least a little behind schedule, and it was always being changed. The simulators couldn't be built to mimic the real LEM until they knew how the real LEM operated. Then, after the simulation hardware was built, there remained the even more difficult task of writing software that would permit the SimSups in the Control Center to put the crews and flight controllers through their paces with realistic, multiple-branch scenarios. In the case of the lunar descent, the simulation software wasn't completed until about two months before G was scheduled to be flown. But that was okay. The White Team was coming off Apollo 9 with "what we thought was a pretty hot hand," Kranz recalled, and they were ready. Two months was plenty for the White Team.

The first simulation of a landing on the moon, on May 29, was in itself a major event within M.S.C. The viewing room was filled with senior NASA officials. Low and Gilruth and Kraft were in the back row of the MOCR. It was not the real thing, but it was exciting nonetheless. That morning, after all the years of preparation, they were going to practice landing on the moon.

Down in the Trench at the Guido console sat Steve Bales. Bales, like most of the others in the front room, was, at age twenty-six, a veteran in

---

* The lunar landing facility at Langley was the creation of one of the old Langley hands who had stayed behind, William Hewitt Phillips, the engineer who had been the mentor for Walt Williams, Chris Kraft, and Chuck Mathews (among many others).

terms of experience (he had been in the Flight Dynamics Branch for five years). Bostick, Bales's boss in the Flight Dynamics Division, had watched him carefully over the last year, as had the flight directors. Steve Bales was still a kid in many ways, inclined to oversleep if you didn't get him on the phone and make sure he was in an upright, conscious position a few hours before his shift, given to enthusiasms and excitements. He was not the image of flight-controller cool—he would sit at his console constantly twisting a lock of his hair at the back of his head so that it stuck out like a small pigtail. But Bales was also a crackerjack Guido, quick, reliable, and knowledgeable, and now Bostick had picked him as Guido for the G Mission, along with Jay Greene for FIDO and Chuck Deiterich for Retro.* During the descent, Guido's was an especially important console, because so many of the things that could go wrong involved errors in guidance; on this first simulation of a descent, Bales was as nervous as he had ever been on a real mission.

The first lunar-descent sim, however, turned out to be routine. The SimSup didn't try any funny stuff at all; he just let the flight controllers and Armstrong and Aldrin determine that they could indeed put the LEM down in one piece if nothing went wrong. And so it went for the first few days of sims, "shooting the nominals," they called it—normal burns, no major malfunctions, no disperse trajectories. "But then the simulation team started putting the meat to us," Kranz said, "and to put it bluntly, we started crashing." Or, almost as bad, they aborted when they didn't really have to.

Several novel difficulties made the descent simulations especially harrowing. The voice and data delay was hard to get used to. At the speed of light, it took 1.3 seconds for a transmission from the spacecraft to reach the earth, and another 1.3 seconds for an answer to return. That added up to 2.6 seconds between the time that an astronaut in the lunar module reported a malfunction and his receiving the quickest possible response. The ways in which that 2.6-second delay could screw things up were legion, the White Team found. The difficulty was compounded by the slow computers of the 1960s. The most critical data during the lunar descent involved navigational calculations based on complex radar data. Those calculations took time to perform, so that four or five seconds

---

* Controllers wanted the first descent shift just as badly as astronauts wanted the mission with the first landing—some of them almost obsessively. It took at least one FIDO a long time to recover from being passed over in favor of Greene. It is indicative of the spirit shared by the F.O.D. managers (and managers in Apollo in general) that the men who could have assigned themselves to the descent phase (Charlesworth to Flight, Bostick to FIDO) did not do so, just as Kranz, as head of the Flight Control Division, could have assigned himself to be lead flight director for Apollo 11.

would elapse between the time the data reached earth and the time the Trench and its back rooms could see what was happening on its screens.

Another problem was the "dead-man's zone," a label that official NASA didn't like at all and tried to discourage. But it was called the "dead-man's zone" for a good reason. During a short period toward the end of the descent, the crew had no way of aborting if something went wrong. During that period, only about ten seconds long, the combination of altitude (by now very low) and rate of descent (still fairly high) was such that even if the ascent engine fired immediately after an abort command, the lunar module would crash before the ascent engine could overcome the inertia of the descent. The theory of how to minimize crashes in the dead-man's zone had been worked out, but the White Team found that it was exceedingly difficult to apply during the sims—especially with that 2.6-second delay, which kept astronauts and controllers from staying in synch.

Then there was the difficulty with the PGNS (Primary Guidance and Navigation System, pronounced "pings" because sometimes the acronym was informally spelled PNGS). In the descent, if the PGNS found that the trajectory was deviating from nominal, it was supposed to keep balancing two competing desirable states of affairs: landing at the assigned target, and landing softly. During the simulations, the PGNS, given a choice between the two, had an unnerving tendency to decide to get to the target, come what may. "On occasion," Kranz said, "if you were far enough off in your navigation state, [the PGNS] would point the LEM directly at the target and try to approach it in pretty much the same fashion that you would fire a projectile." This made for an extremely hard landing, the kind that was euphemistically called "dinging" the spacecraft, less euphemistically known as a fatal crash.

Over in Building 2, the people on the top floors began to get nervous. At that time, the flight director's loop and the air-to-ground loop were piped into the offices of senior management, so they could listen to what went on in the MOCR. "I'd get calls from Kraft or George Low into the phone behind the console," Kranz remembered, "that'd say, 'When are you guys gonna get your act together? You've crashed another simulation. What's wrong over there? What's wrong with your team, or your timing, or your judgment, or whatever it is?' " Kranz began to think of the lunar descent as going into a cauldron.

The late start for the G Mission had meant that Bill Tindall's thousand-ring Mission Techniques circus had been kept busy up until the last minute. The simulations kept bringing up new possibilities, which then had to be thrashed out. The Instrumentation Lab people from M.I.T., the people with the main guidance and software contracts, had been especially hard pressed, since the descent and ascent depended on their new and still-untested programs. The flight controllers would go to the meeting with "straw-man" mission rules, ones they knew weren't good enough but which would provide a starting point, and Tindall would stand up there with his chalk, exclaiming and exhorting, and say, "Okay, M.I.T., what do you think of this one?" M.I.T. would say they didn't have any data for making up their minds, and Tindall would get the SimSups, racing against the schedule, to crank a test of the rule into the next simulation to see what happened. Tindall's final Mission Techniques meeting for Apollo 11 was just one week before liftoff. Getting ready for G was a scramble.

Some time during that harried training cycle of 1969, no one knows exactly when, SimSup Jay Honeycutt went to talk to the controllers in the Flight Dynamics back room. Jack Garman, a twenty-four-year-old controller working there—smart as a whip, computer hotshot—happened to mention the "computer alarms" to Honeycutt. There were many such alarms in the computer software, each designated by a number.

The alarms hadn't previously come to the SimSups' attention because they weren't designed primarily for the controllers or the crew. Rather, they were debugging tools for the software engineers, and they were buried "deep in the bowels of the onboard computer program," as Bill Tindall recalled later. "We had gone through years of working out how in the world to fly that mission in excruciating detail, every kind of failure condition, and never, ever, did I even know those alarms existed." And this was Bill Tindall, who was not only running Mission Techniques, but before that had been one of M.S.C.'s lead engineers on the onboard computer and guidance systems.

Garman and the others working the operating system software knew that the alarms existed, but they hadn't paid any attention to them. As Garman put it, "The problems that triggered the alarms were not problems that could reasonably happen during a mission." But there they were nonetheless, linked to the crew's onboard displays, which meant that an alarm during a flight would produce a flashing light on the display and an audible alarm as well, just like the alarms involving engines or environmental systems. This was just the kind of thing that an enterprising SimSup like Honeycutt was on the lookout for.

*   *   *

A seemingly unrelated event occurred in June, about a month before launch, when the M.I.T. people programming the lunar module's onboard computer sent a Crew Procedures Change Sheet down to Flight Crew Operations.* The Change Sheet involved the mode switch for the rendezvous radar.

The lunar module had two independent guidance systems for homing on the command module. One was the PGNS, which was used to bring the LEM back up to the command module from the lunar surface. The other system, AGS (Abort Guidance System), was for guiding an abort should one become necessary. The mode switch for the rendezvous radar used by PGNS had four settings: Off, Auto, Manual, and Slew. The crew had been instructed to set the switch to Manual before beginning lunar descent. But now the M.I.T. people had decided to have the rendezvous radar keep track of where the command module was during the descent, so that it could take over an abort immediately if that ever became necessary. The Crew Procedures Change Sheet specified that before beginning the descent the mode switch for the rendezvous radar should be set to Auto instead.

The associated changes in the software were loaded into the LEM's computer. Then, after further consideration, M.I.T. decided that the alterations introduced too many new procedural considerations too close to the launch date. But the software changes had already been installed, and to go through the elaborate procedures required to change them back again would be time-consuming. Someone suggested that, if they didn't want to use the rendezvous radar for this purpose after all, they should just withhold the radar data that the PGNS needed to do its abort calculations, thereby preventing it from tracking the C.S.M during the descent phase. It was a simple and apparently no-risk fix.

Two mistakes were made. First, no one realized that withholding the information didn't make the computer stop trying to read the rendezvous radar. All it did was give the computer the impossible task of trying to find a match for a meaningless angle whose sine was 0 and whose cosine was 0; and because computers do not know when tasks are impossible, they just keep trying to do what they have been told. The second mistake was in failing to send down another Crew Procedures Change Sheet canceling the previous instruction to set the rendezvous radar mode switch to Auto—no one thought it was necessary, because (or so they

---

* Nothing involving the Apollo hardware or the mission plan—not a switch, not a number, not a time—could be changed without the documentation of a Change Sheet plus formal approval by a Change Board. The incident described here is a classic example of how elusive the ramifications of changes could be, no matter how strict the system for trying to monitor them.

thought) the original change had been disabled. These two mistakes nearly caused the first lunar landing to fail.

On July 5, just eleven days before the launch, the White Team came to the end of its training for the lunar descent. The crew was to be deployed to the Cape the next day, and the July 5 sims would be their last. By this time, the raggedness was gone. The astronauts and the White Team weren't dinging LEMs any more, and they weren't aborting unnecessarily —confidence was high. The Apollo 12 crew was in the simulator—a normal procedure on the last training sessions, to let the prime crew watch another crew in action, and to give the controllers more work, prompting a crew that wasn't as experienced. Everybody was expecting an easy session. It was the custom for the SimSups to ease off on the controllers on the last day of integrated simulations, to throw them some nominal sims with just a few minor glitches. The theory was that crashing the last sim before launch would be bad for morale.

On the next-to-last simulation of the day, the scenario included one of the computer alarms that Honeycutt had discovered. When the alarm went off, the controllers didn't know what to do with it. Steve Bales, sitting at the Guidance console, saw it and didn't have a clue what was happening. The computer seemed to be saying, "I'm overloaded. I can't do all the work I'm supposed to be doing." If that was really true, then they had to abort—which is the call Bales made, at 10,000 feet, still mystified by the alarm.

At the end of the day, flight director Gene Kranz called the Trench together. The Control Center had no mission rules that covered these computer alarms, and Kranz directed them to find out what every one of the alarms was and how it should be dealt with. Time before launch was so short at this point that there was no time for a formal Mission Techniques meeting.

It was a pain in the ass, many of them thought, because there were so many failure modes on a descent that were much more likely to happen. But they couldn't just leave the alarms dangling. Kranz's team got together with the M.I.T. people and began to work through them. As they did, they discovered a troublesome pattern. A program alarm could be triggered by trivial problems that could be ignored altogether. Or it could be triggered by problems that called for an immediate abort. How to decide which was which? It wasn't enough just to memorize what the program alarm numbers stood for, because even within a single number the alarm might signify many different things. "We wrote ourselves little rules,'' Garman remembered, ''like 'If this alarm happens and it only happens once, don't worry about it. If it happens repeatedly, but other indicators are okay, don't worry about it.' '' And of course, if some alarms happen even once, or if other alarms happen repeatedly and the other indicators are not okay, then they should get the LEM the hell out of there. Trying to hold this list of permutations in their heads, the Guidance controllers went back to reviewing the many other more plausible problems they might encounter on the first lunar descent.

4

On Wednesday, July 16, 1969—2,974 days after John F. Kennedy asked the United States to commit itself to a lunar landing, 169 days before the deadline he had set—Apollo 11 was launched.

The Saturn V carrying Neil Armstrong, Buzz Aldrin, and Michael Collins lifted off from Pad 39A at 9:32 A.M., Eastern Daylight Time, the precise moment selected months earlier. It had been just ten years earlier, in the summer of 1959, that Scott Simpkinson and his team of forty-five had arrived at the Cape and set up shop in one corner of Hangar S, that Joe Bobik had sprayed ammonia to kill the mosquitoes so his men could work through the night on Big Joe, and that Jack Kinzler had carted out the Mercury capsule cushioned by mattresses on the back of a flatbed truck.

As it had for Apollos 8, 9, and 10, the Saturn V performed flawlessly. Then, in a maneuver that had become almost routine, the S-IVB sent the command module *Columbia* carrying the lunar module *Eagle* into its translunar trajectory.

At eight o'clock on Sunday morning, Houston time, July 20, 1969, the LEM *Eagle* disappeared behind the moon and Gene Kranz's White Team took over the MOCR to handle the lunar descent.

Kranz had already been to mass early that morning, at the Shrine of the True Cross near his home in Dickinson. He had listened to his Sousa marches. He had brought a brand-new vest of white brocade with silver thread that his wife, Marta, had made for the day's shift. Now, as Glynn Lunney finished the handover process, Kranz carefully removed the vest from the plastic bag he had hung over the back of his console and put it on as fastidiously as a matador adjusting the jacket of his *traje de luces*.

Kranz looked behind him at the viewing room. There, gathered behind the glass, were Paine, Seamans, von Braun, Silverstein, Elms, Petrone, Rees, Debus, Draper, the astronauts for the next missions, and some astronauts from years gone by. In front of the glass, in the fourth row of the MOCR, were Bob Gilruth, Sam Phillips, George Low, George Mueller, Chris Kraft, and Deke Slayton. The rest of the MOCR was packed with off-duty controllers.

Kranz caught sight of Bill Tindall squeezed into the group in the viewing room. Tindall was Kranz's kind of guy—a man who knew what he believed and wasn't afraid to get emotional about it, who told you where he stood. Plus which, as Kranz saw it, Tindall had been sort of a spiritual leader of the MOCR, their mentor in getting ready for this moment. Kranz beckoned to Tindall to come into the MOCR and plug in beside him, an honor that Tindall would never forget.

While the LEM was still behind the moon, Kranz directed that the doors to the MOCR be locked. The level of tension was such that Steve Bales, sitting at the Guido console next to Jay Greene, smiled to himself—was Kranz trying to keep other people out, or the controllers form escaping? In fact, locking the door was only the visible part of putting the MOCR on "battle short" status, in which the circuit breakers in the Control Center were physically prevented form opening under an electrical load. Kranz was prepared to let a circuit burn up rather than risk a power transient that would cause a major Control Center system to drop off line.

Then Kranz switched from Flight's loop to an auxiliary loop that the people in the viewing room couldn't hear. The tape of what he said is gone. As best he can recall, what he told his controllers was this:

"Hey gang, we're really gonna go and land on the moon today. This is no bullshit, we're going to go land on the moon. We're about to do something that no one has ever done. Be aware that there's a lot of stuff that we don't know about the environment that we're ready to walk into, but be aware that I trust you implicitly. But also I'm aware that we're all human. So somewhere along the line, if we have a problem, be aware that I'm here to take the heat for you. I know that we're working in an area of the unknown that has high risk. But we don't even think of tying this game, we think only to win, and I know you guys, if you've even got a few seconds to work your problem, we're gonna win. So let's go have at it, gang, and I'm gonna be taggin' up to you just like we did in the training runs, and forget all the people out there. What we're about to do now, it's just like we do it in training. And after we finish this sonofagun, we're gonna go out and have a beer and we'll say, 'Dammit, we really did something.' "

Kranz had been standing behind his console as he said all this. He sat down, and, feeling a little abashed, switched back to the Flight loop. He'd really gotten into it. But if not now, when?

As Kranz sat and waited to begin the descent, the NASA cameraman filming in the MOCR caught a shot of him. In it, Kranz is leaning forward expectantly, calm and cool. Behind the console, hidden from the camera, Kranz lifted his hand from a sheet of the flight plan and left behind a perfect, soaking-wet image of his palm.

It took time to check out the LEM and to put it into the initial phase of the descent, so the White Team had been on shift six hours when they approached the moment for powered descent initiation. The MOCR had settled into its rhythm by that time, and Kranz sounded relaxed as he went into a final check before P.D.I. Guido—to be called Guidance now that they were talking on the loops—was Steve Bales. FIDO was Jay Greene. Retro was Chuck Deiterich. CapCom was Charlie Duke. TelCom was Don Puddy. Control was Bob Carlton. Surgeon was John Ziegleschmid. Kranz polled the controllers, called "going around the horn."

"Got us locked up there, TelCom?"

"Okay, it's just real weak, Flight," Puddy answered. Getting good radio communication from the moon was still a matter of constantly juggling signals and sources.

"Okay," said Kranz, who wanted a more straightforward answer, "how ya lookin'? All your systems go?"

"That's affirm, Flight."

"How about you, Control?"

"We look good."

"Guidance, you happy?"

"Go with systems," said Bales.

"FIDO, how about you?"

"We're go," said Greene. "We're a little low, Flight, no problem."

"Rog."

For the next twenty-two seconds, Flight's loop was silent. Then Kranz's voice came on again.

"Okay, all flight controllers, thirty seconds to ignition." His voice was low and mellow, falling off lazily on the word "ignition" as if to say that it was just a stroll in the park, no sweat. And the descent to the moon began.

They had been having problems communicating with the LEM. FIDO intermittently lost data from MSFN ("misfin," the Manned Space Flight Network), but they had tracking data from PGNS, AGS, and Doppler radar as well. Also it looked as if the spacecraft was going to land downrange from the planned spot—the mascons had played tricks with the navigation again.* That one was a little worrisome. The velocity was 20 feet per second (f.p.s.) too high the first time Bales saw data after P.D.I. began. "That really scared the hell out of me," Bales recalled— they were already more than halfway to an abort boundary of 35 f.p.s. He could think of innocuous reasons why the velocity might be too high, but he wanted to be sure this wasn't a sign of some deeper failure in the guidance computer. Bales watched the downtrack error closely, but it seemed to remain at 20 f.p.s.

In a way, Kranz welcomed these problems, because they gave him a look at the way the team was responding to them. "I listened to what they said, but I listened more to how they said it by that time." They were saying it just the way he wanted to hear it. The spacecraft yawed on schedule, and the landing radar began acquiring data for the landing. The clock came up on five minutes and thirty seconds into the burn.

The landing radar had found the surface of the moon and begun calculating where the *Eagle* was. The radar recognized that the altitude was too high and began automatically to adjust velocity to converge with the planned trajectory. Bales began to relax.

"Is he accepting it, Guidance?" Kranz asked Bales, inquiring whether the crew had begun using the landing radar data.

Buzz Aldrin's uninflected voice came onto the loop over Kranz's question. "1202," Aldrin said. He was announcing that a light on his panel was lit, an alarm buzzer was sounding, and the program alarm code 1202 was showing on his computer display.

"Stand by," said Bales, putting off an answer to Kranz.

"1202." It was Aldrin again.

"What's a 1202?" an unidentified voice inquired.

"1202, what's that?" asked another.

"12 . . . 1202 alarm," said Gran Paules, sitting beside Bales. During critical portions of a mission, Guidance had a second controller, "Yaw," so named because his function was to monitor the yawing movements of

---

* "Mascons" were unexpected irregularities in the gravitational field of the moon, discovered by the unmanned Lunar Orbiter missions in 1966-67.

the spacecraft. Paules had won that assignment for the descent. He was talking to Flight now, because Steve Bales had suddenly become exceedingly busy. Paules's voice was firm, but obviously worried.

"Affirmative on that 12," CapCom said, confirming that Houston was picking up the same alarm and also implicitly promising advice to come.

These were the moments when flight directors were forced by circumstance to exhibit a capacity for divided attention. Bill Tindall had always found Kranz in particular to possess this ability to an extreme. "It's like he's got two entirely separate minds. Like two people. He's sitting there running these missions, calling these people, punching the buttons to bring up the loops, thinking about what's happening, and at the same time he's writing down in this precise printing of his, like a telegraph operator, exactly what's going on all the time, keeping a total record of the whole goddamned thing. It's just beyond my comprehension how anyone can do anything like that." But a capacity for such divided attention was essential precisely because of what now happened: Just as Kranz heard that a potentially mission-threatening alarm had just sounded, somebody else also had urgent business for him. Chuck Deiterich's voice came over the loop.

"Flight, Retro."

"Go, Retro."

"Throttle down 6 plus 25."

Retro was giving Kranz the time, six minutes and twenty-five seconds into the burn, when the crew should be expecting the descent engine to throttle down to 55 percent power.

"6 plus 25," said Kranz. In the shorthand of the MOCR, Kranz had just acknowledged Retro's message and at the same time given a tacit direction to the CapCom (which the CapCom recognized as being a direction because he knew this was one of the items of information that the crew must have) to pass Deiterich's message along to Armstrong and Aldrin. Even as Kranz confirmed the 6-plus-25, a voice from the LEM was coming in over his words:

"Give us the reading on the 1202 program alarm."

Some abort decisions were made by the astronauts. Other abort decisions were made on the ground. In the case of a program alarm, the instrumentation in the LEM could tell the astronauts only that an alarm had occurred, and its number. Armstrong and Aldrin could know nothing else about whether it was recurring or how serious it was, and they knew that the ground would have to make the call. When *Eagle* said, "Give us the reading," they were asking whether they should abort. It was now up to Steve Bales.

When the report of the alarm had come in, Bales had half risen from his seat, pulled up by a surge of adrenaline, his left index finger furiously twirling a lock of hair. He began talking to his back room, where Jack Garman also had heard the message. They still didn't have any data—it would be another four or five seconds before the computer brought that up on the screen—but Garman was sure right away what the alarm was, and his voice came over the Guido loop quickly in an almost schoolmasterish voice, each word carefully articulated: "It's executive overflow; if it does not occur again, we're fine."

"Yeah," said Gran Paules in his Texas drawl, "it's the same thing we had"—referring to the sim of a few weeks earlier.

Actually, it wasn't the same thing they'd had. The simulated program alarm had been one that was "possible" during a flight, whereas a 1202 alarm was, they thought, "impossible"—purely a debugging alarm, still in the software only by oversight. But because of the simulation they had studied up on even the impossible ones, so the team was ready. Examined clinically, it was a textbook case of how the flight control system was supposed to function. A type of failure had been foreseen and simulated. Based on that experience, rules had been developed. At the moment of crisis, the man in the back room, Jack Garman, had instantly retrieved from his memory the salient information and given it to his man in the front room. It was cut and dried.

Examined less clinically, the decision was not that clear. "I thought I was right," Bales said years later of the call he was about to make. "Well, I *knew* I was right based on the work we'd done previously." And he trusted Garman, who Bales thought was the smartest computer guy he'd ever met. And yet it still wasn't so simple. "I knew I was executing what we'd decided to do previously," Bales continued. "But then you could ask, well, what was the softness or firmness of that decision made previously? And there was some softness in it. No one knew really what was causing the problem. When you don't know what's causing the problem, you are never absolutely one hundred percent sure that the decision you're making is right. It's like trying to diagnose yourself and taking a pill just on symptoms. So I knew I was doing the right thing from what we'd previously discussed, and I knew darned well there was some softness in that decision." He also knew that two lives could depend on his being right. And so when Steve Bales made his call, he stammered slightly, and his voice was high.

"We . . . we're go on that, Flight."

It was nineteen seconds since he had first told Kranz to stand by. Steve Bales, twenty-six years old, with some advice from a twenty-four-year-old, given nineteen seconds to think it over, told Gene Kranz, Chris Kraft, General Phillips, Administrator Paine, President Nixon, and the world—and, not incidentally, Buzz Aldrin and Neil Armstrong—"Ignore the computer and trust me."

"We're go on that alarm?" Kranz wanted to make absolutely sure.

Again Bales stammered. "If . . . if it doesn't recur, we'll be go."

Jack Garman's voice was in Bales's ear again, straightening out a slight misunderstanding: "It's continuous that makes it no-go. If it recurs, we're fine"—meaning, if it recurred with intervals of several seconds in between, they were fine. He continued to articulate each word carefully, slowly, confidently.

"Rog," Kranz replied to Bales. "Did you get the throttle down, CapCom?" Kranz had already put the alarm aside and was now checking to make sure that, in the midst of the excitement, the 6-plus-25 time had been passed up to the crew.

A different computer alarm came back again later, just after Kranz announced—he couldn't keep a note of triumph out of his voice—that they were go for landing at 3,000 feet.

"Understand, go for landing 3,000 feet," said Aldrin. "1201 alarm."

"Roger, 1201 alarm," said CapCom.

"1201 alarm," said Kranz, an implied question in his voice.

Bales had already been talking to his back room: "What alarm, Jack?"

"Same type, we're go."

Bales began to speak even before Kranz finished. "Same type," he said, confidently now, no stammer. "We're go, Flight."

"Okay, we're go," Kranz said to CapCom.

"Same type, we're go," said Charlie Duke to Armstrong and Aldrin.

Listening to his very words echo through the network, Jack Garman, sitting in obscurity in the Flight Dynamics back room, had a feeling of being suddenly, for that brief and crucial moment, truly connected to the astronauts.

For the next minute Garman listened as the crew performed the final maneuvers before landing. The adrenaline rush that had swept him during the program alarms faded, leaving him drained, with a detached sense of watching events in slow motion. It was then that he heard Aldrin in *Eagle* say, "Forty feet, down two and a half. Picking up some dust." Garman was startled out of his trance. Everything else had felt exactly like the simulations until then. But Aldrin had never said "Picking up some dust" before. The image of the dust blowing up around the LEM made it real, and the enormity of it began to sink in.

Armstrong and Aldrin brought *Eagle* to the surface with a fuel tank that was nearly dry. With thirty seconds of fuel left, CapCom Charlie Duke had made the only call that Kranz was permitting at this point of intense concentration, a fuel call—"Thirty seconds," is all Duke said—and from then on it had been agonizing for the MOCR. Perhaps they didn't have even thirty seconds—they calculated fuel remaining based on the average throttle setting during hover, and the crew had started throttling well before hover. If the fuel went dry, the computers would automatically try to abort the landing and fire the ascent stage—a maneuver that, this close to the surface, left no margin for error at all. Every controller in the MOCR knew all this, and watched the clocks. Or most did. An off-duty Guido sitting on the ledge behind the Trench just buried his head in his hands.

"Contact light. Okay. Engines stop." Buzz Aldrin followed this with the string of procedures he and Armstrong were carrying out, "safe-ing" the vehicle.

"We copy you down, *Eagle*," Duke said.

"Houston, Tranquillity base here," Neil Armstrong announced. "The *Eagle* has landed."

"Roger, Tranquillity, we copy you on the ground. You've got a bunch of guys about to turn blue. We're breathing again. Thanks a lot."

Even as CapCom was saying to *Eagle*, "We copy you down," Kranz was already on the loop to the controllers:

"Okay everybody. T1, stand by for T1."

There were circumstances under which the LEM might have to launch immediately from the lunar surface, so the flight controllers had to make stay/no-stay decisions at touchdown + 1 and touchdown + 2 minutes (called simply "T one" and "T two"). But at this point Kranz had to speak loudly, because the viewing room had erupted. They cheered and stamped and applauded, and then the off-duty people in the MOCR broke into cheers as well. And Kranz could no longer fight off the emotions that were sweeping over him. He knew he had to be getting ready for T1 and T2. But he couldn't make himself do it. He couldn't say anything, he couldn't move. Gene Kranz, the crisis over, froze. His left hand was clutching the handle on the front of the console, his right was holding the pencil with which he wrote his tidy little telegraphic notes. Now, in his frustration with himself, he smashed his right arm down on the desk of his console, breaking the pencil and bruising himself from elbow to wrist. That broke the spell, and he could begin again.

"Okay all flight controllers, about forty-five seconds to T1 stay/ no-stay."

He was back to normal. A few seconds later, irritably:

"Okay, keep the chatter down in this room."

At four o'clock that afternoon, the White Team finally came off shift. Steve Bales was told to go along to the press briefing in Building 1. As he emerged from the Control Center, blinking in the harsh sunlight of a Houston July, he saw a group of demonstrators, kids about his age, out on the grass beside the duck pond. They weren't making a fuss; they just were trying to get attention for their cause, which was more food for hungry people, or something like that. It was 1969, but Bales had been preoccupied for the last few years and had never seen demonstrators before, though he had heard about them. They were a great curiosity to him. He gawked surreptitiously at their sandals and the slogans on their T-shirts and their long hair. It is unlikely that they noticed Steve Bales as he hurried by, with his white shirt, his neatly tied tie, his NASA badge, pens in his pocket, square as could be.

CHAPTER

25

"Well, let's light this sumbitch and it better work"

The events of the rest of that day were watched on television by a worldwide audience estimated to be in excess of a billion people. Neil Armstrong and Buzz Aldrin told Flight Director Milt Windler that they wanted to postpone their planned rest period and proceed directly to the E.V.A. Kraft, not surprised that the crew didn't feel sleepy, agreed. Cliff Charlesworth's Green Team, which had trained for the lunar surface activities, was hurriedly brought on shift (not difficult, since most of the team had been hanging around Building 30 all day anyway). At 9:56:15 P.M., Central Daylight Time, July 20, 1969, Neil Armstrong hopped from the bottom rung of the LEM's ladder onto the lunar surface, proclaiming, "That's one small step for a man, one giant leap for mankind."*

Buzz Aldrin joined Armstrong on the surface. Together they spent two hours and forty minutes gathering twenty-one lunar rocks and deploying some simple experiments. It wasn't a long time, but the first landing was

---

* That, at least, is what Armstrong intended to say. Either he forgot the article "a" or a communications glitch blocked it out, because his sentence came back to earth as "That's one small step for man, one giant leap for mankind." Armstrong was nonplussed when he returned and found out how his historic first words had been heard, pointing out that without the "a" the sentence didn't make sense. He was right, but everyone knew what he meant anyway.

356

considered to be primarily a test of equipment and procedures; the scientists would get more attention in the later flights. Then they returned to the *Eagle* to rest fitfully for six hours—the LEM, with its systems gurgling and whirring, was a noisy place to sleep.

When the E.V.A. ended, it was after midnight for most of the United States. While the nation went to sleep at the end of a momentous day, a relatively small group of Apollo engineers were working throughout the night. They had had no time to celebrate the landing or to watch the moon walk. Nor would they be able to rest any time soon. Even as the LEM's stay on the surface seemed to move serenely forward, the support network labored furiously to keep the mission from self-destructing.

1

Every manned space flight experienced what NASA called "anomalies." The signal that misrepresented the condition of John Glenn's heat shield had been a classic anomaly: Something potentially disastrous had happened, but no one knew exactly what that "something" was. Anomalies continued to occur on the Apollo missions, but, like everything else in Apollo, they became more complicated to diagnose.

The ground's first response to an anomaly, in Apollo as in Mercury, had to come from Mission Control. Only the MOCR and its back rooms were equipped for instant reaction. After their initial response, however, the MOCR could turn elsewhere. Sometimes they needed to have a better idea of what had really gone wrong (why did the S.P.S. engine show slightly less than its rated thrust during that course correction burn on Apollo 8?). Sometimes they needed a prediction of what was going to happen in the future (how long could the ailing fuel cell on Gemini V be expected to limp along?). Often, the MOCR needed a way of actually fixing whatever was wrong.

Such was the case during Apollo 11 after the *Eagle* had landed. In quick succession, Eleven was faced with three important anomalies. To resolve them, the MOCR turned to two other rooms in Houston, SPAN and the MER, and to two distant rooms in Bethpage, Long Island, and Cambridge, Massachusetts.

Out the main door of the MOCR and around two corners was the room called SPAN. SPAN stood for "Spacecraft Analysis," but no one referred to it that way—Scott Simpkinson always swore he had worked there for three years before he found out what "SPAN" meant.

SPAN was staffed by a small but extremely senior group of engineers and controllers. Simpkinson, by this time George Low's Assistant Program Manager for Flight Safety, directed SPAN's activities and was on shift as SPAN operations manager during the landing of Apollo 11. Sitting beside him and across the table were two senior engineers from each of the major contractors. The night of the landing, they were the men who knew their respective spacecraft better than anyone else: George Jeffs, North American Rockwell's chief engineer for the command module, and Tom Kelly, Grumman's chief engineer for the lunar module. Also at the table were equally senior representatives from M.I.T., where the guidance and navigation systems had been developed, and from MPAD. Behind them was a row of consoles manned by supervisory controllers for Flight Dynamics, the C.S.M.'s systems, and the LEM's systems.

SPAN was an interface—the jargon is precisely descriptive in this case—which linked three distinct entities: F.O.D., ASPO, and the contractors. True to Apollo's cavalier attitude toward hierarchies, these senior managers were tucked away in a nondescript back room essentially to act as traffic cops and dispatchers.* If any of the contractors or someone from F.O.D. or ASPO saw an anomaly (the screens available in the MOCR were available on the others' C.R.T.s as well), the report went to SPAN, to be assessed and passed on to the appropriate parties. SPAN's staff, therefore, had to have sufficient technical knowledge and be senior enough to deal instantly, without negotiation, with their own organizations and to reach agreements among themselves without need for approval from higher-ups.

SPAN took the incoming information and, as necessary, honed it into a precise statement of the problem, with a question to be answered. Once

---

* The indifference of many Apollo-era managers toward their status was one of NASA's most endearing traits in those years. During one of the later Apollo missions, Jerry Bostick was in SPAN, working the Flight Dynamics position. SPAN, like the MOCR, always had visitors, and one of them, an unassuming gentleman who looked, someone once said, like a carving of a Black Forest elf, was sitting between Bostick and the pneumatic tube (P-tube) that Bostick used for sending messages to the MOCR and back rooms. Finally Bostick said, "Look, I don't know who the hell you are or what you're doing here, but you could make yourself useful and help get the messages out." And so the man took off his coat and sat there the rest of the shift stuffing messages into the P-tube at Bostick's direction. The next week one of Bostick's friends, just returned from a trip to Huntsville, told him a story. Bostick's friend had attended the weekly staff meeting of Eberhard Rees, who by that time had succeeded Wernher von Braun as Marshall's center director. "Dr. Rees," someone asked, "how did you enjoy being over at Houston for the launch?" Rees is said to have replied, "Well, they had me in this little room, and to be quite honest, I felt in the way, it was so crowded. But then this nice young man gave me a job and I really felt useful."

that question had been defined, SPAN went out to search for an answer. If it wished, SPAN could go directly to the manufacturers' plants. During missions, North American in Downey, Grumman on Long Island, and M.I.T. in Cambridge each had rooms with engineers standing by twenty-four hours a day, with access to all of their plants' archives and testing facilities. The dozens of subcontractors scattered around the country also maintained on-duty staffs. If, for example, Grumman wanted to know the testing history of a particular LEM battery manufactured by Eagle Pitcher in Joplin, Missouri, somebody would be standing by in Joplin who could give Grumman the answer.

But whether or not they phoned the contractors, SPAN's first response when confronted with a new anomaly was to write up a "chit" describing the problem and give it to a messenger who carried it two hundred yards across a courtyard to the third floor of Building 45. This was the Mission Evaluation Room—the MER—the domain of Mad Don Arabian.

"The MER was famous," said one of Arabian's deputies, which was both true and false. The public had never heard of it. To them, Houston was synonymous with Mission Control. But it was famous among Apollo design engineers, for whom it was the equivalent of Mission Control. Indeed, many of them considered the MOCR to be comparatively trivial. In the view of these partisans, when a mission had a real problem, the place that solved it was the MER.

Crowded, sometimes noisy, apparently disorganized, the MER was everything the MOCR wasn't. During a crisis, up to a hundred engineers might be crowded around the MER's six rows of folding tables, the kind used for church suppers. Each system and subsystem in the spacecraft was represented by an engineering team. The contractors assigned some of their top technical people to be in this room, as did ASPO.

Dozens of telephones littered the tables, for, as in SPAN, the engineers in the MER were constantly on the line to their own backup rooms, elsewhere on the M.S.C. campus or, for the contractors, around the country. The difference between SPAN and the MER was that SPAN roughly identified the nature of a problem, whereas the MER was supposed to understand it and solve it while the spacecraft was in flight and before the effects of the problem became critical. If, as was sometimes the case, that meant mobilizing thousands of engineers nationwide, so be it.

The MER was not itself a high-tech room ("We didn't need any fancy damn consoles or anything," Arabian proudly declared). Some of the engineers wore headsets and could plug into their choice of loops. Ten television monitors with telemetry data hung overhead along the far wall. One of the MER's staff was assigned to take Polaroid photographs of any particularly interesting screens, and those were the MER's "hard copy." The engineers brought in binoculars so they could make out what the distant numbers were—during a busy part of the flight the MER presented a strange sight, with dozens of men wearing white shirts and ties, sitting in a closed room, holding binoculars to their eyes.

The assets of the MER lay not in the technology of the room, but in the brains of its members. Assembled in that one room was a vast storehouse of knowledge about how the spacecraft worked and how to jury-rig a fix that would pull the mission out of trouble. And no one exemplified the no-sweat, we-can-do-that confidence of the MER more than its leader, Don Arabian.

"Mad Don" was only one of several appellations for him. Variations included "The Mad Man," "The Wild Man," and "The Wild Arabian." His real name was Donald Dionysios (that's the way his father spelled it) Arabian. In 1949, he had gone directly from the University of Rhode Island to work for Langley, where he relished the precise and elegant engineering-for-engineering's-sake that characterized Langley. Then in 1960 he moved over to join the Space Task Group. In 1967, shortly after heading up one of the panels for the 204 Review Board, he was chosen to be head of Houston's Test Division, part of ASPO.

During a flight, Arabian ran the MER. After the flight, he orchestrated the preparation of the sacred text for each mission, the Mission Evaluation Report, and personally spent most of his time on one chapter in it, the "Anomaly Report." This chapter informed F.O.D., the rest of ASPO, and the contractors which ones of the fifty-odd anomalies reported during a typical mission were hardware malfunctions and which were procedural errors by the crew or ground. The Anomaly Report analyzed each of the true anomalies and reported on what had been done to rectify the situation, ending each section with the imperious pronouncement: "This anomaly is closed."

Arabian had a raised platform just inside the doorway of the MER from which he could look out over the long, narrow room. When a new problem came over from SPAN, he would holler in his pronounced, nasal Rhode Island accent, "All right, folks, we gotta problem here we gotta work, let's get goin' on it!" Then, having assembled the people who would work on it for a preliminary assessment, he would prowl back and forth in front of the blackboard, tall and lean, wide white grin, flashing blue eyes, waving the chalk, stopping periodically to scribble on the board, talking rapidly, contemptuous of grammar. Once in a while he would stop suddenly, head cocked, eyes suddenly withdrawn, as he attacked some new line of thought.

Arabian was known as Mad Don partly because of his ideas and partly because of his unrestrained enthusiasm in expressing them. According to an engineer who worked with him, when there was a meeting to hash out a technical issue, "everybody always shuddered when you got around to Arabian, because we knew it was either going to be back to square one or off into left field—you never knew exactly where he was going to come from." But he was too good an engineer for the others not to listen, no matter how crazy his latest notion sounded at first.

Arabian didn't give a damn about what anybody else thought. He had his own ideas about how things ought to be, and he was sure he was right, and he didn't care if everybody else in the room thought differently. "See, I'm one guy a lot of people are afraid of," Arabian said, accurately. "There's a reason for it, because if anybody does anything technically that's not according to physics, that's bullshitting about something, I will forever be death on them. I mean, you'd better be exact, you'd better show technical elegance. It doesn't make any difference how smart you are, you'd better not ever prostitute physics."

Physics was the one entity that Arabian would forbear from antagonizing. Good-natured and generous in his personal relationships, he had pungent, mostly critical, opinions about everything in his professional life. Organization was just a nuisance ("You find out what the rules of the game are, then you determine how to get around the ones that are detrimental to what you want to do"). People were problematic. A few, mainly Low and Kraft, had his boundless admiration. Arabian also liked and freely praised certain other engineers—Faget, Jeffs, people like that, and most of the contractor engineers who were assigned to the MER. But Arabian thought that just about everybody else was in the way. This tended to include his own subordinates in the Test Division ("I never wanted people to work for me. It just takes away from you"), NASA executives from headquarters ("Hubcaps," useless ornamentation), senior executives with the contractors ("Their goal is just to make a buck"), flight controllers ("They spend most all their time coming up with flight rules. Then when something happens, they haven't the foggiest idea what you need to do''), astronauts (''A lot of them aren't too damned swift''), and, perhaps most eloquently, the Cape. This is vintage Arabian on Launch Operations:

> Those thousands of people in the Firing Room at K.S.C. are not needed. We went and landed on the moon. Strange place, never before been there. Didn't know where the landing spot was gonna be. [Imitates the sound of the LEM's engine shutting down.] They got out, they messed around, ate lunch and everything else, and when they're ready to go they kick the rocks out of the way, brush themselves off. [Acts out parts of both astronauts.] They have no white coats on, there's no ground support equipment, there are no consoles or anything else, and one says, ''Okay, you ready to go?''—''Yeah, I'm ready to go.''—''Okay, you put the sandwiches away.'' And when they're ready to go they press a button. They call up Houston and say, ''Houston, we are go for liftoff.'' Houston says, ''You're go.'' If there's something wrong, Houston couldn't have done a damn thing. That was all on their own, two guys, okay? And they did it six times—launched from the moon, strange environment, without any firing rooms or anything else. Then you say to yourself, ''If they did it there, what the hell's goin' on at K.S.C.?''

All of these heterodox opinions were expressed with a big grin, without personal animosity—and with absolute conviction, anywhere, any time.

Arabian's particular disfavor was reserved for anyone who tinkered with hardware once it had been properly designed and built. He was a relentless foe of the kind of testing they did at the Cape, which in his opinion accomplished nothing except to take operational life out of the systems.* He bemoaned the astronauts' preference for manual systems, which he considered unsafe. But to come full circle, he also took potshots at his own Test Division. The Test Division shouldn't exist, Arabian thought. Reliability and efficiency in a piece of equipment had to be designed in; there shouldn't have to be a separate organization to check that kind of thing. Build the hardware right to begin with, and you won't have to bring all these unreliable human beings into the act later.

And there, finally, Arabian's view of the cosmos rested: He had faith in hardware systems and in designs that obeyed the laws of physics, and

* In this discussion, the word "test" is being used for two different meanings. At the Cape, the test conductors tested to see whether the hardware for the next flight was operating according to specs. Arabian's Test Division was charged with testing the equipment relative to the functions it was supposed to fulfill. Or to put it another way, the Cape tested the equipment, while the Test Division tested the design.

a deep distrust of human performance. "See, the brain is very clever," Arabian said to all who would listen. "It can perceive things, it can create things, and all that. But it's the most undependable, unreliable, unpredictable device that exists."*

That was what made Arabian so good for the MER and so important to Apollo. When an ambiguous problem came in, he wouldn't settle for an explanation unless it fit all the conditions. He was always aware of "the brain's" propensity to jump to convenient conclusions. "If something goes wrong, let's say, and there are ten conditions that must be satisfied, and this one thesis satisfies them all precisely, see, except one, okay? then that ain't it. It's not 'almost': You're either there, or you ain't there." Time and again during Apollo, Arabian's cheerful intransigence turned out to be invaluable. Often the situation that the MER had to deal with was indeed a problem for which there seemed to be an excellent explanation that fit all of the facts, all of them but one, and Arabian would refuse to accept it until that last small, unimportant anomalous fact was understood—when, frequently, it became clear that the problem and its solution were quite different than had been previously thought. In Houston, Arabian was *sui generis*—"a legend in his own time," Chris Kraft once wrote. A knowledgeable observer in the F.O.D. that Arabian disdained, thinking about the people at M.S.C. who made the Apollo flights successful, decided that four men ultimately stood out: Kraft, Low, Tindall—and Arabian.

2

The anomaly that caused the highest pulse rates during Apollo 11 occurred a minute after the landing. It was not mentioned in the press coverage or in the official NASA history, probably because nothing came

---

* Arabian included his own brain in this indictment, citing the time he forgot to lower the landing gear in his little Beechcraft as proof of the brain's treachery. Arabian flew himself everywhere during the Apollo years—mostly alone, however, because he tended to make passengers nervous. Simpkinson remembered sitting in the copilot's seat one day when they got out to the end of the runway and discovered that one of the plane's magnetos was out. "Don't worry about it!" Arabian shouted to Simpkinson over the noise of the engine as he proceeded to take off anyway. On another occasion, after some mysterious episodes in which the Beechcraft's engine had been sputtering and backfiring, threatening to quit altogether, Arabian decided that he needed to do some systematic troubleshooting. So he took the little plane up near its ceiling, about 15,000 feet, where he would have, in his words, "plenty of time"—and proceeded to induce the engine problem, testing out alternative explanations. It was perfectly safe, he said—he made sure he was within coasting distance of an airport.

of it, but for a few minutes it scared the living daylights out of the handful of people on the ground who knew what was happening.

On TELMU's screen in the MOCR and on the lunar module systems screen in SPAN, readings showed that pressure and temperature were rising alarmingly in one of the descent stage's fuel lines. After the engine shut down, a blockage had apparently occurred—almost certainly, Grumman's Tom Kelly in SPAN believed, because liquid helium had frozen a slug of fuel left in the pipe. As they watched the screens, the residual heat from the engine (which had been operating at 5,000 degrees Fahrenheit when the *Eagle* had landed) was moving up toward the frozen fuel, and the question bothering Kelly was what would happen when the trapped fuel suddenly got much warmer. "There's no telling what it will do," he told Low, who by this time was on the phone with Kelly. The fuel was unstable when heated, and Kelly saw a real possibility that the frozen fuel would explode like a small hand grenade.

The MER and the Grumman control center at Bethpage had seen the same pressure readings. Men in both rooms were on the phone to each other, trying to estimate what would happen if the fuel exploded. The best guess was that even a small explosion was likely to squirt fuel or oxidizer into the combustion chamber, which was still thousands of degrees hot.

Some of the people at Grumman argued for an immediate abort, launching the ascent stage and leaving the problem behind. But besides ruining the rest of the plans for the first landing, an immediate launch was impossible: By the time the problem occurred, *Columbia* was out of position for a rendezvous. An option favored by Kelly was to burp the descent engine—restart it at 10 percent power for a split second—and relieve the pressure. By now all the phone lines were open—Low in the MOCR, Simpkinson and Kelly in SPAN, Arabian in the MER. Arabian argued against burping the engine, at least until they knew more about the attitude of the LEM on the lunar surface. If *Eagle* was already tilted at an angle, even a slight thrust from the descent engine at one-sixth gravity might push the LEM over on its side.

As they talked, they saw the pressure in the fuel line start down, edge up slightly once more, then fall abruptly and for good. The blockage had melted or the line had ruptured without damaging anything essential. In either case, the crisis which had flared so suddenly was over as quickly as it had begun. It was an object lesson in Kraft's precept about not doing anything if you don't know what to do.

* * *

The second anomaly occurred when Neil Armstrong, trying to maneuver in the LEM with his pressure suit and backpack on, inadvertently backed into the circuit breaker for the ascent engine's arming switch and broke it. This was not a trivial matter, for the ascent engine ignition button wouldn't work until the ascent engine had been armed. Neither, however, was it a crisis. It was a nuisance which provided an example of the versatility of the LEM's design. The *Eagle* didn't have a backup arming switch, but it did have such redundant wiring that, given time, the Grumman engineers would be able to work out a sequence of switches that would reroute signals in such a way as to arm the ascent engine, even without the arming switch. The MER and Bethpage figured out how to do it while the astronauts were walking on the moon. While the astronauts rested, they simulated the new sequence and passed the solution on to SPAN, where it was reviewed and given to the MOCR for transmission to *Eagle*.

The anomaly that resisted solution the longest concerned the program alarm. The people in the MOCR and the MER understood what had triggered the alarm—the computer was failing to keep up with its work—but even after *Eagle* was on the ground nobody in the MOCR, SPAN, the MER, or at M.I.T. knew why. In a few hours, that same computer would be called upon to perform a lunar rendezvous, and the possibility that troubled everyone was that the computer failure would recur in a more serious form.

Determining what had happened took M.I.T.'s teams in Houston and Cambridge nearly the whole twenty-one hours that the *Eagle* was on the moon. Throughout the night and into the morning they burrowed deep into the computer system, searching for the mysterious something that had been eating up computer capacity during the descent. They ran sims of the software, sims of the hardware, and still they were unable to determine the source of the problem.

On the morning of July 21, George Silver, a heavyset man who looked more like a hard-hat construction worker than the software expert he was, walked into the Apollo Room at M.I.T.'s Instrumentation Lab to supervise the ascent phase. Ordinarily, Silver would have been in SPAN or the MER, but he was in the process of taking over the Systems Test Division at the Instrumentation Lab. When Silver arrived that morning, the team that had worked the descent was still there, unshaven and red-eyed, still working on the program alarm mystery. Something was stealing all the computer time, they told Silver.

Examining the printout, Silver was reminded of some simulations they had run on the computer on LM-1, the first lunar module, flown unmanned as part of Apollo 5 eighteen months earlier. At one point during the tests, they had turned off the simulator and "just let the lines hang open." On that occasion, Silver had discovered that the computer kept trying to find angles and sines and cosines that matched whatever the signal might be, however meaningless. The situation with Apollo 11 looked much the same to him. He asked the weary M.I.T. team whether they had checked the rendezvous radar angles. Yes, they had, and the angles were moving pretty fast. Every time they moved even a little bit, Silver reminded them, that was a cycle steal—the computer was recalculating from scratch—and it could eat up as much as 18 percent of the computer's time. What might be causing the rendezvous radar to calculate angles so obsessively? They hadn't by any chance had the rendezvous radar set on "Automatic," had they?

Steve Bales had stayed up long enough to watch the E.V.A. and then collapsed on a bunk in the controller's lounge. When at ten the next morning he took his place in the MOCR to work the ascent phase, he learned that SPAN still hadn't received an explanation of the computer alarm anomaly. Bales was surprised—he wouldn't have thought that any computer problem in the world could have stumped that team of people for that long, once they knew the symptoms. He was disturbed as well. In planning the flights, everyone had always assumed that the ascent would be more difficult than the descent. For a LEM going down, the surface of the moon was hard to miss. For a LEM going up, the C.S.M. was a small and elusive target. The prospect of lifting off with a computer that had been flighty and overburdened on the way down was not reassuring, and Bales was in no frame of mind for more dramatics. He had had enough excitement the preceding afternoon to last him for the rest of his career.

Bales was sitting at his Guidance console in the MOCR when the call came through, just thirty minutes before the scheduled liftoff: Tell the crew to turn the rendezvous radar switch to "Manual," and they won't have any more computer alarms. "I gotta believe you," Bales said, and passed the word on to Glynn Lunney, who was Flight for the ascent. Lunney had to believe it as well. It was one of the last messages sent up to the crew before liftoff from the moon: Please set the rendezvous radar switch to "Manual."

3

L.O.R. had been considered so unrealistic in 1961 because of the maneuver that was now about to take place for the first time: a launch-to-rendezvous by a spacecraft 240,000 miles from home.

The first reason why that prospect had been so forbidding is that spacecraft cannot move sideways any great distance. They can move up, down, forward, or backward, but for practical purposes they are locked into the plane in which they have been launched—sideways movement against that momentum is extremely costly in propellants. Therefore the first constraint of space rendezvous is that the planes of the two rendezvousing spacecraft be virtually identical. This requires, obviously, great navigational precision during powered flight to orbit.

The next problem is timing. Because spacecraft carry limited propellants, it is not enough that the two spacecraft be in the same plane; they must also be reasonably close to each other when the second spacecraft reaches orbit, which means that the second spacecraft must be launched within a narrow time interval. If both spacecraft are launched precisely from the equator (of earth or moon) into an orbit directly above the equator, then the timing is fairly simple: Once each orbit, the second spacecraft has a brief period of time when it can launch and be in position to rendezvous. But if the plane of both spacecraft is anything except precisely equatorial, then the timing of launch is complicated by the fact that the earth and the moon rotate—the source of a phrase familiar to television viewers of launches at Cape Canaveral, the "launch window." A spacecraft which can launch to rendezvous at time X cannot necessarily launch to rendezvous the next time the orbiting target spacecraft comes around; it might have to wait hours or even days, depending on the orbits involved. On a lunar mission, both spacecraft had tightly budgeted resources, and couldn't afford to wait long if a launch window were missed.

Even after the second craft has lifted off at the right time into the right trajectory, the pair of rendezvousing spacecraft face a number of additional obstacles. If a spacecraft gets into orbit and finds itself in the identical plane as its target but just half a mile back, the pilot cannot simply step on the accelerator to catch up, because increasing his speed will increase the height of his orbit. The same principle holds true whether he is above, below, behind, or ahead of his target: He cannot point the spacecraft in the direction he wants to go and turn on the engine. Unless he is within a few feet of his target, the astronaut must rely on precise tracking data combined with sophisticated computer programs to tell him what direction to point, the duration of the burn, and what throttle setting to use.

These are just a few of the reasons why, in 1961, space rendezvous was so daunting. Since then, rendezvous had been gradually demystified. In August 1965, the first rendezvous in history had been planned for Gemini V. Gemini V would release a pod with a small radar transponder, let it drift away, and then find it again. But early in its flight, Gemini V developed a problem with its fuel cells and the rendezvous attempt was delayed. By the time the problem was solved, the pod had drifted too far away to be found again. Rather than waste months of training, Houston and the astronauts forgot the pod and rendezvoused with a point in space.* Bill Tindall cheerfully consoled the men and one woman of Ed Lineberry's Rendezvous Section by telling them that, technically, it amounted to the same thing.†

Four months later, in December 1965, Gemini VI rendezvoused not with a phantom, but with another manned spacecraft, Gemini VII. This was followed by an unbroken sequence of successful rendezvous in Gemini VIII, IX, X, XI, and XII. The Gemini flights performed co-elliptical rendezvous and equi-period rendezvous; they rendezvoused from above and they rendezvoused on the first apogee. These were followed by the successful rendezvous of the LEM and C.S.M. in earth orbit on Apollo 9, and in lunar orbit on Apollo 10.

Thus on July 21, 1969, rendezvous itself was not the fearsome thing it

* When at first it appeared that there would be no rendezvous at all on Gemini V, the bitterly disappointed members of the little Rendezvous Section did the only sensible thing and retired to the Flintlock. At about ten o'clock that night, Bill Tindall called them there with the wonderful news: The fuel cells were functioning again and they could do a rendezvous—albeit with a point in space—after all. Tindall had it all worked out, and they should get back to Building 30 right away. By that time, most of the members of the Section couldn't even navigate to the Flintlock parking lot, let alone find their way to Building 30. As the story is told, the surviving members of the Section who straggled back to Building 30 that night proved not only that rendezvous could be done, but that you didn't even need to be sober.

† This is the first woman—her name was Cathy Osgood—we have mentioned whose role was something other than wife or secretary. There were just a handful of others (Rita Rapp, who ran the nutritional program for the astronauts, was apparently the highest ranking). Women engineers were exceedingly rare either in NASA or on the contractors' staffs.

had seemed at the beginning of the space program. The main concern now was that the ascent engine start. To that end, the ascent engine on the LEM had been made as simple as a rocket engine could be, with only two moving parts in the entire assembly. It had a magnificently reliable record in the ground tests.

Nonetheless, the first liftoff from the moon was a tense moment for everyone, just as Eight's entry into the first lunar orbit had been. For Mike Collins, circling the moon alone in *Columbia,* the lunar liftoff was the most nerve-wracking moment in his long flying career: During the six months since he had been selected to be the command module pilot, his secret terror had been that he might have to come back to earth alone. For Glynn Lunney, waiting at the flight director's console, this moment was an odd mix of emotions. As at liftoff for any flight, he felt tense ("trying to hold my breath for eight minutes"). But in an odd way, the ascent from the moon was not as draining on the flight director as a launch from earth or any other phase of flight where there was a choice of aborting or going ahead. With the lunar ascent, "there were no decisions to make, so in that sense it was easy. You just say, 'Well, let's light this sumbitch and it better work.' "

It did. After a stay of twenty-one hours, *Eagle*'s guillotines slashed the connections linking the descent and ascent stages, the ascent engine went instantly from a cold start to full power, the computer did all its work without lighting up any master alarms, and within three hours the *Eagle* was keeping station with *Columbia,* ready for an uneventful docking.

4

At dawn on Thursday, July 24, 1969, Apollo 11 splashed down in the south Pacific close to the Navy aircraft carrier *Hornet.* In the MOCR, which had passed control to the recovery team in the adjacent room when the parachutes deployed, a television image of *Columbia* bobbing in the water was projected up on the right-hand screen. The MOCR was jammed with people, waving small flags and smoking the traditional splashdown cigars. The large twenty-by-ten-foot screen, which for the past eight days had been showing trajectories to the moon and lunar-orbital tracks, went black. In a moment, it lit again, with these words:

I believe that this nation should commit itself to achieving the goal, before this decade is out, of landing a man on the moon and returning him safely to earth.

John F. Kennedy to Congress, May 1961

Since *Columbia* had splashed down, the right-hand screen had been displaying the mission patch of Apollo 11. Now these words appeared above it:

Task accomplished.

July 1969.

CHAPTER

26

"I think we need to do a little more all-weather testing"

Walter Kapryan, "Kappy," looked unhappily through the windows of the Firing Room at the skies over the Cape. It was Friday, November 14, 1969, the launch of Apollo 12 was scheduled for 11:22 that morning, and Kapryan was directing his first launch since taking over the Launch Operations Directorate from Rocco Petrone. Kapryan, a Langley man and an early member of the Space Task Group, had been deputy director of Launch Operations before his promotion. Petrone, who had moved into Sam Phillips's job as Apollo program manager when Phillips returned to the Air Force, had left Kapryan a daunting record to live up to. Petrone had launched every Apollo flight thus far at precisely the planned second—two Saturn IBs and six Saturn Vs: eight flawless launch schedules in a row.

The weather outside was iffy. Yesterday, it had been plainly unacceptable, with thunderstorms passing through the Cape all day. The skies had cleared during the night and it had looked as if the launch was safe, but now the morning had brought more clouds and intermittent rain. The pressing question was whether the current conditions fell within the weather rules for a launch.

371

Kapryan kept planes and weather balloons at varying altitudes above the Cape throughout the morning. He wasn't worried about the rain—short of torrential downpours, the Saturn V wasn't affected by rain—but either high winds or lightning would scrub the launch. All morning, however, the word coming back to the Firing Room was the same: Winds were light at all altitudes. The Sweeney meters they used for measuring lightning potential showed no problem. The countdown continued.

By eleven o'clock, Kapryan was still edgy. They were not in violation of the weather rules, but the cloud cover was at less than a thousand feet and the rain was getting worse; it just didn't feel like a good day to launch a Saturn V. But none of Kapryan's options was without risk. The Saturn was fully fueled and (after a quick fix of a minor fuel cell problem) working perfectly. The spacecraft was working perfectly. The astronauts—an all-Navy crew consisting of Pete Conrad, Alan Bean, and Dick Gordon—were already on board. To offload the propellants, reload them, and recycle the countdown was a major job that would introduce a variety of hazards. The spacecraft and launch vehicle were never going to be more ready than they were now. The next time, there might be a small, not-quite-in-violation problem with the vehicle that would worry Kapryan a lot more than the weather was worrying him today.

To increase the pressure, the President of the United States, Richard M. Nixon, was sitting in the small V.I.P. room, a triangular glass enclosure which jutted out over the Firing Room like the prow of a ship. Kapryan knew he couldn't let Nixon's presence affect his decision, but having the President watching from up there, waiting to see the advertised launch, didn't make Kapryan's job any easier. Kapryan silently asked himself, as he would in the last minutes before every launch he directed, What am I doing up here? Isn't there a better way to make a living?

At the rows of consoles below him, the countdown was as smooth as he could wish. Kapryan decided: As long as the rain didn't get worse and the wind didn't rise and there was no lightning, the rules said Launch, and that's what he was going to do.

1

In the MOCR at Houston, it was also a first-time day for Gerry Griffin. Griffin was part of the third generation of flight directors, brought in along with Milt Windler and Pete Frank in 1967. He was the first of his group to be named lead flight director and the first to be Flight on a launch phase. Still baby-faced at age thirty-four, an Aggie from Texas A&M with a high-pitched Texas twang, Griffin knew from the loops at the Cape that they were having weather problems in Florida and had doubted that they were going to launch. As the countdown continued toward T − 0, he was pleasantly surprised. In accordance with MOCR custom, he had no television image on any of his console's screens, but he assumed that the weather must be okay—otherwise, the Cape wouldn't go ahead.

At 11:22, right on schedule, the Saturn V fired up and rose from Pad 39A. The commander of Apollo 12, Pete Conrad, was an exuberant, playful man who felt no compulsion to pretend that he wasn't excited. His voice came over the loops full of joy. "This baby's really goin'!" he reported. "That's a *lovely* liftoff!" Only thirty-six seconds into the mission, Conrad's gleeful commentary was interrupted by loud, continuous static that drowned out all communication. Twenty-six seconds later, the static finally faded and Conrad came on the air again, his voice strong and clear, but a little breathless:

"Okay, we just lost the platform, gang. I don't know what happened here. We had everything in the world drop out."

In the MOCR, all Gerry Griffin knew was that the air-to-ground loop was filled with static and that neither he nor any of his controllers were getting any data from the spacecraft on their consoles. This did not necessarily constitute an emergency. Because of the vagaries of the ground stations and the many kinds of electromagnetic interference with which the communications system had to contend, it was not unusual to lose data or even voice contact with the spacecraft for a few seconds. Still, Griffin didn't like losing communication during the launch phase, always a worrisome time anyway. Then came Conrad's voice, reporting that the guidance platform was no longer functioning, which meant that the spacecraft now had no way of knowing where it was or what its attitude might be.

CapCom Gerry Carr acknowledged with a calm "Roger," which, in the absence of any data from the spacecraft, was the only thing he could do. Data were still coming in from the launch vehicle—the systems in the Saturn were completely independent from those in the C.S.M.—and the Saturn V was still pumping away, gaining altitude, maintaining the correct trajectory. But Carr didn't need to tell that to the astronauts, who were still being pushed back in their seats by the 3-g force of the S-IC stage.

Then came Conrad's next message: "I got three fuel cell lights, an A.C. bus light, a fuel cell disconnect, A.C. bus overload, 1 and 2, main bus A and B out." Conrad's voice was calm but strained. He was reporting that, for all practical purposes, the spacecraft was inoperative: all electrical power was down except for the emergency batteries which ordinarily were used only for entry.

Conrad had barely finished his sentence before Griffin was on the loop to his EECOM, John Aaron. "When things start going to worms," Aaron would point out later, the EECOM was the most likely person to know what was going on. Only something in the electrical system EECOM monitored could be causing the multitude of difficulties that Twelve was encountering. "How's it looking, EECOM?" Griffin asked. He fully expected Aaron to come back with a recommendation to abort.

John Aaron had intended to be a math and science teacher and raise Herefords. But upon graduation from Oklahoma's Southwestern State College in 1964 with a degree in engineering, he had applied to NASA on a whim, and a few weeks later someone in Houston had telegraphed an offer of a huge salary—or so it had seemed to an Oklahoma farm boy. Aaron hadn't known for sure what they were doing down in Houston, but he decided he'd go down and do whatever it was for a couple of years and then return to his Herefords. Since then, he had come to delight in the work, and he had already acquired a reputation within F.O.D. as one of the sharpest of the EECOMs. What he was about to do would make him a local legend.

Like other controllers, Aaron had set up his panel of buttons to light for various malfunctions and, ordinarily, he would have gone first to the lights to try to identify what had happened. But this time it was hopeless: "The whole place just lit up," Aaron recalled. "I mean, *all* the lights came on. So instead of being aids to tell you what went wrong, the lights were absolutely no help at all." He turned quickly to his screens.

The screens hadn't gone blank, nor had the numbers turned to zeros. Rather, it was as if the screen had suddenly gone from English to an unknown language. The hundred-odd parameters on the two screens still showed values, but values that didn't make any sense. Aaron, like everyone else in the MOCR, had no idea what had happened. Unlike everyone else in the MOCR, however, Aaron saw a pattern in the numbers, and the pattern was somehow familiar. Now all he had to do was remember why.

*   *   *

About a year earlier, Aaron had been sitting in the MOCR at midnight as part of a small team led by Glynn Lunney. They were "watching" a test at K.S.C.—just one more way of familiarizing themselves with their systems. The test was proceeding normally when the parameters on Aaron's screens suddenly changed to a strange pattern—not zeros, but an incomprehensible set of values. Then the numbers returned to normal.

The morning after the test, Aaron retrieved a hard copy of the anomalous screens from the computer and took it back to his office— there was no particular reason for doing so, just Aaron's uncommon curiosity. Aaron couldn't make any sense of the numbers he had seen. As persistent as he was curious, Aaron got Lunney to call the Cape and find out what had happened. The Cape wasn't pleased to have Houston call and demand to know how they had screwed up, but they disclosed nonetheless that a test conductor had accidentally dropped the power system on the C.S.M. to unusually low voltages.

Aaron went looking for one of the instrumentation specialists at M.S.C., trying to find out why the screen had reacted in such a peculiar way under low voltage. They spent hours on it. Finally, the instrumentation guy zeroed in on the signal-condition equipment, S.C.E., a box of electronics that performed an obscure role in translating the information from the sensors into the signals that went to the displays in the spacecraft and on the ground. It had a primary and an auxiliary position. In the primary position, where it was ordinarily set, it had a sensor that would turn the S.C.E. off under low voltage. In the auxiliary position, the S.C.E. would attempt to run even under low-voltage conditions. "You know," the instrumentation guy told Aaron, "that signal-conditioning equipment had tripped off because you were in primary. Now, if you'd gone to auxiliary, you would have wiped this circuit out and you would have got your readings back." Aaron thought that was interesting.

It is part of flight-controller etiquette to credit the back room reflexively. Whenever flight controllers are reminiscing about a memorable call, the formula is always, "Ol' Bill had been working with the boys in the back room, and they came up with . . ." Steve Bales, who had done that with Jack Garman on Apollo 11, put it best when he compared the men in the back and front rooms to two mountain climbers roped together. And virtually without exception, the assumption that the back room was involved in a major call is correct. In the case of John Aaron on Apollo

12, it is not. There was no time. When Griffin asked Aaron, "How's it looking?" Aaron was just starting to call his back room.

"Is that the S.C.E.?" he asked, already sure of the answer.

"Boy, I don't know, John," a worried voice came back, "It sure looks—"

Griffin, getting no answer to his first call to EECOM (he could not hear Aaron's exchange with the back room over his loop), tried again, needing an answer quickly: "EECOM, what do you see?"

Aaron cut off his back room and punched up the Flight loop. "Flight, EECOM. Try S.C.E. to Aux."

Griffin was surprised. In the first place, he was ready to call an abort, and was already preparing himself for that irrevocable step. In the second place, he had no idea what "S.C.E." referred to. Never in any of the simulations or the Mission Techniques had that switch been mentioned. Griffin wasn't sure he'd heard Aaron right, and in fact he hadn't.

"Say again. S.C.E. to 'Off'?"

"Aux," corrected Aaron. The MOCR's was truly the most economical language in the world.

Griffin played it back to Aaron, needing to be sure: "S.C.E. to Aux."

As he confirmed, Aaron loosened up, using two whole words: "Auxiliary, Flight."

Griffin still had no idea what Aaron was talking about, but once again trust made the system work.

"S.C.E. to Aux, CapCom," Griffin said.

CapCom sat immediately in front and to the left of the flight director. Carr turned his head and looked back up at Griffin with a "What the hell is that?" expression on his face, but his was not to question why. In this situation, CapCom had two responsibilities: to communicate clearly and to radiate confidence. Carr did both, his voice sounding as if he were relaying a standard procedure that would make everything okay: "Apollo 12, Houston. Try S.C.E. to Auxiliary. Over."

"What panel, EECOM?" Griffin asked. He was asking Aaron on which instrument panel in the spacecraft this hitherto unknown switch was located—if Carr, an astronaut, didn't seem to know what S.C.E. referred to, it was entirely possible that Conrad and his crew didn't either.

Pete Conrad, riding on top of a Saturn V in a spacecraft whose alarm panel was lit up like a pinball machine, seemed as mystified by the instruction as Carr and Griffin had been.

"N.C.E. to Auxiliary," he said dutifully.

"*S*.C.E., *S*.C.E., to Auxiliary," CapCom repeated. This time the crew heard right. Al Bean, the lunar module pilot, knew where the S.C.E. switch was, and clicked it to the position labeled "Auxiliary."

Griffin was still worried that the crew didn't know how to put S.C.E. to Aux and asked again, "What panel, EECOM?" But by this time everything had changed.

"We got it back, Flight," Aaron said laconically, meaning that the data—interpretable data—had come back up on his screen. "Looks good." Now that he had data, Aaron could also deduce that the fuel cells that powered the C.S.M. had, for reasons unknown, been disconnected.

A minute and fifty seconds had passed since launch; thirty seconds since Conrad had first reported his status. Sitting in the viewing room, Bill Tindall looked at the people sitting around him, listening to the flight director and air-to-ground loops. The V.I.P.s—headquarters people, wives, politicians, celebrities—were laughing and talking, relaxing now that the tension of liftoff was over. Nothing they had seen in the room beyond the glass, nothing they had heard in the voices of the controllers speaking their incomprehensible language, had hinted to them that anything was wrong. One of the senior officials from F.O.D., who like Bill Tindall knew something was badly wrong, tried to quiet the others so he could hear.

"My first thought was, I can't believe this is happening to me," Gerry Griffin recalled. "It wasn't panic at all, more sort of a feeling of 'My gosh, what's goin' on here?' " Every other flight director who had launched a Saturn V had had an easy time of it, comparatively speaking—except for Charlesworth, who had lost the two engines on Apollo 6. But that had been on the second stage of an unmanned vehicle.

Griffin glanced at one of his console's screens. The five F-1s on the first stage, close to the end of their two-and-a-half-minute life, were still burning unperturbed. He decided that even if they still had to abort, there was no sense in hurrying; at least this way they were gaining altitude, and with altitude Griffin also gained time and options if he had to bring the crew down. He went around the horn, polling his controllers to see whether they were go for staging. As soon as he had completed that, EECOM was on the loop again.

"Try to put the fuel cells back on the line and if not, tie bat [battery] Charlie to Main A and B," said Aaron.*

Carr passed the word to the crew. It was only seconds later when EECOM reported again to Flight that Fuel Cell 1 and Fuel Cell 2 were on line. Fuel Cell 3 would come up again within seconds. The entry batteries could be taken off line. The spacecraft was alive again.

Half a minute later, after the S-IC had staged and the S-II had taken over, Pete Conrad came back on the air.

"Okay, now we've straightened out our problems here," he reported cheerily. "I don't know what happened; I'm not sure we didn't get hit by lightning." The crew had seen a bright white light at the time their problems began.

A few seconds later, the irrepressible Conrad was back. "I think we need to do a little more all-weather testing," he said confidentially.

"Amen," replied CapCom Carr. "Amen."

The onboard tapes reveal that Conrad, Gordon, and Bean laughed the rest of the way into orbit, making jokes about the way every alarm in the spacecraft had gone off at once.

Down below, Griffin had heard Conrad's speculation about the lightning. Dang it, I bet that's what happened, Griffin thought to himself.†

"The one thing we missed was the Ben Franklin situation," Petrone mused. "Somehow that had never entered into our discussion." What they had done, they realized later, was to launch a 363-foot lightning rod, with the equivalent of a copper wire in the form of a trail of ionized gases running all the way down to the ground. Even though there was no lightning in the vicinity before launch, Apollo 12 could create its own. And that is exactly what it did, discharging the cloud into which it had entered. A few of the soaked observers at the viewing area could see it even from four miles away, a long flash of lightning down the exhaust plume to the steel of the umbilical tower.

Actually, Don Arabian's anomalies team later determined that *Yankee*

---

* "Main A- and B" referred to Main Bus A and Main Bus B. A "bus" (a term which will recur in the story of Apollo 13) may be thought of as a set of wall plugs. The fuel cells generate electricity and feed it into the buses; the various pieces of equipment in the spacecraft draw electricity out of the buses.

† Texas Aggie jokes were to Texas what Polish jokes were to the rest of the country. Even before Apollo 12 left earth orbit, people in F.O.D. were going around saying, "Did you hear the one about the Aggie who launched a Saturn V in a thunderstorm . . .?"—grossly unfair to poor Gerry Griffin, who had nothing to do with the launch decision.

Clipper—the Apollo 12's command module—was hit twice by lightning, once 36.5 seconds after launch at an altitude of 6,000 feet, when it discharged the cloud it was flying through, and again 16 seconds later, at an altitude of about 14,000 feet, when it triggered a cloud-to-cloud bolt. Each time, the lightning acted on *Yankee Clipper* in two ways, as Arabian described to a visitor (accompanied by a vivid reenactment of the crew's reaction and the sound of the alarms). First, the lightning itself, a jolt of 60,000 to 100,000 amperes, flowed through the metal exterior of the Apollo stack from top to bottom. The only damage it did, however, was to destroy some external instrumentation for measuring temperatures and R.C.S. reserves, none of which was critical. Because of the way the Apollo stack was electrically grounded, the direct charge did not penetrate the spacecraft itself. In addition, though, the lightning induced electromagnetic fields within the stack. Induced voltages and currents, powerful but not nearly as devastating as a direct jolt, raced through the electrical circuitry and, among other things, knocked the fuel cells off line and caused the guidance platform to begin tumbling. As it turned out, the shocks were not powerful enough to destroy the circuits.*

2

Once in orbit, the crew was preoccupied first of all with getting the guidance platform realigned.

The platform was at the heart of the Inertial Measuring Unit (I.M.U.), a metal sphere about the size of a beach ball located beneath the center couch of the command module. Within the I.M.U.'s casing were three nested spheres called "gimbals." The gimbals, each controlled by a gyroscope, functioned like a gimbal in a boat that lets a table surface remain unaffected when the boat heels. At the center of the three gimbals was the "platform"—the table surface—which had been aligned before

---

* The reason Apollo 12 was able to get to orbit was that the guidance system for the Saturn V, buried within the Instrumentation Unit (I.U.) at the top of the S-IVB stage, was unaffected by the lightning. If its platform had tumbled, the Saturn would have gone out of control within a few seconds. Part of the reason the spacecraft was so affected by the lightning while the Saturn was not involved the spacecraft's greater exposure—it was positioned like the tip of a lightning rod—and part of it was luck, as Arabian emphatically pointed out. Neither the spacecraft nor the launch vehicle had been designed with lightning in mind. In the case of the I.U., induced currents reached the guidance system's circuits but the computer software kept the platform from tumbling. In addition, the incident on Apollo 12 dramatically vindicated the decision early in the 1960s to have separate guidance systems for the spacecraft and the launch vehicle.

launch. From the moment of launch, each movement in the spacecraft was measured by accelerometers and registered in the guidance system, which "added" that movement onto the previous movements it had recorded. In other words, the guidance system always knew where it was relative to some specific zero point. When the main power buses went out during launch, the gimbals lost their balance, as it were, and the result was that the platform physically tumbled.

The crew of Apollo 12 had to realign the platform through star sightings taken on board *Yankee Clipper*, a task that was never easy but was particularly difficult when the spacecraft was not (as for sightings during a nominal mission) in deep space with lots of time. The crew, already busy checking out the spacecraft, had somehow to work in the star sightings during the half of each orbit when they were in the earth's shadow. With much effort, and at nearly the last opportunity, Command Module Pilot Dick Gordon succeeded, and the onboard guidance system could take over from the I.U. in the S-IVB stage that had been doing all the work up until then. This left the Apollo's managers still confronting the larger question: Should Twelve proceed to the moon?

As EECOM, the man responsible for the environmental and electrical systems, Aaron was once again the man on the spot. This time it wasn't just Gerry Griffin coming to him, but Chris Kraft, descending from the fourth row. "Now, young man," Aaron remembered Kraft saying to him, "you've got an hour and a half to figure out whether that spacecraft's ready to go to the moon or not." They could take no longer than that, because they couldn't let the spacecraft continue in earth orbit longer than three revolutions. On each revolution, the earth was changing position under them; after three orbits, the ground stations would be unable to track the burn, emergency recovery vessels would be in the wrong parts of the world's oceans, and the spacecraft itself would be out of position for the lunar trajectory needed to conduct the mission.

"[Kraft] really put it on me," Aaron said. By this time, they were certain that the cause of Apollo 12's problems had been lightning. But how does one determine in a hour and a half that a spacecraft which has been struck by lightning is sound enough to go to the moon? It was not the kind of exercise for which the flight control team had ever trained. Griffin and Aaron didn't have time to invent a procedure, so they decided to section out the piece of the flight-maneuver checklist that the crew would ordinarily have used before burning into lunar orbit. Their logic was that no maneuver in the flight was more critical than the L.O.I. burn—if the spacecraft was in good enough shape to pass the L.O.I.

checklist, it was good enough to proceed out of earth orbit. They worked through the checklist with Conrad's crew, testing the propulsion systems, gimbal systems, gyros, computers, counting down to the last few seconds before actually firing the S.P.S. Everything worked.

During that same hour and a half, the MOCR's support network was working furiously, concentrated in Building 45 and at the North American Rockwell plant in Downey. (Bethpage, Long Island, where Grumman was located, couldn't do much except speculate: The LEM was still housed in the SLA, inaccessible.) Watching the data in the MER, Arabian reached the conclusion that the spacecraft looked fine except for the destroyed instrumentation, which had no bearing on the mission, and there was no reason not to continue. For Arabian, it was just one more instance where you examine the physics of the situation: "You do whatever the data say. All the systems seemed all right, so go ahead." The MER passed its recommendation along. To Bill Tindall, watching the data, it looked as if they had "a normal spacecraft," once they had reset the platform. But, it was later pointed out to him, a large and undesigned-for electrical charge had run through every circuit in *Yankee Clipper*. "Aaah—who cares?" Tindall replied, with a big laugh.

Griffin recalled that Kraft had tried to take some of the pressure off—"Don't forget that we don't *have* to go to the moon today," he had said to Griffin. But the determinations that flowed to the flight director's console all said the same thing. "We kept clicking off that checklist," Griffin recalled, "and when we got to the end, we all kind of said, 'Well, there's this unknown about a few things—we don't know where all that stray electricity may have run around in the cabin—but everything we can check looks okay. Is there any reason not to go?' And we looked at each other and said, 'Hell no, let's go.' " Griffin turned around to Management Row and told Kraft that's what they were going to do.

It was a unique test for senior management who were close enough to the day-to-day operations to have a reasonable technical grasp of the situation—specifically, the directors of Flight Operations (Kraft), Flight Crew Operations (Slayton), ASPO (McDivitt), and the Apollo Program itself (Petrone). They had never overridden a flight director's decision during a mission, but never before had a strategic choice of this nature presented itself. There was an argument for terminating the mission no matter what the data said. The unknown in Griffin's mind was in theirs as well. The spacecraft hadn't been designed for the treatment it had gotten during launch. Why take a chance?

Kraft was in Houston; Slayton, McDivitt, and Petrone were at the Cape. They had been conferring over the phone lines. Apparillty there was no argument and no agonizing. There was one awful possibility that they couldn't check out with telemetry: Conceivably, the electricity had blown the pyrotechnics that operated the entry parachute system. But that was extremely unlikely, and—thinking back on it Petrone shrugged fatalistically—if that had happened, what difference would it have made to their decision about going to the moon? The crew wouldn't have come back alive no matter what they did.

"We had a very short time to discuss it," Petrone recalled. "But all of our minds were saying the same thing. You're up! The big part of the mission had been accomplished, and everything was working, you've got the table lined up, everything's pumping, everything looks good on board. Houston on the ground says, 'Everything looks good here.' So we all said, 'Hey—obviously.' No big discussion. Wasn't needed. Once you've got all those other parameters falling in line. . . . Let's go." So when Griffin told Kraft that he was going to go for T.L.I., Kraft just nodded.

But suppose the situation had been posed as a scenario on the ground, when writing mission rules. Under those circumstances, would Tindall's Mission Techniques meetings have come up with a rule saying that if the spacecraft is hit by lightning and the electrical system goes down and the platform tumbles our Standard Operating Procedure will be to conduct an hour-and-a-half check and, if nothing seems wrong, continue? Well, said an Apollo veteran, now a senior NASA manager, maybe not. Or suppose, twenty years later, after the *Challenger* accident, that a comparable situation were to occur with the shuttle. Would NASA go ahead with the equivalent of a translunar injection? He laughed aloud at that one. Not a chance. But this is now, he said. That was then.

3

When Apollo 11's *Eagle* had landed on the Sea of Tranquillity four months earlier, no one knew at first where it was. "We were all scrounched around trying to figure out where the hell it was," Tindall recalled. "Kraft was there, and Sam Phillips was there, and they were talking to the scientists in the back room," trying to pick up clues from Armstrong and Aldrin's description of the terrain. "It was a comedy." Phillips turned to Tindall, whose Mission Techniques determined the landing procedures. "On the next mission," Phillips said firmly, "I want a pinpoint landing." "So help me God," Tindall recalled, "that's what he said. I thought it was impossible."

Tindall convened his lunar landing group. The task looked impossible, because the moon had proved to be unmanageably lumpy, in a gravitational sense. "Mascons," uneven concentrations of the lunar mass, constantly introduced unexpected deviations into the Apollo lunar orbits. Their effects were too complex to be fully modeled.

After they had wrestled with the problem for a few meetings and rejected a few unworkable ideas, Emil Schiesser, a soft-spoken, intense young man who was one of MPAD's experts on deep-space navigation, introduced an idea. As the spacecraft came out from behind the moon, it was heading toward the tracking stations on earth. And, as everyone knew, this created a Doppler effect that was used for tracking the vehicle.*

As the LEM continued on its circular orbit, it was no longer approaching the earth's tracking stations, but was moving at an increasing angle relative to them until, as the LEM headed back around the moon again, it was moving directly away from the earth. During its entire period, measurements of the LEM's Doppler effect from the earth showed a predictable pattern.

Now came Schiesser's imaginative leap: Suppose that we forget about trying to model perfectly the effects of the mascons, he suggested, and concentrate instead on modeling what the shifts in frequencies should look like during the course of a landing at point X, from the time that the LEM appears around the edge of the moon until touchdown. With this predicted pattern of frequencies in front of us, we can watch what the actual frequencies are, and calculate the difference. Then we can use the difference between the predicted and the actual frequencies to decide how far off target we are. It was, Tindall reflected, "astounding"—simple and obvious after you heard it, as elegant solutions seem always to be.

No matter what the source of navigational errors—mascons, venting of the spacecraft, changes in trajectory from firing of the R.C.S. thrusters, or an imprecise burn—Schiesser had given them a way to determine

* The Doppler effect is the apparent change in the frequency of a light wave or a sound wave—or, as in the Apollo case, a radio wave—caused by a change in the relative position of the source of the wave with respect to the observer. It is the effect that a person standing beside a railroad track hears in the changing pitch of the whistle of a train while it is approaching, passing, and then moving away.

precisely how much they needed to change the planned course of a descending LEM. Now what they needed was a way for the LEM to use that information to achieve the pinpoint landing that Phillips wanted. One option was to update the "state vector" in the onboard computer— the information required to compute the vehicle's precise position and velocity. But a state vector is expressed in seven pieces of information, and asking the lunar module pilot to input seven new nouns, each of them about eight digits long, during the middle of a lunar landing was far too unwieldy.*

The group came up with an alternative. ("I didn't, of course," Tindall said quickly; "I never did anything. I just got the people together to do these things.") Don't try to tell the LEM where it really is, went this bright idea. Just tell it that the landing site has moved. If they found that the LEM was coming down 800 feet short of the landing site, they would tell the computer to land 800 feet farther down-range than it had planned. To implement this procedure required only one number—"Noun 69," they decided to call this piece of crucial targeting data.

MPAD's mathematicians went off to model the Doppler readings they needed for Apollo 12's target; the computer people went off to modify the onboard computer's software; the flight operations people went off to devise crew procedures and begin simulations. Elegant as Schiesser's procedure was in theory, it was far from simple to put into practice. The earth did not stand still while the tracking stations measured the Doppler effect; it rotated, and the effect of the earth's rotation had to be taken into account. The tracking stations handed off from one to another as they rotated out of position, and each of them was at a different latitude; the effects of the angle at which they received the signals from the LEM had to be incorporated. The time measurements had to be so precise that only a nuclear clock was sufficiently accurate. And all of these changes had to be completed and practiced during the interval between the July landing of Apollo 11 and the November launch of Apollo 12.

At the Mission Techniques meetings, FIDO Dave Reed, who would be handing the lunar descent for Twelve, talked with Pete Conrad about the target strategy. Conrad was supposed to land his LEM close enough to Surveyor III, an unmanned probe that had soft-landed on the moon three years earlier, to enable him and Al Bean to walk over to it and bring pieces back with them. The object was to obtain information about

---

* The astronauts instructed the computer by means of numbers representing "nouns" (data points) and "verbs" (procedures for using the data).

the effects of exposure to the lunar environment over a period of years. Since Apollo 12 did not have a battery-driven Lunar Rover (which would not be available until Apollo 15), the maximum landing error that would still permit Conrad to reach the Surveyor was on the order of 2,000 feet. Even if the navigation on Twelve was markedly improved, it seemed implausible that they could get that close on the second manned landing—Armstrong had been more than four miles off target. So the conversations between Reed and Conrad had an air of unreality about them.

"Where do you want me to put you down?" Reed asked Conrad. The first time, Conrad told Reed to put him on the far side of the crater in which Surveyor would be resting. They ran a couple of simulations that way, with Conrad in the LEM simulator watching an image of Houston's plaster-of-Paris lunar surface as he came in over a configuration of three craters they called "Snowman," which would tell him he was on the right track. Then one day Conrad told Reed he didn't want to walk that far. Put me on the near edge of the crater, Conrad said. So Reed changed the parameters in the software. Watching Reed calculate these minute changes, apparently in all seriousness—the crater itself was only 600 feet across—Conrad finally told Reed. "You can't hit it anyhow! Target me for the Surveyor." "You got it, babe," Reed said, and he proceeded to enter coordinates that would land the LEM precisely on top of the Surveyor if Schiesser's bright idea worked.

At midnight Houston time, November 18, four days after *Yankee Clipper*'s encounter with lightning, Emil Schiesser stood behind two Martin contractor personnel at their consoles in a corner of the first floor of the Control Center, near the computers. As the LEM *Intrepid* appeared at the edge of the moon, the screens began to fill with the tracking data they had been waiting for. The three of them then began filling out their cheat sheets, more formally known as Procedures Sheets—they looked something like tax forms—copying numbers from the screen. Then, as Conrad and Bean streaked across the face of the moon under powered descent, the three of them began figuring out the value for Noun 69—by hand. The Control Center's computers didn't know how to do something as simple as multiply two numbers, Schiesser said, and they hadn't bothered to bring in a mechanical calculator. They scratched out their calculations, passed the number to the Trench, who gave it to Flight, who told CapCom to transmit it to the crew.

*   *   *

As *Intrepid* descended through 7,000 feet toward the lunar surface, it pitched to an upright position and Pete Conrad and Al Bean got their first look at the approach to the landing site. In front of them, right where it was supposed to be, was Snowman. "Hey, there it is!" announced Conrad. "Sonofagun, right down the middle of the road!" They had a chance, at least, of getting close to the Surveyor. As the LEM continued past Snowman, Conrad began to realize just how close they were going to come: "Hey, it [*Intrepid*] started right for the middle of the crater," he cried. "Look out there! I just can't believe it! Amazing! Fantastic!"

Dave Reed had still been a little off. *Intrepid* wouldn't have actually landed on top of Surveyor if Conrad had let it alone—it would have been perhaps as much as 150 feet away. That was close enough to shower the Surveyor with dust—too close. Conrad took manual control, steering the LEM across the crater to a safe distance—on the far rim after all.

A few years later, after the four LEMs following *Intrepid* had each descended to within a few yards of their targets, Neil Armstrong was giving an interview to a historian of the manned space program. The interviewer, thinking of astronauts and senior managers, asked Armstrong: After all the years he had spent at M.S.C., who did he think stood out in talent and ability? Neil Armstrong grinned. "Emil Schiesser!" he replied. "I'd vote for Emil every time."

CHAPTER

27

"You really need to understand that the C.S.M. is dying"

When Apollo 13 lifted off on April 11, 1970, manned space flight was one day shy of the ninth anniversary of Gagarin's flight. In the intervening years, the actual flights had been unexpectedly safe. Thirty-seven times, men had sat atop rockets and been blasted off into space; thirty-six times, they had returned safely. Only once—if one discounted rumors of unacknowledged Russian catastrophes—had a flight resulted in a fatality: In April 1967, Vladimir Komarov of the Soviet Union had perished after his parachutes failed to deploy properly during an emergency entry. Of the thirty-six crews that had returned safely, the Americans had launched twenty-two—six in Mercury, ten in Gemini, six in Apollo.

So perhaps it was time for something to go wrong. Or perhaps Providence was rebuking the Apollo engineers for the sin of hubris. Or perhaps the superstitious were right, and NASA was asking for it. Flight number 13? Launched at the thirteenth minute of the thirteenth hour of the day, Houston time?* What else could one expect two days into the flight, when April 13 arrived, but that something awful would happen?

* People who set launch times aren't thinking about superstitions, NASA officials said later.

But while Apollo 13 was objectively a failed mission, it was something else altogether to the people who were involved in it. For Glynn Lunney, thinking back to the long development of operations that had begun with such halting steps in 1959, "Apollo 13 was the crowning achievement"—not just for the Control Center people, but for the people flying and all the people on the ground. It was crowning, because it was ultimate—no crisis that manned space flight faced in the future could be much worse, Lunney thought, and still survivable—"We were as close as you get to the edge and are still able to pull back."

1

The crisis on Thirteen began on the third evening of the flight. The command module *Odyssey* and the lunar module *Aquarius* were outward bound, 205,000 miles away from the earth. Jim Lovell, veteran of two Gemini flights and Apollo 8, at that time the most experienced American astronaut, was commanding the flight. Fred Haise, the lunar module pilot, was making his first space flight, as was Jack Swigert, the command module pilot. Swigert was in a unique situation. As part of the backup crew for Thirteen, he had been pressed into service just three days before launch when it was discovered that astronaut Charlie Duke, who had been working with the prime crew, had contracted German measles. Lovell and Haise were immune, but the prime CMP, Ken Mattingly, was not.* Lovell fought to keep Mattingly on the crew—even if he got the measles, they would be on the way home in trans-earth coast when it happened, Lovell argued, and measles weren't that bad anyway—but Administrator Paine finally sided with the doctors and insisted that Swigert replace him. Swigert had done well during the training that he and Lovell had packed in during the two days before launch, and he got along fine with Lovell and Haise. Still, he was anxious that he do his job just right. So far, the only problem had been getting good readings on the quantity of hydrogen left in one of his storage tanks.

At nine o'clock that night, Houston time, Lovell and the rest of the crew completed a television broadcast. None of the three networks had carried the show—by April 1970, flights to the moon were old hat, as

---

* Mattingly, who didn't get the measles, was command module pilot on Apollo 16. "CMP," by the way, was pronounced "simp" by the controllers. The lunar module pilot, LMP, was "limp," and the commander, getting a little respect, was C.D.R.

were ill-lit pictures of whiskery astronauts floating around in their spacecraft. The main audience had been the people at M.S.C., watching on the television monitors scattered around the Center. In the MOCR, flight director Gene Kranz had permitted the O&P officer to throw the television image onto one of the screens on the front wall. With the spacecraft safely on course, the LEM docked with the command module, and nothing much to do for twenty hours until L.O.I., this was one of the laziest shifts during a lunar flight.

Seymour (Sy) Liebergot, a Californian, almost elderly by MOCR standards (thirty-four years old), was the White Team's EECOM for Thirteen. As the television show ended, he got onto the loop to Flight to ask for a "cryo stir."

"Cryos" referred to the cryogenics: two tanks each of liquid oxygen and liquid hydrogen which produced the spacecraft's electricity, oxygen supplies, and water. Chilled to their liquid state because that was the only way to store a sufficient amount for the lunar journey, the O_2 and the H_2 (as the controllers referred to them) were fed into three fuel cells, which converted them to electricity and, as a by-product, drinking water. The electricity passed into two main electrical buses, A and B, from which the equipment on the spacecraft drew power.

The system was so simple and so redundant that it seemed foolproof. Each of the two tanks of oxygen and hydrogen was adequate to last the entire mission if something were to happen to the other one. Any two of the three fuel cells could meet all the spacecraft's needs, and just one of them could bring the spacecraft home. There were two main buses, either one of which could if necessary distribute power to all the spacecraft's equipment. Each of these redundant systems had (almost perfectly) the best of both worlds: They were independent insofar as potential failures were concerned, but linked so that they could take over one another's duties. Piping, valving, wiring—everything was designed so that problems could be bypassed. Short of being hit by a meteoroid or something equally improbable, it was thought that the Apollo spacecraft could never run short of water, oxygen, or power.

Liebergot had decided this would be a good time to help Swigert get a good quantity reading on the H_2 tank that had been giving him problems. To do that, the astronauts would turn on two small fans in each of the four cryo tanks. The gases tended to settle into layers with different temperatures and densities, and stirring with the fans was necessary to produce the homogeneous tank temperature necessary for an accurate reading. It was a routine procedure which they ran once a day. Tonight's was an extra one prompted by the balky H_2 tank.* At 9:07 P.M., CapCom Jack Lousma passed the request for a cryo stir to the crew.

Three men were working in EECOM's back room that shift: Dick Brown, the electrical power systems (E.P.S.) specialist; George Bliss, an environmental control system (E.C.S.) specialist; and Larry Sheaks, another E.C.S. specialist. It was not customary for the astronauts to tell the ground precisely when they flipped the switches that turned on the fans, so Liebergot and his two E.C.S. men began watching the H_2 quantities carefully as soon as the request was passed up. The numbers in question were located at the bottom-right corner of a screen called "CSM ECS CRYO TAB."

At 9:08, the conversation on the EECOM loop, which until then had been quiet and desultory, suddenly changed. In the background, on the air-to-ground loop, Swigert could be heard saying, "Okay, Houston, we've had a problem." Immediately thereafter, Larry Sheaks cried out indignantly, "What's the matter with the data, EECOM?!" "We got more'n a problem," chimed in Dick Brown. Liebergot himself was now looking at a second screen on his console, called "CSM EPS HIGH DENSITY," which showed the status of the electrical system. "Okay, listen you guys," Liebergot said to his back room, "we've lost Fuel Cell 1 and 2 pressure." George Bliss ran his eyes over the cryo screen, and then reported the rest of it: "We lost O_2 Tank 2 pressure. And temperature." The crisis had begun.

1 minute. To the left of Liebergot and up a row was Gene Kranz. At 9:07, he and his White Team were entering the last hour of their eight-hour shift. The conversation over the flight director's loop had been leisurely, as he and Guidance discussed whether to uplink some data into the onboard computer directly from the ground, or to read it off to the crew and let them enter the information themselves. Kranz was listening to someone else when Swigert said "we've had a problem," but he clearly heard Guido Will Fenner report a few seconds later: "We've had a hardware restart. I don't know what it was."

"Okay," said Kranz. "G.N.C., you want to look at it and see if you see any problems?" A hardware restart indicated some unusual event that the computer had detected and was checking. Then Lovell's voice stopped Kranz short:

* If Liebergot had not ordered the extra cryo stir, the accident would not have been prevented; just postponed to the next day.

"Houston, we've had a problem. We've had a Main B Bus under-volt." An "undervolt" meant a substantial reduction of power into Bus B, jeopardizing the equipment running off it. Kranz went to his EECOM: "You see an AC Bus undervolt there, EECOM?"

"Negative, Flight." This was true—Liebergot had not seen the undervolt—but his answer was not entirely forthcoming. Liebergot was already looking at several other serious problems, and he wanted a few seconds to sort them out.

Kranz persisted: "I believe the crew reported it." Even as he spoke, Brown was telling Liebergot from the back room that he had seen an undervolt as well.

"Okay, Flight," Liebergot amended. "We got some instrumentation problems. Let me add them up."

Liebergot wasn't the only person who thought he had instrumentation problems. All around the MOCR, controllers were seeing strange indications on their screens. INCO (the instrumentation and communications systems officer, formerly known as TelCom) reported to Flight that the high-gain antenna had switched to high beam. Kranz, who was as baffled as everyone else by what was happening, told INCO to try to pin down the exact time of the change, hoping that it would give him some clues. Then G.N.C. Buck Willoughby was on Flight's loop, telling him that the spacecraft was changing attitude in unexpected ways. The helium valves on some of the R.C.S. (reaction and control system) jets that controlled the spacecraft's attitude were acting as if they were closed. They should be open. Kranz fielded this strange collection of problems, wondering what in the hell had happened up there.

When Liebergot told Kranz that he would add up the instrumentation problems, he did not mention that he was also looking at a loss of pressure in two of the three fuel cells. This was true to flight-controller custom: You didn't go to Flight with a problem until you were reasonably confident that it was a problem. In this case, the probability that two independent fuel cells would go bad at the same moment had to be somewhere in the vicinity of six or seven decimal places. They could, however, appear to go bad at the same time because of glitches in the electrical system, and that, Liebergot thought, was surely what had happened.

At this point, Liebergot was assuming that the reading for O_2 Tank 2 pressure was another instrumentation problem caused by the same electrical difficulties that had produced the appearance of low fuel cell pressure. The tanks, metal spheres twenty-six inches in diameter, contained nothing but liquid oxygen and a few simple and reliable pieces of equipment—a heating element, two fans, and a few sensors. Nothing much could happen to them. And even if something had gone wrong with O_2 Tank 2, that still left O_2 Tank 1, which would automatically feed all three fuel cells and maintain full power even if Tank 2 quit working. If Liebergot wanted to treat the oxygen tank as the cause, he had to hypothesize a failure that (a) had disabled Tank 2, (b) left Tank 1 still functioning, but nonetheless (c) somehow disabled two fuel cells. That didn't make sense. The spacecraft wasn't built that way. It was doubtful that such a failure could even be simulated—what could a SimSup use as a cause? The alternative and much more plausible explanation was that Liebergot was seeing a problem in the electrical system. Perhaps it was a pure instrumentation phenomenon, and the readings were lying to him. Perhaps the problem was producing interference in the power flows, in which case they would find some way to bypass the errant circuitry.

Liebergot's instrumentation theory was instantly reinforced. The Main Bus B undervolt warning light on his panel blinked off. Swigert reported the good news from *Odyssey*: The voltage was looking good. Still, he added, "We had a pretty large bang associated with the caution and warning there." Shortly thereafter, Lovell reported that the jolt must have "rocked the sensor" for the O_2 quantity in Tank 2. It had been down around 20 percent, but now was "full-scale high" again.

And so it went for the first few minutes—conflicting data, a stream of seemingly unrelated problems. Around the MOCR and in the back rooms, controllers tried to put the pieces together, and they just didn't fit.

John Aaron was at home, shaving off the day's stubble after spending a long shift in the Control Center. Arnie Aldrich, chief of the C.S.M. Systems Branch in F.O.D., called on the phone from SPAN. "John, there's something funny going on here," he said to Aaron. "A lot of the guys think it's instrumentation problems and flaky readouts." Aldrich had a long cord on his phone, so Aaron asked him to walk along the rows of consoles and read him some data. Aldrich went to the SPAN consoles and read off some numbers. Aaron would later realize that being away from the MOCR, not glued to his own console, made it easier for him to see. In fact, standing in the quiet of his bedroom with shaving cream on his face, it was all quite clear. "Arnie, that's not an instrumentation problem," he said to Aldrich. "It's a real problem."

$*$ $*$ $*$

5 minutes: The difference between EECOM's situation and everyone else's was that the other guys' problems were fairly minor. EECOM's problems were either trivial (if it was bad instrumentation) or life threatening (if the electrical and life-support systems of the C.S.M. were going dead). Like Bales on Eleven and Aaron on Twelve, Liebergot was the man sitting out in the open, the man with the best opportunity to make the mistake that would, in another MOCR euphemism, "blow the whole mission." The difference between Liebergot's situation and Bales's in Eleven or Aaron's in Twelve was that Bales and Aaron had had the option of stopping everything immediately and coming home—Armstrong could have called off the landing, Conrad could have fired the escape tower. Liebergot was stuck. If it wasn't an instrumentation problem, this crisis had no easy way out. There might be no way out at all.

Kranz inquired of Liebergot how things were going. Liebergot replied that the crew were "flipping their fuel cells around"—trying to reconfigure them. Kranz was impatient: "Well, let's get some recommendation here, Sy, if you got any better ideas."

Just then Swigert reported the inexplicable and disconcerting news that they were getting an undervolt on Main Bus *A*, the one that until now had still been working. "Sy," Kranz said, the exasperation in his voice now unmistakable, "what do you want to do? Hold your own and—" Kranz broke off to listen to the spacecraft, which was adding that Main B was "reading zip," distributing no power at all. Still nothing from Liebergot. "Sy," Kranz said, "have you got a sick-sensor-type problem there or what?"

Liebergot had been busy talking to his back room. He now came onto Flight's loop and recommended that the crew try to reconnect the fuel cells which, Liebergot hoped, had been thrown off line by the mysterious jolt. CapCom passed the instruction up to the crew while Liebergot returned to the EECOM loop which connected him to his back room. Remembering that O_2 pressure reading, Liebergot called to Sheaks. "Larry, you don't believe that O_2 Tank 1 [sic] pressure, do you?" Sheaks came back confidently. "No, no. Surge tank's good. Manifold's good, E.C.S. is *good*." Reassured, Liebergot went back to the screens and to a discussion of the electrical problem with Dick Brown.

Odyssey came back on the line. "Okay, Houston, I tried to reset and Fuel Cell 1 and 3 are both showing gray flags. But they are both showing zip on the flows."

"I copy, Flight," Liebergot said unhappily, indicating that he had heard the crew's transmission. This had to be an instrumentation problem. They couldn't possibly be having a simultaneous failure in two independent fuel cells—especially since, he had just been reassured, the oxygen was doing fine. And yet every time he got a new reading on the fuel cells, their condition seemed to be deteriorating.

"Okay, what do you want to do?" Kranz asked.

Liebergot suppressed an impulse to reply, "I want to go home." He passed along a new configuration of linkups between fuel cells and buses that would reveal this problem for what he hoped it was: some blown circuitry that they could work around.

10 minutes. Sitting in the back room, pulling up screens, Dick Brown was beginning to wonder whether it was instrumentation after all. He called to the front room.

"Let's throw a battery on Bus B and Bus A until we psyche it out," he said to Liebergot. "We're getting undervolts." Brown's recommendation meant that the command module would be running off battery power. Liebergot suggested the less radical change of limiting the battery to Bus B and leaving the fuel cells on Bus A. "I want to psyche out what those fuel cells are doing here," Brown repeated. He confessed his growing fear: "We might have a pressure problem in the fuel cells—it looks like two fuel cells simultaneously."

"That can't be!" Liebergot protested.

"I can't believe that right off the bat," Brown agreed. "But they're not feeding currents."

John Aaron's performance on Twelve crossed Liebergot's mind. Aaron had seen that unbelievable mess of parameters on his screen and immediately he had come back to Flight—Do this! Do that!—and everything had been straightened out in minutes. Seconds, even. Liebergot felt as if it were taking him forever to figure this thing out. He was disappointed in himself, and at the same time had a sense of being at the edge of an abyss.

Liebergot pushed these thoughts to the back of his mind and, one hand clutching one of the metal handles to the electronics drawer on his console, began the first of a series of attempts over the next hour to resuscitate the electrical power. Occasionally he had to remind himself to swallow.

12 minutes. Guido Will Fenner reported to Kranz that the spacecraft's attitude was still changing, when it shouldn't be. "He ought to stop it,"

Fenner told Kranz, "he" meaning the crew. Fenner was thinking about one of the perpetual concerns during the Apollo flights, gimbal lock.

All of the spacecraft's course corrections, orbital insertions, and entry maneuvers depended on knowing not only precisely where the spacecraft was in space, but also the precise attitude of the spacecraft. Whenever the crew performed a maneuver, they had to know, within fractions of degrees, in which direction the nozzle of the engine was pointing. The guidance platform, encased within its three gimbals, provided that information. Unfortunately, there were certain circumstances under which the three gimbals could not perform their function of keeping the guidance platform steady. In particular, if the spacecraft were to get into a position where two of the three gimbals in the guidance system were lined up in the same plane, the gimbals would be unable to let the platform swing free.* This was "gimbal lock," and would result in the platform losing its alignment—in effect, becoming lost.

A fourth gimbal would have taken care of the problem—no matter what crazy position the spacecraft got itself into, the platform would have remained steady as long as the guidance system remained powered up. Gemini had used a four-gimbal system. But a fourth gimbal would have been heavy. It would have added substantially to the bulk of the system, because it would have had to fit around the other three. And the Instrumentation Lab at M.I.T., which designed the system, was comfortable with three gimbals—that's what the Lab had used for Polaris. Three gimbals were enough for any maneuver the spacecraft might be required to make, as long as everyone paid attention to what was happening. After a long battle with the astronauts (Jim McDivitt threatened not to fly in a spacecraft without a fourth gimbal), a three-gimbal system was used.

In the event that the spacecraft did accidentally go into gimbal lock, it would be possible to realign a platform from scratch, but, as on Apollo 12, it was a complicated, tedious job. And most certainly, Fenner wanted to avoid gimbal lock with a spacecraft that already seemed to have more problems than it could handle. Now, he was warning Kranz that the spacecraft was moving unpredictably. Pushed by an unidentified force, *Odyssey* was vulnerable to gimbal lock.

*   *   *

* This discussion is highly simplified. Readers who work with mathematics may find it convenient to think of the gimbal lock problem as analogous to an indeterminate solution for a set of equations.

14 minutes. Lovell reported seeing something venting from the service module—a gas of some sort. He didn't say so, but in his own mind he was pretty sure it was oxygen—the O_2 Tank 2 pressure was reading zero and there was a big sheet of what looked like white smoke out his window. It added up. He was also pretty sure that it was only a matter of time until the C.S.M. went dead. For Lovell, the moment when he looked out the window and saw the venting was the moment when he stopped being disappointed at losing the lunar landing and started wondering how they were going to get home. For CapCom Jack Lousma, a fellow astronaut, Lovell's report of venting was the most chilling moment of the flight. The problem could not be just instrumentation or an electrical screwup. Something violent and destructive had happened to the service module more than 200,000 miles away from home.

Kranz understood the implications too, and knew that everyone else in the room had heard Lovell's calm report. He decided it was time to give a little speech to the controllers. "Okay, now, let's everybody keep cool," he began. "We got the LEM still attached, the LEM spacecraft's good, so if we need to get back home we got a LEM to do a good portion of it with." He went through the priorities: Don't blow the command module's internal batteries, don't do anything to blow the remaining Main Bus. "Let's solve the problem," he concluded, "but let's not make it any worse by *guessin'*." Kranz spat out the word as if it were an epithet.

A few seconds after Kranz had finished, Brown called Liebergot on the EECOM loop. He sounded curiously formal, but that was because he had something momentous to say: "EECOM, this is E.P.S. I think we ought to start powering down." Powering down meant that they would start taking some equipment off line, making it temporarily inaccessible to the crew. It was not something one did to a spacecraft in flight unless there was a compelling reason, but Liebergot immediately agreed—he had been coming to the same conclusion independently. "Let's get the power-down list, here," Brown said. For the first time that night, Brown's voice shook. For the next minute, the only sound on the EECOM loop was the rustle of pages as each of them looked through his copy of the emergency power-down checklist.

It was then that Liebergot looked up at his screen, and suffered yet another shock: The pressure on his other O_2 tank was falling.

In preparing for disasters during flight, the world of flight control began with certain verities. One was that structural materials like lines and bulkheads had a reliability not just of .9999 or even .999999, but of

1. Most of them had margins that protected them well beyond the worst-case design requirements. A second verity was that while two of the mechanisms of the spacecraft might plausibly fail simultaneously, it was not plausible that two redundant elements within the same system would fail simultaneously. This confidence was based on simple mathematics. With thousands of parts in the spacecraft, even very small probabilities of failure in the individual components added up to a reasonably large probability that some two of those thousands of mechanisms would fail at the same time. But if there are two completely independent oxygen tanks and each of them has a reliability of .9999 (for example), then the chance that both of the oxygen tanks will fail is $.0001^2$, or one in 100 million. The likelihood of a dual failure within any of the handful of major systems—electrical, oxygen, water, propulsion, and the like—was therefore exceedingly small.

Until the moment that he saw the pressure on O_2 Tank 1 falling, Liebergot (if he believed his unbelievable screens) had been looking at a double failure in the fuel cells plus an oxygen tank failure. Now, if he accepted the newest information before him, he was confronted with a quadruple failure, a pair of double failures within systems. As Swigert would write later, "If somebody had thrown that at us in the simulator, we'd have said, 'Come on, you're not being realistic.' "

Around the corner from the MOCR, in SPAN, Scott Simpkinson watched the data with growing concern. Like everyone else, he had been hoping that it was an instrumentation problem. Now someone told him that the pressure in the second O_2 tank was dropping. "At that point," he said later, "we began to get everybody in the world working on it."

18 minutes. Kranz had just finished warning the crew through CapCom to maneuver out of a new threat of gimbal lock when EECOM got back to Flight with his recommendation. "Flight." Liebergot cleared his throat nervously. "I think the best thing we can do right now is start a power-down."

Kranz tried to make the best of it. "You want to power-down, let us look at the T.M. [telemetry] and all that good stuff, and then come back up [power-up again]."

"That's right." Liebergot played along. They could still hope that this was something to be explained and fixed.

*   *   *

Chris Kraft, recently promoted to deputy director of M.S.C., was in the shower when he got a call from George Abbey. Kraft dressed without taking time to dry himself and drove to M.S.C. at speeds up to 90 m.p.h. He walked into the MOCR, and over to Kranz's console. Kranz glanced up at the man he called Teacher. "We are in deep shit," he said.

32 minutes. CapCom Jack Lousma, feeling a little apologetic that the MOCR didn't seem to be able to pull a rabbit out of the hat tonight, called up to the crew. "Thirteen," he said, "we've got lots and lots of people working on this. We'll get you some dope as soon as we have it, and you'll be the first to know." There was a pause of several seconds, and then back from *Odyssey* came Jim Lovell's voice, ironic and as arid as the 200,000 miles of vacuum it had crossed:

"Oh. Thank you."

It was classy, it was cool, and it reminded his listeners: This episode might be a strain on you folks in the MOCR, and that's too bad, but three of us up here have a unique interest in how it all comes out.

As for Kranz, a strange change in his style was taking place. During the first few minutes after the problem was reported, he had been brusque, even querulous as people were slow to explain to him what was going on. His little speech to the controllers had been intended to reassure, but his delivery had had a hard edge to it. Now, as the crisis deepened, his tone was getting more and more casual, sometimes almost gentle, as he went from one controller's problems to another's.

Aaron arrived at the Control Center. He walked slowly through the MOCR and the back rooms, stopping periodically to plug in at the consoles. "You can visualize what had happened," he reminisced. "Everyone was glued to his own tube and he was digging into his own area deeper and deeper and deeper." Their disbelief in what was happening, Aaron thought, was induced to some degree by their training. The SimSups' favorite trick was to confuse the controllers with instrumentation problems and then "while you were off chasing the flaky readouts they would drop the hammer on you with some real big failure and see how well you could back out of that." The simulations always had a way out; surely this situation must have a way out as well. To Aaron, it was as if the controllers were saying, "We're not going to give up, we're not going to give up, we can't give up, we've never had this happen before." He began preaching to them in his Oklahoma twang. "You guys are wasting your time," he said to them. "You're convincing yourself this is some kind of funny instrumentation problem. You really need to understand that the C.S.M. is dying.''

38 minutes. Liebergot watched the pressure in his remaining O$_2$ tank slowly fall. George Bliss called from the back room to tell him that now the pressure in the surge tank was falling as well—the C.S.M.'s systems had automatically tapped into it to try to make up for the deficits elsewhere. This was not permissible. The surge tank was a small bottle of gaseous oxygen on board the command module which was normally used for entry. Ordinarily, it could be refilled any time by one of the O$_2$ tanks. But if they were gone, the surge tank had to be preserved at all costs. Liebergot called Kranz, asking him to instruct the crew to isolate the surge tank.

Kranz was still focused on trying to save as much of the mission as he could, which at this point meant trying to restore pressure in O$_2$ Tank 1 and keep at least one fuel cell working. "Why that?" he said sharply. "I don't understand that, Sy." If he wanted to keep the fuel cell going, why pull off one source of oxygen?

"We want to save the surge tank which we will need for entry," said Liebergot. There was a brief pause.

"Okay. I'm with you. I'm with you." Kranz's voice was resigned. "CapCom, let's also isolate the surge tank."

The term for what Kranz had been doing since he had heard about the anomaly was "downmoding," moving from one set of options to another, more restricted set. As he did so, he was trying to disturb the standard configurations as little as possible—Kranz, like Kraft, had formulated some precepts of flight control, and one of his favorites was "Tread lightly lest ye step in shit." He had been trying, he recalled, to "move very easily and progressively through this thing—because you did not want to close out any of your options for any of the systems, because you never knew which ones you were going to get." At this point, the options were dwindling to a precarious handful.

42 minutes. In an attempt to keep O$_2$ Tank 1 alive, Liebergot instructed *Odyssey* to use manual heaters to increase the pressure. The heaters started, and Liebergot watched the pressure reading. It should have been going up as the heat was applied. It wasn't. He called Dick Brown, who was watching the same screen in the back room.

"E.P.S. EECOM. It looks grim."

"Yes it does," Brown replied.

"We got the five amps [of power, to heat the heaters], and no pressure increase."

"Just went down."

Liebergot sighed. "It's going down," he told Brown. "We're losing it." Liebergot called Flight and asked that the crew turn on the fans in O_2 Tank 1. That didn't help either. The pressure continued to fall.

46 minutes.

"Flight, EECOM."

"Go ahead."

"The pressure in O_2 Tank 1 is all the way down to 297. You'd better think about getting in the LEM and using the LEM systems. I'm going to have to power way down. I don't know if I'm going to be able to save the O_2 for the [remaining] fuel cell." Liebergot, like Aaron, now knew that the command module was dying. If that was true, the only way the crew could get back was by getting into the LEM and surviving on its oxygen, water, and power.

Kranz pushed a little—the heaters weren't working, perhaps; maybe there were some circuit breakers to be checked, some way to get those heaters going and bring up the O_2 pressure, some way to supply oxygen to the fuel cell.

"We saw the current," Liebergot said bleakly, but Kranz persisted, so Liebergot gave him some circuit breakers to check on panel 226.

Kranz was determined to try every possibility, but he was not oblivious to what Liebergot was telling him. A minute later, Kranz called TELMU with a brief message: "I want you to get some guys figuring out minimum power in the LEM to sustain life."

In the middle of the warm Houston night, the cars began pulling into the empty parking lots outside Buildings 45 and 30. They wheeled across the lots heedless of signs and painted arrows. Occasionally rubber squealed as someone braked abruptly into a space. When they got out, the drivers half-ran, calling to each other with hurried exchanges of whatever fragmentary information they had.

Ed Fendell, now an INCO, was one of them. He had been chatting with Gerry Griffin over a beer after their softball game when he got a call from the INCO on duty. By the time he arrived, the lots were already filling up. To hell with that, Fendell thought—there was an open spot right next to Building 30. In his rush, he locked his keys in the car, which sat there for the next five days—in the space clearly marked as reserved for the director of Flight Operations.

Over in Building 45, Sid Jones, one of the MER's shift leaders, had been presiding over a half-empty MER when Swigert reported that Apollo 13 had a problem. It took him only a few minutes to decide that they were going to need all the help they could get. He called Arabian, who had left just an hour earlier, and some of the key systems people. Owen Morris, ASPO's chief of engineering for the lunar module, got back quickly. It was easy to see that people were upset by what was happening, but things hadn't settled down to the point that much was getting done. Morris called the LEM people up to his desk beside the dais and started handing out assignments. He was also on the phone with Bethpage.

51 minutes. "Sy," George Bliss told Liebergot from the back room, "it looks like from the leak rate that we've got one hour and fifty-four minutes left until we're down to 100 p.s.i. in Tank 1. Which is about the end of it for the fuel cells."

As often happened, Flight was asking a question while a controller was talking to the back room. In this instance, Kranz was saying, "I see that juice is still going down there, EECOM. You got any more suggestions?" Liebergot hadn't heard. He called Flight to pass along Bliss's bad news.

"Flight, EECOM."

Kranz repeated his question: "Any more suggestions in trying to pump up O_2 Tank 1 pressure?"

Throughout the crisis, Liebergot had been reasonably successful in suppressing his emotions, keeping his voice calm and businesslike. For one moment, in one word, his unhappiness showed. "No," Liebergot said miserably, a brief wail. Then his voice was back to normal: "Flight, we're going to hit 100 p.s.i. in an hour and fifty-four minutes. That's the end, right there."

In less than two hours, the command module would be lifeless. Kranz acknowledged.

54 minutes. In SPAN, Scott Simpkinson was puzzling over the loss of oxygen. One last possibility occurred to him—it involved a radical cure, but this was a radical problem. He got himself patched through to talk with Liebergot.

"Hey, did we close the reactant valves on those two fuel cells that's down?'' Simpkinson asked. ''We could be losing cryo through there.''

There was one reactant valve leading into each fuel cell. Simpkinson was pointing out a source of leakage that Liebergot hadn't considered. But the solution—closing the reactant valve—was final. Once its reactant valve had been closed, a fuel cell could not be reactivated. Still, it was a new idea. Liebergot broached it to Brown, who was dubious. ''I'm not sure this problem's not back up in the tank somewhere,'' he said. ''And if we can get that O_2 straightened out it might straighten out the—''

Liebergot interrupted. ''I don't think we're going to get it straightened out, Dick.''

''I . . . I don't either,'' Brown finally answered. There was a long silence. And then, for the first time since they had been thrown into the middle of this mystery, they began to see how it might have happened. ''There's a possibility that there could be a leak between that reactant valve and the rate sensor,'' Brown conceded. ''If there is, then shutting it off may help us. But that'd be a dual failure, not a single failure.'' And a dual failure still seemed as improbable to both of them as it ever had.

EECOM pondered that for a moment. Finally Liebergot said, ''It'd have to be in the cryo tanks for them both [both fuel cells] to go.''

''Somewhere back there,'' Brown agreed.

But, still, that left the problem of explaining why both O_2 tanks were failing. Then the nickel dropped. ''Yeah, that's right, it's common,'' Liebergot said excitedly. ''Dick, if you blasted a hole downstream between the rate transducer and the reactant valve, that'd be, that, that . . .''—Liebergot could barely get the words out, it suddenly seemed so obvious—''that's manifold right there, all three cells.'' He was very close to the truth.

At that same moment, Kranz was on the loop with FIDO. The spacecraft was rapidly nearing the outer limits of the time at which it could perform a ''direct abort,'' returning to earth without first looping around the moon. But a direct abort meant using the S.P.S. engine on a very sick service module. Kranz told FIDO to concentrate exclusively on returns around the moon, using the LEM's engines. ''Unless we get a heck of a lot smarter, I think we are wasting our time planning on using the S.P.S.''

Arabian had arrived at the MER about half an hour after the problem had first been reported. It was at times like this, one observer reflected, that

Mad Don was at his best. Arabian's extravagances were muted but his quickness and insightfulness remained. Also, he didn't have any trouble making up his mind under pressure—"He is very decisive," the observer said, "and when you're under a lot of stress that's very helpful." Arabian began by pulling the power people and thermal people up to the blackboard, where they mapped out the minimum voltages under which different pieces of *Odyssey*'s equipment could operate. After half an hour, they had worked up a basic configuration. Arabian didn't wait for it to get written down; he and a few of his team trotted over to Building 30 to talk to the F.O.D. people themselves.

69 minutes. The crew had closed the reactant valve on Fuel Cell 3, and it hadn't helped—nor had Liebergot and Brown really expected it to. They were now increasingly convinced that something drastic had happened to the manifold leading into all three cells. The time was 10:17. Kranz had just informed all the controllers that they should begin to hand over to Glynn Lunney's Black Team. Liebergot stood up shakily from the console to turn it over to Clint Burton. It had been a long time since he had last told himself to swallow, and he found he could barely talk. "I was so relieved to be relieved," he would say many years later, when he was finally able to smile about it.

2

The White Team filed downstairs to Room 210 on the second floor of the Control Center, the Data Room, where they could stretch out the analogue strip charts on tables. Room 210 was to be the command post for the rest of the mission, and Kranz's first command decision was that the White Team would not take another regular shift in the MOCR. Milt Windler's Maroon Team, Gerry Griffin's Gold, and Glynn Lunney's Black would stand the regular shifts while the White Team devoted itself to planning the burn to bring the crew home and the events leading to entry, which would involve procedures that essentially were without precedent.

Liebergot and the other systems people began to examine the "d-log," the data log, a second-by-second dump of all the telemetry that had been received just prior to, during, and just after the anomaly. And it was then that Sy Liebergot began to understand the basics, though it would be many more weeks before anyone would fully understand the night's events. In the end, NASA would find, this is what had happened*:

In October 1968, when the O_2 Tank 2 used in Apollo 13 was at North American, it was dropped. It was only a two-inch drop, and no one could detect any damage, but it seems likely that the jolt loosened the "fill tube" which put liquid oxygen into the tank.

In March 1970, three weeks before the flight, Apollo 13 underwent its Countdown Demonstration Test which, like all C.D.D.T.s, involved loading all the cryos. When the test was over, O_2 Tank 2 was still 92 percent full, and it wouldn't detank normally—probably because of the loose fill tube. Because a problem in the fill tube would have no effect on the tank's operation during flight, the malfunction was not thought to be relevant to flight safety.

After three unsuccessful attempts to empty the tank, it was decided to boil off the oxygen by using the internal heater and fan. This was considered to be the best procedure because it reproduced the way the system would work during flight: heating the liquid oxygen, raising its pressure, converting it to gas, and expelling it through the valves and pipes into the fuel cells where, in flight, it would react with the hydrogen. So they turned on the tank's heater.

A technician working the night shift on Pad 39A was assigned to keep an eye on the tank temperature gauge and make sure that it did not go over 85 degrees Fahrenheit. It was not really necessary that a human serve this function, because a safety switch inside the tank would cut off the heaters if the temperature went beyond the safety limit. And, in reality, the safety margin built into the system meant that the temperatures could go considerably higher than 85 degrees without doing any damage. But the precautions were part of NASA's way of ensuring that nothing would go wrong.

After some time, the technician noticed that the temperature had risen to 85 degrees, but all he had been told was that anything in excess of 85 degrees was a problem, so he let the heater run—about eight hours, in all. No one had told him that the gauge's limit was 85 degrees. That's as high as it could measure. Thus the technician could not tell that temperatures

* After the crew had safely returned, Paine set up an investigative board similar to the one established after the 204 fire. Scott Simpkinson oversaw M.S.C.'s part of the investigation. Don Arabian headed the team of engineers who, in a tour de force of engineering detective work, tracked down the convoluted history of O_2 Tank 2. As the finishing touch, Arabian took an identical O_2 tank, subjected it to exactly the sequence of events that he believed to have caused the Apollo 13 anomaly, and produced exactly the same result.

inside the tank were actually rising toward a peak of approximately 1,000 degrees Fahrenheit, because the safety switch had failed.

It had failed because of one small but crucial lapse in communication. Eight years earlier, in 1962, North American had awarded Beech Aircraft a subcontract to build the cryo tanks for the service module. The subcontract specified that the assembly was to use 28-volt D.C. power. Beech Aircraft in turn gave a small switch manufacturer a subcontract to supply the thermostatic safety switches, similarly specifying 28 volts. In 1965, North American instructed Beech to change the tank so that it could use a 65-volt D.C. power supply, the type that would be used at K.S.C. during checkout. Beech did so, neglecting, however, to inform its subcontractor to change the power specification for their thermostatic safety switches. No one from Beech, North American, or NASA ever noticed this omission.

On all the Apollo flights up through Twelve, the switches had not had to open. When the tanks were pressurized with cryogens hundreds of degrees below zero, the switches remained cool and closed. When, for the first time in the history of the cryo tanks, the temperature in the tanks rose high enough to trigger the switch—as O_2 Tank 2 emptied—the switch was instantaneously fused shut by the 65-volt surge of power that it had not been designed to handle. For the eight hours that the heaters remained on, the Teflon insulation on the wires inside the cryo tank baked and cracked open, exposing bare wires.

On the evening of April 13, when Liebergot ordered the cryo stir, some minute shift in the position of two of those bare wires resulted in an electrical short circuit, which in turn ignited the Teflon, heating the liquid oxygen. About sixteen seconds later, the pressure in O_2 Tank 2 began to rise. The Teflon materials burned up toward the dome of the tank, where a larger amount of Teflon was concentrated, and the fire within the tank, fed by the liquid oxygen it was heating, grew fierce. In the final four seconds of this sequence, the pressure exceeded the limits of the tank and blew its dome off, jarring the shelf above it with a force of 86 g's for about eleven microseconds, and slamming shut the reactant valves on Fuel Cell 1 and Fuel Cell 3. Then the Teflon insulation between the inner and outer shells of the tank caught fire, as did the Mylar lining in the interior of the service module. The resulting gases blew out one of the panels in the service module. That explosion also probably broke a small line that fed a pressure sensor on the outside of O_2 Tank 1, opening a small leak.

Once the service module panel blew out, the vacuum of space extinguished the fire. All the crew knew was that something had made a loud bang and jolted the spacecraft. Actually, the crew of Thirteen was exceedingly lucky. The explosive force could have broken the tension ties holding the command module to the service module, depriving *Odyssey* of the precious few hours the remaining fuel cell gave them. It could have cracked the heat shield. Or the wiring in the tank could have short-circuited later in the mission, after the LEM had already been used for the lunar landing. If the explosion had occurred while the LEM was actually on the surface, the result would have been especially ghoulish, with two healthy astronauts stranded on the moon while the command module and its pilot died overhead.

The rise in pressure in O_2 Tank 2 that preceded the explosion had been deceptively normal. First the pressure rose for twenty-four seconds, then it stabilized. Even if Liebergot, Sheaks, or Bliss had noticed the rise in O_2 pressure during this period, it would not necessarily have caught his attention. A rise in pressure was, in and of itself, part of a normal fluctuation. For the next fifteen seconds, the pressure held steady. Then it rose again for another thirty seconds. Liebergot, Sheaks, and Bliss were still concentrating on the quantities shown in the tanks of hydrogen—the readings for which they had requested the cryo stir in the first place. The readings for the O_2 tanks were also at the bottom of the screen, just to the left of the ones for the hydrogen.

At this moment, there was no light to blink on and alert the controllers to the changing pressure. The EECOMs had set up a yellow warning light on the EECOM console to indicate a change in pressure in the cryo tanks, but it was a "pay-attention" light that went on whenever the pressures were moving markedly up or down, which they did frequently, and on this occasion the warning light was already lit because of changing pressure in the hydrogen tank.

So none of the three noticed the numbers for O_2 Tank 2 during four particularly crucial seconds. At 55 hours, 54 minutes, and 44 seconds into the mission, the pressure stood at 996 p.s.i.—high, but still within the accepted limits. One second later, it peaked at 1,008 p.s.i. By 55:54:48, it had fallen to 19 p.s.i. During those same four seconds, the temperature in the tank went from − 329 to + 84 degrees Fahrenheit. If one of them had seen the pressure continue on through the outer limits, then plunge, he would have been able to deduce that O_2 Tank 2 had exploded. It would have been a comparatively small leap to hypothesize that the explosion had damaged the piping from the second O$_2$ tank as well. And once they had been able to conceive of something having occurred that disabled both oxygen tanks, then the problems reported in the electrical system would have made sense.

Sy Liebergot had, in his estimation, failed to live up to the controller's code. Two numbers among the fifty-four parameters on his CRYO TAB screen had silently changed—or more precisely, many of the numbers had silently changed, but two of them had changed anomalously—and, at a time when he had reason to be looking elsewhere, Liebergot had not noticed. His two E.C.S. men in the back room hadn't noticed either, but that didn't make Liebergot feel any better. Every night for the two weeks after the accident, he was awakened by a nightmare. In the dream, he was in the MOCR, the *Odyssey*'s crew reported that they had a problem, he looked at the screen only to see a mass of meaningless numbers, and then he relived the next anguishing hour.

After two weeks of the nightmare, the dream changed. In the new dream, which he had only once, Liebergot looked at the screen before the bang and saw the pressure rising. He realized that the rise was abnormally fast. "Flight," he called to Kranz, "pressure's going up in O$_2$ Tank 2. Think we have a failed-on heater. Tell the crew to turn off the heater." The crew turned off the heater, and the pressure kept going up. Liebergot said, "Flight, have them pull the breakers on panel 226." The crew pulled the breakers. Then the tank blew, and Liebergot saw the pressure drop and told Flight exactly what had happened. And then everything else occurred just as it had in reality—for noticing would have saved some time in figuring out what had gone wrong, but it could not have prevented the crippling of Apollo 13.

The nightmares stopped. Except for that, however, Sy Liebergot still didn't feel much better. A flight controller was supposed to notice everything, all the time.

CHAPTER

28

"You don't have time to worry"

Glynn Lunney had been in the Control Center even before the explosion, reading the flight log in the MOCR and poking around the back rooms as was his custom before taking a shift. Fourteen minutes into the crisis, when Lovell reported that something was venting, Kranz told his assistant flight director to get hold of Lunney and let him know what was going on. From then until the time he took over, Lunney watched from a seat beside Kranz, just as the rest of the Black Team's controllers watched from chairs pulled up next to their counterparts. "Sort of a dawning was going on," Lunney said, "as Gene worked the problem with the guys and I sat there and listened to it. . . . It wasn't a thing where immediately you knew, 'Boy, we've blown the tanks and we gotta get this thing back from the moon and power up the LEM.' " The Control Center reacted cautiously, Lunney said, "and properly so. You don't want to jump off the deep end."

1

Sixty-nine minutes after the explosion, when Lunney took the console, he was determined to keep the C.S.M. alive. *Odyssey* was partially powered

408

down, but all its crucial life-support systems were still operating off the one remaining fuel cell. If one of the oxygen tanks and one of the fuel cells could be kept on line, they could sustain that configuration all the way back to earth. For the first minutes of his shift, Lunney concentrated on that possibility. Repeatedly, he pushed Clint Burton, the Black Team's EECOM, to come up with new ways of dealing with the failing fuel cells and the dropping oxygen pressure. First of all, was Burton sure that the dropping pressure in O_2 Tank 2 was legitimate? Yes, Burton was. What confirming information did he have? The temperature was rising, as it would if the pressure were falling.

Lunney was persistent because the next step they were contemplating was shutting off the reactant valve in Fuel Cell 1, as they had done already in Fuel Cell 3. If they shut it off and then came up with an Aaron-like solution that suddenly got the O_2 pressures back up, the door would still be closed on two-thirds of the C.S.M.'s power supply. It was like shooting a lame horse if you were stranded in the middle of a desert. It might be the smart thing to do, but it was awfully final. Lunney, like Kranz before him, had no way of knowing that the explosion had instantaneously closed the reactant valves on both fuel cells 1 and 3.

At ten minutes into his shift, seventy-nine minutes after the explosion, Lunney was close to exhausting the alternatives.

"You're ready for that now, sure, absolutely, EECOM?"

"That's it, Flight."

"It [the oxygen pressure] is still going down and it's not possible that the thing is sorta bottoming out, is it?"

"Well, the rate is slower, but we have less pressure too, so we would expect it to be a little bit slower."

"You are sure then, you want to close it?"

"Seems to me we have no choice, Flight."

"Well . . ."

Burton, under this onslaught, polled his back room one last time. They all agreed.

"We're go on that, Flight."

"Okay, that's your best judgment, we think we ought to close that off, huh?"

"That's affirmative."

Lunney finally acquiesced. "Okay. Fuel Cell 1 reactants coming off."

It was uncharacteristic behavior by Lunney—"stalling," he would later call it. "Just to be sure. Because it was clear that we were at the ragged edge of being able to get this thing back. . . . That whole night,

I had a sense of containing events as best we could so as not to make a serious mistake and let it get worse.''

Throughout the long evening, Swigert's voice never sounded so despondent as it did a minute later when he reported from *Odyssey*: "Fuel Cell 1 is closed."

Though he continued to seek a way to hold on to the C.S.M., Lunney was now thinking about the LEM, *Aquarius,* with its independent guidance, control, and environmental systems. He called TELMU on the loop.

"TELMU, Flight."

"Go, Flight."

"Is there anything simple that we can refer the crew to, to get them thinking about using the LEM here?'' Lunney asked. ''Or have you got anything in the checklist paperwork to describe to them what your intentions are?''

"Negative," TELMU replied. "There is nothing documented in [the crew's contingency procedures]. We're thinking about using the LEM as a lifeboat. We have some procedures here on the ground, though."

"I'm sure you do," said Lunney. "What do they amount to? Flying with the tunnel open?''

"Rog. Just a LEM low-power load supplying power to the C.S.M.''

At this point, the MOCR's thinking about the lifeboat procedure was still collectively confused. Theoretically, there were two ways to do it: Either the crew could power up the LEM and live in it, or the LEM's batteries could be used to keep the C.S.M.'s systems going. Instinctively, TELMU and some others in the MOCR had thought first that evening about using *Aquarius* to power *Odyssey*. When Lunney asked, TELMU was still off on that tangent. But using *Aquarius* for that purpose wasn't really feasible. The C.S.M. was set up to supply power to the LEM, not the other way around. The MER would soon figure out how to reverse the process, but there was no off-the-shelf procedure. Furthermore, *Odyssey*'s systems were power-hungry; *Aquarius*'s batteries couldn't possibly feed them for long enough. Lunney had not thought through all of this in detail, but he was suspicious.

"Supplying power to the C.S.M.?'' he asked TELMU dubiously.

"Yes," said TELMU, "about five amps."

"Ah . . . to what?'' asked Lunney, meaning: What did they think they were going to hook up with? What subsystems in the C.S.M. would the power be used for? Lunney asked EECOM to get to work on answers.

Personally, Lunney began thinking about shutting down *Odyssey* altogether and sending the crew to live in the LEM.

Lunney had come on shift with a schedule at the back of his mind. While watching Kranz and listening to the loops, he had heard Liebergot report that they had an hour and fifty-four minutes of oxygen pressure left. That had been less than twenty minutes before Lunney took over the flight director's console. Thus Lunney should have been able to figure on squeezing at least an hour and a half of power out of O_2 Tank 1 and Fuel Cell 2 after he came on shift. But it was a night during which the flight director was destined to get only bad news. Eighteen minutes into his shift, EECOM informed him that he had forty minutes before the pressure in Tank 1 hit 100 p.s.i., functionally zero. The pressure was dropping faster than before.

Lunney tried one last time to save the C.S.M. "Clint, let me ask you, is there anything you want to do trying to pump up the other tank? Or are you satisfied that both of these tanks are going down, and we're past helping them, even with batteries? That's what I'm getting at. I'm just trying to be sure that you've gone through everything and you don't have any other tricks up your sleeve."

"Sure don't, Flight," Burton said.

Since the initial report of a problem, as first Kranz and then Lunney had been forced into a steadily shrinking circle of options, the flight control team had kept searching for something that would enable them to avoid the last, drastic step of giving up on the command module and using the LEM's systems. Now Lunney had exhausted the alternatives, and he turned his full attention to getting *Aquarius* ready to serve as a lifeboat. By this time the notion of using the LEM to power the C.S.M. had been discarded.* The plan was to have the crew live in *Aquarius* until just before entry, when they would move back into the C.S.M. and power up with the C.S.M.'s three internal batteries that were its source of power for the last phase of the flight.†

Lunney was embarking on an exceedingly tricky project that neither he

---

* The tape of the flight director's loop does not reveal when or how this change in plans took place, because much of the discussion at this point was being conducted off the loops as Lunney conferred with people crowding around his console. At about this time, word was passed to the MOCR that the electrical umbilical linking the LEM and the command module could carry only a 17-ampere load, too little to power the command module's systems.

† The batteries were designed to last for half an hour in a fully powered configuration—the time between jettison of the service module and the command module's splashdown—with a generous reserve.

nor any of his team had ever practiced: to close down a command module in such a way that it could be brought back to life some days later, and, at the same time, power up a lunar module without any help from the command module. And this novel, complex procedure, filled with possibilities for irretrievable error, was to be accomplished—or so Lunney now thought—with less than forty minutes of power from Fuel Cell 2 remaining.

When CapCom Jack Lousma had told the crew a few minutes earlier that the ground was starting to think about using the LEM as a lifeboat, the crew was glad to hear it. "That's something we're thinking about too," Lovell said. They'd already been talking it over, and the first thing the crew proposed to do was run a quick P52 (program number 52) and get the LEM's platform aligned.

It was an important decision. If the crew was to get home, they had to have an aligned, functioning guidance platform. But an I.M.U. was also an extravagant consumer of electrical power and water (for cooling purposes), both of which were now in precariously short supply. *Odyssey* and *Aquarius* each had platforms that were not in any way linked with each other. The Thirteen crew was proposing to power up *Aquarius*'s I.M.U., run a P52 on *Odyssey*'s computer to give them the C.S.M.'s alignment data, and then punch those numbers into *Aquarius*'s system. The question was: Should they do it? The next time that *Aquarius* would have to have the platform operating could be as long as twenty hours away, when, after they came out from behind the moon, the LEM's descent engine would establish a trajectory for home. Furthermore, the platform could be aligned from scratch, through star sightings taken by the crew. The Apollo 13 crew had been trained in that procedure, and the Apollo 12 crew had demonstrated that it could be done in a real emergency. Therefore, one option was to leave the platform powered down until a few hours before it had to be used.

Lunney and Control and Guido discussed it over the loops and agreed that they ought to power down *Odyssey*'s I.M.U. and align *Aquarius*'s I.M.U. later, using the Alignment Optical Telescope (A.O.T.). Lunney told *Odyssey* to hold off on the P52. Tom Stafford, who had commanded Apollo 10 a year earlier, was in the MOCR listening to this exchange. As soon as it was over he approached Lunney for a private conversation. They were on the wrong track, Stafford said. Maneuvering the docked spacecraft with the LEM's thrusters to get manual star sightings was going to be next to impossible. It was essential they get an alignment now, from the C.S.M., while it still had power. Stafford's was the knowledge of a man who had flown the LEM, and Lunney respected it.

Clint Burton had more bad news.

"Flight, EECOM. Flight, EECOM."

"Go ahead."

"Okay, got an update on the time," said Burton. "It looks like we got about eighteen minutes until we get down to 100 p.s.i., and that's the cutoff point."

It had been four minutes since EECOM had told Flight that they had forty minutes left. The leak was getting faster, and Lunney's margin, if forty minutes had given him any margin at all, had disappeared.

After listening to Stafford, Lunney went back to his controllers and asked that they look at the trade-off—could they stand the power drain if they turned on the LEM's platform now? They would work on it, they replied: "Stand by one, Flight." "All right," Lunney muttered. Then, emphatically: "Pronto."

The exchanges over Flight's loop were now continuous bursts of queries and instructions to the flight controllers around the room, marked every few minutes by another update on the time remaining before the command module went dead. Like a variety-show juggler with a dozen plates spinning on the top of sticks, Lunney went from one decision to another, keeping the plates spinning, sometimes getting back to the one at the other end of the row only when it seemed to be wobbling on its last revolution. Throughout it all, Lunney, aged thirty-three, maintained the mildly distracted air of an experienced parent getting many small children ready for school.*

It seemed to the astronauts that the ground was taking an awfully long time to recognize what was obvious to them, that they had to get into the LEM immediately. But they couldn't simply climb in and shut the door. Powering up the LEM was much more complicated than throwing a power switch, and activating the LEM's many systems was much more

---

* At one of the most hectic points, Lunney even took a few seconds to make a small ironic joke that no one else got. TELMU had forgotten about using the LEM to power the C.S.M. and now, ten minutes later, was focusing exclusively on powering up the LEM's own systems. Lunney asked TELMU whether they needed to send any power at all from the LEM to the C.S.M., if just to keep the I.M.U. heated. "I'm not sure we can do that," TELMU told Flight, as if surprised Flight could even consider such a thing. "Can't do that," Lunney acknowledged, amusement in his voice. "That was a long time ago."

complicated than flicking a few switches to "On." Fred Haise had three onboard checklists for powering up *Aquarius* and activating its systems under various circumstances, none of which applied to the circumstances he faced now. Furthermore, the shortest of the established checklists would take two hours to complete. TELMU's team had been busy cobbling together a way of powering up *Aquarius* without any cooperation from *Odyssey* and devising for the scheme a drastically shortened checklist for activation. Finally CapCom Lousma was able to tell the crew that he had a procedure for powering up.

"Okay," said Swigert. "That sounds like good news." He did not need to add "It's just in time"—everyone knew that already.

Lunney was having trouble getting things to happen. He had been persuaded that the crew had to run a P52 in the command module to get a course alignment for the LEM's platform, and Guido and Control now agreed. But Guido was struggling simultaneously with a completely separate issue that had come to preoccupy him. His back room was posing the hair-raising possibility that if they turned off the command module's guidance system and let it sit in the cold—surely subfreezing; how much colder than that no one could predict exactly—it wouldn't work when the time came for entry. The guidance people, along with teams in the MER and at M.I.T. in Cambridge, were trying to calculate whether that fear was warranted, and, if so, whether they could afford to run the I.M.U.'s heaters for the duration of the flight.

Three times Lunney had tried to focus Guido's attention on the status of the alignment presently residing in *Odyssey*'s computer. Now Guido came onto the loop, saying that the crew ought to do a P52.

"That's what I'm asking you!" said Lunney exasperatedly. "Do you have [a good alignment] now? We don't have much time. Do you have a good one now? As far as you know?"

"A good alignment?" Guido asked, still absorbed in deciding whether they should keep the I.M.U. in the command module heated.

"Yeah, that's what I'm asking you," said Lunney. "You have a good alignment? I'm not worried about tenths of a degree either."

"Well, it oughta be that good, Flight," said Guido, aggrieved— ordinarily, it should be good within hundredths of a degree.

"Okay, yeah," said Lunney wearily. Just then EECOM came back to him.

"We need to open up the surge tank," Burton said. "That manifold pressure is dropping." No longer was Burton's back room worried about having enough oxygen for power, but about whether the crew would have enough oxygen in the command module to breathe.

Lunney grunted softly, as if someone had just hit him in the stomach—which is how in fact he described that moment years later. It was like a blow, he said, and then a hole into which his stomach and the rest of him were starting to slide. It was the one time during the entire night when for a moment Lunney backed away from the rush of events and "had a sense of 'Holy Christ, it's this bad. It has really happened this way.' " The surge tank was the last reserve of oxygen which had to be protected for the entry, and Burton wanted him to use part of it, lest the crew asphyxiate now.

"Okay," Lunney replied after a pause. "Wouldn't you rather pump that up in the LEM?" He didn't want to tap the C.S.M.'s surge tank.

"Well," said Burton, "we got to get into the LEM first, Flight."

"CapCom, get 'em going in the LEM," Lunney said quickly. "We gotta get oxygen on in the LEM."

Lousma passed the word on to the crew and turned to Lunney. Years later, Lousma himself would not remember, but Lunney felt that the CapCom—"Bless his stout Marine heart"—had seen his sinking feeling in his face. Off the loop, Lousma reminded Lunney to see how far TELMU had gotten in preparing his abbreviated activation checklist. Maybe they could use that rather than ad-hoc the steps. "That one comment at that one point got me right back out of the hole and on track," Lunney recalled. "After that you're busy, it's just blowin'-and-goin' time, and you don't have time to worry."

The power from the remaining fuel cell in *Odyssey* was now good only minute to minute. Since the LEM still wasn't prepared to take over, it was agreed between the ground and Swigert that he should switch to one of the C.S.M.'s internal batteries just before the fuel cell failed. While they didn't want to do this a moment sooner than necessary, they couldn't afford to delay it a moment too long, lest they lose all power temporarily—and, with it, control of the spacecraft and the course alignment stored in the I.M.U.

The twenty minutes that followed were perhaps the most rushed, the most confused, and in retrospect the most glorious of the four-day crisis. The Black Team was racing to complete in minutes a procedure that usually took hours, under the threat of imminent disaster if they failed and under the more subtle threat of a delayed disaster if they got the LEM started but didn't shut down the C.S.M. properly.

In the controllers' memories, the years have tended to smooth over the rough spots. Many of the controllers recall it as having been a piece of cake. Lunney himself conceded that "it was an exciting time," something of a blur in his memory, but orderly. "I look back on it and reflect upon the kind of teamwork and lack of panic," he said. "[There was] no sense of 'My God, the sky is falling.' Probably each individual had the same few seconds I did, but there was no sense of it in the Control Center. Listen to [the tapes] and it all sounds tickety-boo, like the guys had been training for it all their lives." In reality, the tapes reveal that Lunney's memory was two-thirds right. Teamwork, yes. Lack of panic, yes. Tickety-boo, no.

The MOCR was jammed with people by this time. It wasn't like the big moments in other flights, when the off-shift controllers would come in and watch quietly. Now, each console was surrounded by a knot of men trying to put together new procedures, amend old ones, and interpret the data, talking to back rooms that were also crowded with reinforcements. It was so noisy that Lunney had to take time out to ask for quiet. The attention of the people talking on the loops was unusually fragmented, and the result was occasionally missed signals, confusion, and contradiction.

EECOM's back room had been wrong about needing the surge tank; Burton came on a minute after he had asked for it to tell Flight to "disregard the surge tank request." Lunney had time only to acknowledge, "All right," because he was struggling to get the LEM activated.

A few minutes later, Burton was back with a request to "get the crew to stuff the power-down."

"Okay," said Lunney cautiously, "but we wanted to keep [*Odyssey*'s I.M.U.] up for a little while, I thought you just said, with Bat A, to get a course align in the LEM."

Burton realized his mistake—when he said, "stuff the power-down," he hadn't meant to include the I.M.U. "Well," he said apologetically, "maybe there's a bit of confusion here."

"Yeah, you don't want to turn I.M.U. off until we get a course align in the LEM, right?"

"That's affirm, Flight."

The guidance experts remained absorbed in trying to analyze whether they could turn off the heaters in the I.M.U. Lunney, worried about the power drain, checked in with G.N.C.

"G.N.C. I haven't heard from you on the heaters yet on the I.M.U."

"Rog, Flight," G.N.C. replied. "Preliminary look at it looks like we could pull those circuit breakers, power-down the heaters. We're still trying to get a handle on it."

Thinking how much of the command module's battery power had to be saved for the entry phase, Lunney observed, "We may well have no choice."

"That's a fact," G.N.C. conceded.

Lunney was struck by a thought. "How many amps do they take?" he asked, meaning amperes per hour.

"Ten."

In their current situation, this was an enormous amount of power, and to be of any use it would have to be maintained throughout the rest of the flight.

"Aww!" Lunney exclaimed. "Ten amps!?"

"Rog," the reply came back, sheepishly, as G.N.C. suddenly realized that his team had been missing the point.

"I'm looking at . . ."—Lunney did a quick mental calculation—"fifty, sixty, eighty hours." (Another nervous "Rog.") "That's academic!" (A hurried "I understand.")

A few minutes later, G.N.C. came back with a correction. If they kept heat on the I.M.U. but turned off the gyros and the rest of the apparatus, they would need only .8 amp per hour, and they recommended leaving the heater up. Lunney pointed out briskly that .8 amp per hour still worked out to 64 amp-hours. "I can't afford that."

They would assume that the I.M.U. could survive without a heater.

"We only almost lost it once," Lunney said. It occurred during the final stages of the move from *Odyssey* to *Aquarius,* when two separate processes were going on simultaneously. In *Aquarius,* Lovell and Haise were moving through their improvised activation checklist, bringing the LEM's systems on line. For the past ten minutes, they had been preoccupied with getting a good course alignment, which they had done in record time. Jack Swigert was still in *Odyssey,* powering down the systems one by one.

The LEM now had a computer that was up and running, and a good set of angles punched into the guidance system. In their relief, the reaction of the men in the MOCR was to assume that now the LEM was operational and that *Odyssey* could safely render itself helpless. The problem was that the activation routines were being relayed up to the two spacecraft in chunks. Among these chunks was the one which included the steps necessary for pressurizing the reaction and control system

(R.C.S.) in *Aquarius* and powering up the "balls"—short for "eight-balls," referring to the displays that let Lovell and Haise know the spacecraft's attitude. But the actual rate at which the crew in *Aquarius* was accomplishing these steps, and the actual order in which they were being performed, were in doubt.

Reflecting on it afterward, Lunney said that at that point he should have stopped and established a clear-cut understanding of the obvious: Whatever else you do, don't power-down the command module's platform and its R.C.S. until the LEM's control systems are up. But he didn't, and the two procedures got out of synch with the rest. *Odyssey*'s platform and R.C.S. were powered down, *Aquarius*'s R.C.S. and "eight-balls" were not yet ready to take over. As they listened to the traffic, the mix-up slowly became apparent.

"Okay, we haven't gotten ourselves in a position here, where, uh, we have no attitude control in either vehicle, have we?" said Lousma uncertainly.

"Uhh . . . I'm . . ." Lunney sounded almost abashed. "I'm waiting to see when we get attitude control in the LEM. Would you ask them to call us when they have attitude control in the LEM and then we'll power the inverters, et cetera, down in the C.S.M.?"

Lousma asked *Aquarius* to inform him when they had attitude control. Haise replied that they were still working on the pressurization of the R.C.S. system—which meant they still didn't.

"Hey, Flight," Lousma said plaintively, "they don't have attitude control in the . . . and we don't have it in the C.M., uh, C.M.S., uh . . ." Lousma was a little flustered himself.

"Well, they're trying to get it up, right?" Lunney said.

"Yeah, they are," replied Lousma, but he was openly unhappy.

The spacecraft did not go to gimbal lock while it drifted—the gap in control was only about two and a half minutes long—nor would it have been catastrophic if it had, Lunney reflected. They would have put the crew to sleep, worked up some procedures, and done a manual course alignment when the crew was rested. Nonetheless, Lunney was always irritated with himself for having allowed it to happen—"It turned out to be not a real problem, but I was really pissed at myself because it was—well, it was kind of the only mistake we made that night."

At nine minutes until midnight, Jack Swigert closed down Fuel Cell 2, completing the shutdown of *Odyssey*. The LEM was now fully operational, with an alignment stored in its guidance system. It had been two hours and forty-three minutes since Swigert had first reported a problem, about an hour and twenty minutes since the MOCR had first realized the crew would have to use the LEM. Right after Apollo 13, John Aaron recalled, F.O.D. prepared special checklists for activating the LEM as a lifeboat, and during the sims for every mission that followed, the flight control team was thrown at least one massive Thirteen-type failure on a translunar coast. But no matter how good the new checklists were, and no matter how much they trained and scrambled, Aaron said, "we never did it that quick again."

Swigert drifted through the docking tunnel into *Aquarius*. This was alien territory for Swigert, who as a command module pilot had never spent any training time in the LEM simulator. He looked down at Lovell and Haise from the hatch. "It's up to you now," he said.

2

Lunney's team remained on shift for another seven hours. During that time, Building 30 became, as it would be for the rest of the mission, the end of a gigantic funnel of information. Throughout the first few hours of the crisis, the Control Center had had to make most of the decisions itself—events were flowing too quickly to bring in more than the most urgently required information from the outside world. Now expertise of every sort was flowing through the network. When Simpkinson said that SPAN put "the whole world" to work on the problem, he was barely exaggerating.

In Houston, it was no longer just the MOCR, the MER, and SPAN which were filling with people. Bill Tindall, who had recently been promoted to deputy director of Flight Operations, convened a semicontinuous Mission Techniques meeting to concentrate on ways to use the LEM's navigational system for tasks it had never been designed to accomplish. The astronauts congregated in Building 5 where the flight simulators were installed. Stan Faber, chief of the Flight Simulation Branch, began running a nonstop show, with crews following one another in the simulator, testing out each new procedure that was required to keep Thirteen going, until just before splashdown—in all, seventeen astronauts participated at various times during the flight. Former astronaut Jim McDivitt, the new manager of ASPO, was taking all of ASPO and splitting it up into small problem-solving teams—a power team, a cooling team, a trajectory team. "SPAN" was no longer a little room in Building 30; for practical purposes, one SPAN engineer recalled, it expanded to include the whole Center.

Joe Mechelay of the Test Division arrived to take his shift in the MER on Tuesday morning and found "mayhem." Usually, only one or two of the systems teams would be busy at any given time; now they all were. The room was jammed not just with people, but with the schematics and books of specifications that were spread out everywhere. Despite the appearance of chaos, however, the MER was by this time well organized. For each system, it had been established what needed to be done and which were the most crucial items. Arabian himself devoted his attention to whichever problem was at the top of the priority list, shifting to a new group as soon as the old problem seemed to be under control.

The scene was much the same at the contractors. When Tom Kelly got to the Bethpage plant at 3 A.M., he found five or six hundred people already congregated there. At M.I.T., Downey, and at smaller places like Hamilton Standard in Windsor Locks, Connecticut, which had built the LEM's environmental control system, staff engineers showed up unbidden in the middle of the night to see what they could do to help. Few went home. The next three and a half days would find engineers draped anywhere on their company's premises—desk tops, chairs, conference tables—catching a catnap. In some cases, the managers would get together and decide that nothing was being accomplished by this kind of devotion and that it would really be better if people went home for a few hours and slept in a comfortable bed. But most of the engineers would unobtrusively refuse to obey, unwilling to be away for even that long.

Within hours, Apollo 13 also became a story that engaged the attention of the outside world as nothing else about the space program had done except the first lunar landing. When a Grumman engineer hurrying to Bethpage was stopped for speeding and explained where he was going, the police gave him a high-speed escort down the Long Island Expressway, sirens screaming. France was just the first of thirteen foreign governments—the Soviet Union among them—that offered the United States use of their naval vessels for emergency rescue service. In the next days, Pope Paul prayed with an audience of 10,000 for the astronauts' safe return. Rabbis in Jerusalem offered special prayers at the Wailing Wall. Throughout Europe, bulletins on the crew's condition interrupted news programs. The U.S. embassy in London compared the British reaction to that which had followed Kennedy's assassination. The front pages of papers from Oslo to Bangkok were covered with news of Thirteen.

The Manned Spacecraft Center itself was in the eye of the media hurricane. Hundreds of journalists descended on Houston after the accident (until then, interest in Thirteen had been tepid), but they were confined to the area around the public affairs offices. The only people who saw much of them were the flight directors who briefed the press after every shift. None of the other people at M.S.C. who were working the problem had time to notice how preoccupied the world had become with what they were doing. Later, when FIDO Dave Reed watched a film that NASA had made about the flight, he found himself most deeply affected by the picture of a young Asian woman watching Apollo 13 coverage on a store-window television somewhere on the other side of the world from Houston. Her face was a study in concern. He hadn't known.

3

Inside the Control Center, the pace during the rest of the Black Team's shift slowed after the C.S.M. had been powered down. The remainder of the night was occupied with the work necessary to configure *Aquarius* for the long trip home, and with deciding how to put the spacecraft back onto a free return to the earth.* Finally, after almost nine years of manned space flight, the Retros were going to have a chance to use their expertise in aborts.

Down in Room 210, Chuck Deiterich, the lead Retro for Apollo 13, led the discussion. At the time of the accident, the spacecraft was close enough to the earth for a "direct abort" that would have brought it back to earth without going around the moon first. But as Kranz had quickly realized when he heard about the venting, something disruptive had happened back in the service module. They had to assume that the big S.P.S. engine was unusable. This left two widely divergent options, Deiterich explained. They could wait for another eighteen hours until the spacecraft had swung around the moon, fire *Aquarius*'s descent engine nearly to exhaustion, and get the crew back on earth in two and a half

---

* "Free return" refers to an outward-bound trajectory that would use the moon's gravity to swing the spacecraft around the moon and back to a safe landing on earth. At the time of the accident, Apollo 13 was not on a free-return trajectory.

days. Or they could perform a brief burn right away to put the spacecraft onto a return trajectory that would return the crew in four days.

Getting the men back quickly was obviously attractive, but it had serious drawbacks as well. Using up *Aquarius*'s fuel was not the kind of action that they wanted to take immediately if they could buy themselves time. It would leave few options if they were to make a mistake or if something else went wrong. Deiterich argued for a short burn now to get the spacecraft on a free return; they could make a second burn later, after they had taken the time to think about what kind of burn they wanted. There were objections to that strategy as well—two burns meant roughly twice as much drain on the spacecraft's electrical and water supplies as one burn—but the flight control philosophy on this point was deeply ingrained: Don't cut off an option unless you absolutely have to. Kranz approved Deiterich's approach and Lunney concurred.

A burn by an Apollo spacecraft has been compared to coming about in a sailing ship—it takes the same kind of extensive preparation. In the case of the first burn of the docked LEM and C.S.M., it required more than an hour after the burn decision had been made to calculate the trajectory of Thirteen (which had been changed significantly by the explosion and venting) and to check out the peculiarities posed by a docked burn. The DPS engine, which could be gimbaled in varying directions, had to be lined up exactly with the center of mass of the docked assembly that it would be pushing. Then there were small frights to contend with—the astronauts, conducting a final check before the burn, discovered that a switch that could eject the LEM's descent stage with its DPS engine was ''On.'' It was quickly turned to ''Off.''

Finally, at 2:43 in the morning, Lovell pushed the ignition button and the DPS engine ran at low throttle for thirty seconds, putting the spacecraft into a trajectory that, even without a second burn, would bring it down in the Indian Ocean not quite four days later. Lovell was relieved. He wasn't completely confident that the burn provided them a survivable entry, but at least the spacecraft would intercept the earth's atmosphere. In his mind, this was much better than the alternative they had just avoided—orbiting the earth indefinitely, in a lonely revolution with an apogee of 240,000 miles and a perigee of 3,000 miles, a ''perpetual monument to the space program.''

When the burn had been completed, CapCom passed up some advice from the astronauts in Building 5 on how to maneuver a spacecraft in this unholy configuration. *Aquarius*'s crew also put the spacecraft into a rough, manually controlled version of passive thermal control (P.T.C.) to prevent overheating of the spacecraft's sunward parts. It was a nasty job to do without sleep, without practice, after six hours of a life-threatening crisis that had drained the crew of its reserves, but somehow they accomplished it. Jim Lovell sent Fred Haise off to the rapidly chilling command module to get some sleep.

At 7:13 Tuesday morning, ten hours after coming on duty, Glynn Lunney's Black Team turned their consoles over to Gerry Griffin's Gold Team. Though there had been many other memorable shifts in that room—Gene Kranz's White Team had just finished one of them the preceding evening—no flight control team had ever been asked to make so many life-and-death decisions, find so many improvised fixes, and sustain their nerve for so long. Glynn Lunney's Black Team left the MOCR still unsure whether they had saved the crew of Thirteen. They had at least preserved them.

29

"I hope the guys who thought this up knew what they were doing"

The first time anyone had ever thought about using the LEM as a lifeboat was in 1961, even before the L.O.R. decision had been made. Two of Grumman's designers, Tom Kelly and Al Munier, were in the preliminary phases of the design when Munier had a thought. Suppose something happened on the way out to the moon, Kelly remembered him musing. Could this little machine they were designing bring the command module back? Kelly and Munier worked through a few calculations. "Nothing had been built yet," Kelly recalled, "so changing the specifications really didn't cost anything." Planning for the lifeboat option "gave you more fuel to land [on the moon] with, it let you stay on the moon a little bit longer, and it gave you a rescue option." In writing up the specs, Grumman increased some of the quantities on the consumables, and Kelly kept the lifeboat idea in the back of his mind.

In 1963, the possibility of using the LEM as a lifeboat came up at North American in the course of building the command module. North American inquired of NASA whether the LEM would have a thrusting

424

capability sufficient to maneuver it with the command module attached. This intrigued Bob Piland, then acting head of ASPO, who responded with a request that North American look into the possibility of using the LEM to provide a backup in case the S.P.S. failed "in translunar flight." Eventually it was determined that the LEM could function as the source of engine power for a combined LEM and C.S.M. and, if necessary, push a crippled command module out of lunar orbit and back home.

In the Flight Operations Directorate, the lifeboat option had come up in various forms. Glynn Lunney remembered that when he was head of the Flight Dynamics Branch in the early 1960s, his guys had started to talk about what they called a "lifeboat" procedure. It wasn't part of a systematic strategy, Lunney recalled; just another example of their "try-everything way"—"Hey, you know, we've got another engine on the LEM, we ought to know how to use the LEM engine [for docked burns]. We don't know what for, but who knows?" They worked out some procedures and filed them away. The EECOMs had gotten some practice in lifeboat procedures as well. Once during a simulation of lunar orbit, a glitch in the cabin pressure showed up on the EECOM's screen just before loss of signal as the spacecraft went around. The EECOM didn't take the hint. When the command module came back around into radio contact twenty minutes later, the flight control team abruptly discovered that the spacecraft cabin pressure had gone to zero and the astronauts were already buttoned up in their suits to stay alive. The flight controllers used the LEM's oxygen supply to repressurize the C.S.M.

The White Team in particular had gotten some relevant lifeboat experience out of Apollo 9. During an Apollo 9 simulation of "mini-maxi double-bubble rendezvous," which involved a wide separation between the LEM and C.S.M., the SimSup failed the LEM engines, forcing the crew and the flight control team to conduct a rescue in which the command module maneuvered into a rendezvous with the LEM instead of the other way around. Since the orbital characteristics of the two craft meant that the rescue would require thirty hours, and the LEM for Apollo 9 was only an eighteen-hour spacecraft, the White Team had to work out procedures to keep the LEM's astronauts alive. During the Apollo 9 flight itself, the White Team gained another useful bit of experience when the lunar module conducted DPS burns with the command module attached, to test the ability of the LEM to control both spacecraft.

Such experiences led to a few procedures that could be pulled off the shelf during Apollo 13. Still, the flight control team had never developed procedures for an accident of the magnitude of Thirteen's. They had

imagined losing one of the two oxygen tanks, or a battery, or even the
S.P.S. engine. But never had the simulations postulated that the C.S.M.
would be completely dead.*

1

The problem that preoccupied the MER and the contractors from the time
of the accident through the rest of the mission was whether *Aquarius* had
enough water, oxygen, and electricity to sustain life until Apollo 13 could
get home.† It was known as the "consumables problem."

Aquarius had been designed to take Lovell and Haise to the lunar
surface (less than five hours), remain there for another thirty-three hours,
then rendezvous with the command module (about two hours), and still
have a five-hour reserve—a lifetime of forty-five hours. Counting from
the time the LEM was activated during Apollo 13, the return could take
as little as seventy-seven hours or as much as a hundred, depending on
what kind of additional burns were decided upon. And those decisions,
filled with trade-offs and hazards, depended in turn largely on the
consumables situation. The specialists on the LEM quickly came to a few
basic conclusions.

Oxygen was not a problem. To support a lunar landing, the LEM was
designed on the assumption that its door would be opened when the
astronauts went outside for their E.V.A., venting all of the cabin's
oxygen. For Apollo 13, two E.V.A.s had been planned, meaning that the
LEM had to carry enough oxygen to repressurize the cabin twice, plus
provide enough oxygen for the time that the crew was in the cabin. With
that much and with the safety margin built into the system, *Aquarius* had
plenty of oxygen for the trip home even with all three astronauts drawing
on it. Because the LEM used batteries instead of fuel cells, oxygen didn't
figure in the calculations about power supply.

In contrast, water was a huge problem. The electronics in the
spacecraft generated heat which was carried off by glycol circulating
through the systems. The warmed glycol was recooled by running it

---

* It was assumed that anything that knocked out both oxygen tanks would also have destroyed
the spacecraft. One result of the Apollo 13 experience was a change in simulation policy. FIDO Dave
Reed summarized the new ground rule: "They can throw anything at us they want and we won't
object."

† They didn't have to worry about food. People don't die from starvation in four days, and
anyway the C.S.M.'s food supply was still accessible, even though much of it was virtually inedible
without hot water to mix it with.

through tubes encased in ice. The ice was made by the cold of space from water supplied by the LEM. As the glycol ran through the pipes, the ice vaporized and boiled away. Every system that remained powered up used water.

Aquarius carried 338 pounds of water in its tanks. During the first hours after the crew moved in, the LEM was consuming water at the rate of 6.3 pounds per hour. Arithmetic immediately revealed that, at that rate of consumption, *Aquarius* would have no water at all after fifty-four more hours—twenty-three fewer hours than the fastest possible return to earth. The astronauts would survive without water for the extra twenty-three hours, but the equipment wouldn't.

Electrical power was also in short supply. The descent stage of *Aquarius* had four batteries; the ascent stage had two. From the six batteries together, and after subtracting the power that had already been expended in checking out the LEM, the planners on the ground could count on fewer than 2,000 ampere-hours of electricity. Ordinarily, the LEM used 50 amps per hour. Two thousand divided by fifty gave only forty hours. Moreover, they were going to be making extraordinary demands on *Aquarius*. As the men in the MER and at Bethpage calculated the maneuvers *Aquarius* would have to make and the power it would take from the LEM to power up the partially depleted entry batteries in *Odyssey*, it was decided that *Aquarius* would somehow have to use a maximum of 15 amps per hour when it was not maneuvering. Until the day of the accident, the LEM's designers had calculated that the LEM's minimal configuration used 20 amps per hour.

To conserve water and power, the engineers in the MER went systematically through the spacecraft, looking at every component to see whether it could be turned off completely. If not, they calculated the lowest voltage at which any given piece of equipment would operate and the minimal configuration in which a system would continue to perform the functions it absolutely had to perform. Forget the specs, forget the rule books, forget the operations manuals, Arabian announced. Go back to the physics of the design that no one knows better than you, the knowledge that's not written down anywhere, and decide what's the least we can really get away with.

For example, the DPS engine contained an Abort Sensor Assembly (A.S.A.) that monitored its performance. The A.S.A. included small heaters that kept it at 70 degrees Fahrenheit. What would happen if the heaters were turned down so that the fluid in the A.S.A.'s gyroscopes was maintained at a few degrees above freezing? Would the A.S.A. still work? There was no time for anyone in the MER or at Grumman to conduct tests; the answer had to be provided on the basis of the inadequate data at hand. In the case of the A.S.A., the engineers decided that it should be able to function under those near-freezing conditions, and so the crew was told to turn the heaters down. Everyone hoped for the best.

As such questions spread through the support network, often ending in the laps of the people who had originally designed the equipment, cautious engineers were typically faced with a choice between two incompatible demands. Each was sensitive to the critical nature of his own piece of the puzzle (for example, it could be disastrous if the A.S.A. failed to detect a malfunction in the DPS during the next burn). The temptation to insist that his piece of equipment be operated within the tolerances for which it had been designed was powerful. And yet each also understood that somehow the voltages and the heat sources in the lunar module had to be cut drastically. Small dramas were played out all over the country as designers passed on the word, usually with too little data to be sure they were right, that their babies would continue to function under the unprecedented conditions that Houston was proposing. This process of redefining the LEM's capability continued throughout the flight, as the reserves dwindled and the conditions within the spacecraft kept changing. Years later, John Strakosch, a Grumman engineer at Bethpage, would say it was the most intensely concentrated work he'd ever done—an effort that in his memory seemed to last eighteen or twenty hours. His handwritten log of his activities reveals that it lasted for what amounted to three unbroken days.

Bit by bit, the LEM was powered down to 15 amps per hour, and the astronauts, wearing thin clothing designed for a long trip in a confined space at 70 degrees, began to get cold as the temperature dropped below 60 degrees and kept going down.

In the process of inventorying the consumables during the first night, someone discovered that, as things stood, the astronauts would asphyxiate from carbon dioxide buildup before they got home. In both the LEM and the command module, carbon dioxide was removed by circulating the air through canisters of lithium hydroxide. The problem was that *Aquarius* had only two such canisters, not nearly enough to last the journey home. *Odyssey* had plenty of canisters, but they were the wrong size and shape to fit the LEM's equipment. The problem was passed along to the Crew Systems Division: Figure out some way to use the C.S.M.'s lithium hydroxide canisters in the LEM. Unless they came up with an answer, none of the rest of the planning was going to make any difference.

2

At eight o'clock on Tuesday morning, less than twelve hours after the explosion, a meeting of NASA's most senior managers, from the Houston directorate chiefs up through administrator Thomas Paine, convened in the viewing room behind the third-floor MOCR. Gerry Griffin's Gold Team was now on shift, its members occasionally looking back curiously at the all-star assemblage behind the glass.

Chris Kraft led the discussion. The question before the group was how quickly to bring the crew back. It had already been decided to conduct the next burn at two hours after pericynthion (the spacecraft's closest approach to the back of the moon), making it known as the P.C. + 2 burn.

Kraft explained the options. They could, if they chose, bring the crew back in less than a day and a half after that with a long burn at full throttle (the "fast" burn), if they were willing to bring *Odyssey* down in the Atlantic where NASA had no recovery ships. Kraft didn't spend much time on this option, for it was only a few hours faster than the second option, which, in less than thirty-nine hours after the burn, would put the spacecraft in the southwest Pacific, the prime recovery area. The third option was a shorter, lower-power burn (the "slow" burn) that would return the spacecraft to the prime recovery area sixty-three hours after the burn, or twenty-four hours later than the second option.*

Given the concern about the consumables and the risks inherent in keeping the crew in a disabled spacecraft any longer than necessary, the case for the fast burn—in its second version—seemed so compelling that the astronauts over in Building 5 were already practicing it. However, Kraft wanted the slow burn even at the cost of the extra twenty-four hours. So did his lead Retro and his lead FIDO, Chuck Deiterich and Dave Reed. Kraft called the two of them into the viewing room so that the others could hear why, firsthand.

The fast burn would take virtually all the propellants that the LEM had, they explained, with little energy left for "tweaking" the trajectory if it

---

* They had no intermediate choices if they wanted to land in the prime recovery area. A spacecraft returning from the moon had limited maneuvering range. If one wanted to put the spacecraft in a particular spot along the potential landing track, it was available only once per earth rotation.

were in error. It didn't take much of an error in the burn to make tweaking necessary—at 240,000 miles away, an error of a tenth of a foot per second in a burn could compound in such a way that the spacecraft would miss the earth altogether.

The fast burn would also require that the heavy service module be discarded beforehand. "They don't have a rearview mirror on that spacecraft," Reed pointed out. "We don't know what's happened back there." Why take the damaged service module off any sooner than they had to? More important, the service module covered *Odyssey*'s heat shield and protected it from direct exposure to the heat (on the sun side) and cold (on the shadow side) of outer space. No one had ever tested what would happen to the heat shield if it were exposed to the thermal conditions of outer space for forty hours before it had to work. No one knew how much effect it would have on letting heat out of, or cold into, the command module itself. Deiterich and Reed didn't want to use Apollo 13 to find out.

As for the extra twenty-four hours that the crew would have to spend in space: By this time, Tuesday morning, people were confident that the consumables could be stretched to last two and a half more days. The main problem still outstanding, the shortage of lithium hydroxide, wouldn't be avoided by the fast return—the LEM would run out before the crew got home, no matter which burn they chose. And the slow burn still kept options open. If some additional problem developed on the way home, they could always use the remaining propellants to speed up the spacecraft and make an emergency landing in the nearest available ocean. As he talked, Reed was gratified to see Bob Gilruth, still center director at Houston and the grand old man of manned space flight, nodding in agreement with each of his points.

Kraft had reached his own opinion after thrashing out the options at a meeting of the flight directors a few hours earlier. When Deiterich and Reed had finished, Kraft summarized the options, leaving no doubt that he wanted the slower burn.

For Glynn Lunney, standing alongside Kraft, the reaction of the men in the viewing room was inspiring. The option Kraft was recommending had the obvious disadvantage of leaving the crew at risk for twenty-four hours longer than the alternatives, but there was no quibbling. Lunney: "These men who we all grew up thinking were kings, absolute tops in our business, sat in this room and said to us, 'Look, we're here to support you guys, and we just want to hear what's going on and we want you to know that if there's anything at all that you need, you just let us know.' " Then and in the days to come, Lunney recalled, "they never second-guessed, never reversed anything, never even came close to that. No cover-your-ass. The whole set of interactions was one of the most supportive and reassuring you can imagine." The gathering quickly ratified Kraft's choice: The Apollo 13 crew would come home more slowly, but, they hoped, more surely.

3

At that point, one day into the crisis, Chris Kraft and the flight directors who attended the press briefings were confidently telling the world that the astronauts were going to get back safely. Asked to reminisce years later, the flight controllers, almost to a man, said they were never in any doubt about whether they could get the crew back once it had been determined that the consumables would stretch. It is curious but true that none of them seemed to doubt that a way would be devised to use *Odyssey*'s lithium hydroxide supplies to stretch the LEM's—somebody would figure something out, they assumed.

The optimism in the press briefings may have been influenced by public relations, and the confidence in the recollections may have been influenced by the passage of time. Yet it does appear that the flight controllers really were confident throughout. Gene Kranz explained it as a matter of training: "I don't think anybody was pessimistic. Not at least on the White Team. . . . The training was such that, by the time you finished the process, you had the confidence that, given a few minutes, you could solve any problem. That's all there was to it. And it didn't matter what the size, what the magnitude, what the origin of the problem was. The fact was that you could solve any problem that came up." Don Arabian in the MER, at least as cocky as any flight controller, felt the same way—after the first half hour, he knew that "Yeah, we can get 'em back." It was no big deal, he said.

People who were neither controllers nor Arabian tended to be less sanguine. On Tuesday evening, a day after the explosion, journalist Robert Sherrod encountered George Low sitting in the front row of the viewing room and asked him what the chances were that they would get the crew back. Low, who knew that Sherrod was writing a book and not for the next day's newspapers, looked at him bleakly and said, "Fair." The problem, he explained, was that all the systems were now committed.

There was no margin for anything else to go wrong. He refused to put a percentage on the odds. A little later, Sherrod mentioned Low's response to two M.S.C. officials. They thought Low sounded optimistic compared to people they had been talking with—"Lots of guys around here wouldn't [give] you a fifty-fifty chance," one said. The next day, Sherrod encountered ASPO engineers Aaron Cohen, one of M.S.C.'s leading experts on the C.S.M., and Owen Morris, one of its leading experts on the LEM. Both were still skeptical about success. "It's so far to go," Cohen said. Years later, when he was about to become center director at Houston, Cohen had not forgotten his anxiety during Thirteen. Even in retrospect, it was to him a very close thing.

Despite the controllers' optimism, which to some degree was part of their job description, the design engineers were right about how closely the margins had been shaved. A senior engineer in the Test Division, speaking just a few days after Apollo 13 had landed, said that when he heard the news about the accident he assumed the Apollo 13 crew were goners. "If you had asked [before the accident] what would happen if we lost both oxygen tanks fifty-eight hours into the mission, we'd have said, 'Well, you can kiss those guys goodbye.' "

4

Each of the major participants in Apollo 13 remembered a different moment that, to him, represented ultimate crisis. For Lunney it was the final few minutes of the transfer from *Odyssey* to *Aquarius*. For Gerry Griffin and Jim Lovell, the "biggest heart-stopper" (in Lovell's words) occurred while they were preparing for the P.C. + 2 burn.

The technical problems of piloting that Lovell and Haise faced—Swigert, as command module pilot, wasn't trained to fly the LEM—were enormous. *Aquarius* was about the same length as the C.S.M., but it was little more than half the weight, and the LEM's biggest engine, the DPS, had less than half the thrust of the C.S.M.'s engine. Using *Aquarius* to guide the C.S.M. was like using a small car to push a limousine, but in three dimensions, with requirements for precise adjustment. It had already been a problem during the first night, when Lovell and Haise had had to maneuver the assembly into a passive thermal control (P.T.C.) mode, the slow, turning-on-a-spit spin that Joe Shea had suggested five years earlier. It became an even bigger problem on Monday afternoon, when *Aquarius* had to get a star sighting to prepare for the P.C. + 2 burn.

Checking the alignment with star sightings was standard procedure on every flight. Even when they were functioning normally, guidance platforms tended to drift a little. Before each major maneuver, the astronauts went through a straightforward and largely automated procedure. The astronauts would select an appropriate star from their list and ask the spacecraft to find it for them. The spacecraft, consulting the I.M.U., would orient itself so that, if the platform was properly aligned, the star in question would be centered in the cross hairs of the crew's sextant, the A.O.T. (Alignment Optical Telescope). If the star wasn't exactly centered, the astronaut looking through the A.O.T. made the necessary adjustment to center it, and the computer entered the adjustment into the I.M.U., which then corrected itself.

When the crew of Thirteen tried to do this with the alignment it had borrowed from the C.S.M., they found that the debris from the explosion in the service module still accompanied them, reflecting light and creating a swarm of false stars indistinguishable from the real ones. They reported the problem to the ground. "A genius in Mission Control," Lovell wrote in his account of Apollo 13, suggested using the sun instead (it was Ken Russell, the lead Guido). Not even the particles from the debris could block out a target as large as the sun.

Deiterich was unhappy with the prospect—the sun was too big for the kind of accuracy that star sightings would have given them. But no one had any better ideas, so Gerry Griffin passed along the procedure to Lovell and Haise, who tried to implement it. What followed was an example of the kind of obstacles that the crew and the ground had to surmount throughout the flight.

The LEM simply was not designed to do certain kinds of things, and getting a manual star sighting with thirty-two tons of C.S.M. hanging on to it was one of them. Whereas the star-sighting telescope in the command module was mounted so that it swiveled, the simpler one in the lunar module did not. To get a star sight to check *Aquarius*'s alignment before the big burn, Lovell would have to guide the whole cumbersome LEM/C.S.M. assembly manually and then hold it steady for long enough to get a fix.

By Tuesday afternoon in Houston, Lovell and Haise had gotten only a few hours of sleep and their energy reserves had been drained by more than eighteen hours of crisis. Now, without fail, they had to carry off this extremely delicate maneuver. As Lovell made their preparations to begin,

Flight Director Gerry Griffin was so nervous that, later, he could barely make out his own handwriting in the flight log.

The ground had worked out the attitude that Lovell would have to reach. The astronauts manning the flight simulator over in Building 5 had practiced the procedures. Now it was Lovell's turn, to work the spacecraft painstakingly into the prescribed position. He was grateful for his experience on Gemini XII, when he had had to maneuver Gemini XII with the dead mass of an Agena rocket hanging on to its front, though that did not make the present job less demanding—"It was a complete learning curve, right now," as he described it. Later, listening to a tape of the air-to-ground loop toward the end of the maneuver, Lovell was reminded of the song from *My Fair Lady*:

"Okay," said Lovell. "We got it. I think we got it. What diameter was it?"

"Yes," Haise confirmed. "It's coming back in. Just a second." The spacecraft had slipped out of position.

"Yes, yaw's coming back in. Just about it."

"Yaw is in . . ." At this point Lovell was struggling to hold the spacecraft steady, balancing a vehicle that kept insisting on moving.

"What have you got?"

"Upper right corner of the sun . . ."

"We've got it!"

In the MOCR, the Gold Team cheered and pounded on their desks. Griffin himself would always feel a chill up his back when he recalled the moment. The crew had a valid alignment and knew how to position the vehicle for the P.C. + 2 burn. For Griffin, this was the moment when he became confident that the crew was coming home alive.

For Gene Kranz, "the only dicey thing that occurred throughout all of Apollo 13" came at the conclusion of the P.C. + 2 burn.

The burn itself had been an anxious event. When the White Team came into the MOCR at four o'clock Tuesday afternoon to take the first of the two shifts Kranz had assigned to it for the rest of the flight (the other would be the entry itself), many of its controllers had been working continuously for thirty hours. Astronaut Joe Engle was still in the simulator giving the team a final verification of the DPS burn procedures. The fallback procedures to use in the case of a delayed or incomplete burn were still being worked out. And *Aquarius* was eating up electricity at the rate of 40 amps per hour, jeopardizing its slender margin for the trip home. From the fourth row, Chris Kraft watched the energy consumption with increasing apprehension and occasionally communicated his concern to Kranz. But there wasn't anything Kranz could do at the moment—the LEM had to prepare itself for the burn, and that required most of its systems to be powered up.

When the time for the burn arrived, CapCom Vance Brand was so keyed up that he was watching the wrong electronic clock among the several which were counting down to various events; he called "Mark," which was supposed to come exactly forty seconds before ignition, three minutes early. The mistake was caught and corrected. Lovell ignited the engine on schedule—it had to be done manually, for the computer was not programmed to handle a docked burn—and slowly brought the engine up through three stages to full thrust, where it continued to burn for four minutes and cut off within 130 milliseconds of the prescribed time. Thirteen was on its way back to a landing in the Pacific's prime recovery area.

Kranz's problem began as soon as the burn was completed, when three formidable gentlemen, each with something on his mind, converged on him at the flight director's console. Kranz later agreed with Lunney that, for the most part, senior management played an unassuming and helpful role throughout the crisis. But, he added, "all those folks were giving us our head while we were fighting the problem and when they had pretty much no option to intervene." With the P.C. + 2 burn behind them, he suddenly discovered that "now management wanted to sit down and discuss philosophy thereafter"—"management" meaning Max Faget, Deke Slayton, and Chris Kraft.

For one of the first times since he had played through the options for dealing with John Glenn's heat shield problem, Max Faget was deeply involved with an ongoing flight. Having examined the characteristics of the underpowered, unwieldy, damaged spacecraft that was heading homeward, he was adamant that Kranz get the spacecraft back into passive thermal control immediately. Otherwise they were in danger of overheating systems that were in the sun and freezing up propellant lines that were in the sun's shadow. Faget was worried that unless they got into P.T.C., they would be unable to power up the command module's systems when the time came for entry.

But getting the spacecraft into P.T.C. could take as long as two hours. Deke Slayton, director of Flight Crew Operations, wanted to get the crew to sleep, and he was supported by Chuck Berry, the head of M.S.C.'s medical office. The crew was dead on their feet; they wouldn't be able to execute the delicate maneuvering for a P.T.C. now; they had to get some rest.

For his part, Chris Kraft was still worried about power consumption, as he had been since he had stood behind Gene Kranz on Monday night and watched the first hours of the crisis. Now that the burn was over, Kranz should get down to his 15-amps-per-hour-ration right away and let P.T.C. wait.

As each argued his point of view, others got into the act. The TELMUs, also worried about the power profile, were on Kraft's side. The astronauts in Building 5, knowing the mistakes that you could make when you were cold and exhausted, were on Slayton's side. It was a real uproar, Kranz remembered. As it happened, Kranz was one of the few who agreed with Faget.

"The Flight Director may choose to take any necessary action," the rules said. It was too bad if the crew was tired, Kranz told the others, but they were going to feel a hell of a lot worse if they tried to power up the C.S.M. for entry and found out it didn't work. They weren't going to go to sleep; they weren't going to power-down; they were going to put the spacecraft into passive thermal control.

The P.T.C. maneuver the night before had been a rough-and-ready approximation. The spacecraft had gotten into more or less the right position, and every ninety minutes Lovell or Haise had fired thrusters to give the spacecraft a turn. Now, however, the LEM engineers and the astronauts in the simulator had worked out a way to get Apollo 13 into a clean, automatic, sustained P.T.C., if Lovell and Haise—still subsisting on a few hours of sleep—could carry out the instructions that were relayed up to them.

The first attempt didn't work. Instead of spinning smoothly on its axis, the spacecraft developed a wobble—a "coning angle." "And they're tired," Kranz recalled of the crew, "I mean, they are really starting to drag at this time." That, Kranz remembered, thinking of the eminent crowd surrounding the flight director's console, was when "*everybody* was sort of glowering." They wanted to get the crew to sleep, they wanted to get powered down, and "I'm still after passive thermal control."

They didn't call him General Savage for nothing. "Okay, we're gonna go do the thing again," Kranz announced, and that's what they did. Fortunately (probably for Kranz as well as the crew), this time the crew did it perfectly. "So we set up the passive thermal control, and we got powered down and put the crew to sleep, and everybody got happy," Kranz concluded. "But that was what I'd say was the only significant crisis that erupted."

It all depended upon your point of view.

5

An issue that had not been stressed during the decision between a slow and a fast burn, but which was as important as any of the others, was that the Control Center badly needed the extra twenty-four hours to prepare for entry. Kraft had asked Kranz about it before he went into the meeting in the viewing room Tuesday morning, and Kranz had been blunt, telling Kraft that "there was no way" the team could be ready for entry unless they opted for the slow return. This was why Kranz had taken the White Team "off line," as he put it, to concentrate on the entry. The entry for Apollo 13 would require a completely novel checklist, one which took into account a host of anomalous factors: The batteries in the C.S.M. would have to function for a matter of hours instead of the usual half hour or forty-five minutes; the LEM was going to execute preparatory maneuvers that were ordinarily done with the S.P.S. engine; the crew would be powering up a command module that had been cold and wet for more than three days. Once the White Team had compiled this checklist, they would have to couch it in such a way that its hundreds of steps could be copied and understood by an exhausted crew.

Because the White Team spent all of Tuesday preparing for the P.C. + 2 burn, it wasn't until Wednesday morning that they began concentrated work on the entry. They were supposed to come to the job fresh; Kranz had insisted that they go home and get some rest after the burn. But some didn't obey (Kranz himself was a bad example, subsisting on his catnaps in the viewing room), and others went home only to end up working on their taxes (Wednesday was April 15). So as the White Team assembled Wednesday morning, they were a disheveled and weary crew, with a little more than two days left before splashdown. Under ordinary circumstances, preparing an entry checklist for a lunar mission was a three-month job.

On that same Wednesday morning, the lithium hydroxide in *Aquarius* was depleted and the carbon dioxide levels approached the danger point.

By that time, however, a solution had been passed up to the crew. Working nonstop since the early hours of the explosion, members of the Crew Systems Division had invented a box that could hold the C.S.M.'s lithium hydroxide canisters and be connected to a hose in the LEM ordinarily used to suck air out of space suits. The box was built out of storage bags, tape, and the stiff plastic covers from the crew's checklist book. The remaining problem, once the people in Crew Systems had figured out a solution, was explaining to the crew what they had done. Hours had been spent carefully working out verbal instructions for building a contraption which the crew had never seen. (Which would be more accurately understood: Cut a piece of tape "thirty inches long" or "a good arm's length"?) CapCom read up the instructions, Haise assembled the materials, and Swigert and Lovell managed to put it together before the carbon dioxide level became intolerable. Within a few hours, the level was back to normal.

Down in Room 210, a kind of collaborative choreography was under way. Kranz had assigned Arnie Aldrich to put together the integrated checklist that would jettison the service module, transfer the crew from the *Aquarius* back into *Odyssey*, jettison *Aquarius*, and position *Odyssey* for the entry. John O'Neill was in charge of translating the checklist into formats that could be used by the crew, making sure that the astronauts in the simulators had verified that they worked. John Aaron was the "power manager," responsible for deciding whether the checklist that Aldrich was developing fit within the constraints of electricity and water. Kranz himself refereed arguments, set deadlines, enforced decisions, and kept management informed.

The White Team, now augmented by controllers who had not been working the Apollo 13 mission and calling itself a "tiger team" in the time-honored aviation tradition, split itself into small working groups that held meetings in nooks and crannies throughout the Control Center. Every three or four hours they would reconvene in Room 210. Aldrich and O'Neill would act as scribes, assembling the new contributions from each group into the right sequence. Then Aaron would analyze whether the procedures were affordable.

Aaron's job was seemingly impossible. Not only did the entry batteries have to last three times as long as normal, one of the three batteries had been half depleted during the changeover to *Aquarius*.

Aaron worked backwards, making a budget in the manner of a family that doesn't really have enough money to do something but is going to do it anyway. Then he called together the flight controllers and said, "Here's the plan, what do you think of that?" When they exploded with protestations and predictions of catastrophe, Aaron and Kelly made slight adjustments—still far short of everyone's minimal requirements—and went back for another try. "We were in the rationing business," Aaron said. Then, once they had a scheme that worked, at least on paper, they were "the castle guard for the plan, handing out electrons."

Some of the decisions that Aaron had to make were draconian. A G.N.C. asked Aaron for permission to turn on the Primary Guidance and Navigation System, PGNS, during the entry. Denied. Then how about the secondary system, consisting of six gyroscopes? Nope, Aaron didn't think they could afford even that. This left the G.N.C. with the prospect of a tertiary system, which did nothing but measure g forces. If the g-forces buildup varied from the prescribed rate, the astronauts were supposed to adjust the tilt of the C.S.M.—"riding down the g's," it was called, a terrifyingly primitive method. Aaron compromised: In addition to the meter registering g's, G.N.C. could power-up three of the six gyroscopes on the secondary system. That's the kind of power squeeze they all faced.

Even being ruthless wasn't enough. After many iterations, the tiger team had completed a checklist that they thought would work. It also left a 16 amp-hours margin, the minimum that the recovery people insisted upon in case the *Odyssey* landed in the "Stable 2," or upside-down, position in the ocean. This stringent budget was taken over to the flight simulators, where Ken Mattingly, the astronaut who would have been on Apollo 13 if he hadn't been exposed to the measles, tried it out. To Mattingly's surprise, the hastily prepared checklist worked without a hitch. But instead of landing with a 16-amp-hour surplus, Mattingly ended with a 10-amp-hour deficit.

There was no place to cut. Aaron was already holding most of the systems to budgets that the controllers had originally said they couldn't possibly live with. Bill Peters, the lead TELMU, came to the rescue. He was now confident that *Aquarius* would have a small reserve of battery power, and the MER had figured out a way to pass power from the LEM to the C.S.M.* Aaron could gain 20 amp-hours by using *Aquarius* to charge *Odyssey*'s half-depleted Battery A.

The creation of the checklist was a controlled frenzy in which solutions were developed, rejected, and replaced in such rapid-fire sequence that it

---

* The spacecraft's design provided for power to flow the other way, to top off the LEM's batteries from the C.S.M. The MER had found a way to do it backwards.

was difficult at any one moment to be sure how far they had to go. When the Apollo 13 crew inquired about progress, CapCom Joe Kerwin assured them, deadpan, that the checklist would be ready "by Saturday or Sunday at the latest." Splashdown was Friday.

An Apollo spacecraft returning to earth from the moon did not aim directly at the earth, a passage which no heat shield could survive. Rather, it aimed at the leading edge of the earth so that, as the spacecraft sped by, it would be caught by the earth's atmosphere and gravity. If the spacecraft was too far from the leading edge, it would continue past the earth into an elongated earth orbit. If it was too close to the leading edge, it would enter the atmosphere at a steep angle and burn up. The area that was neither too high nor too low was about a degree and a half wide—ten miles wide at the point of entry—and was called the "entry corridor." From Apollo 8 onward, one of the chief functions of the Trench was to make certain that the returning spacecraft was in the middle of it.

As early as Wednesday morning, lead Retro Deiterich and lead FIDO Reed were puzzled and concerned by Thirteen's errant trajectory. For unknown reasons, the trajectory was "shallowing," moving toward the high side of the corridor. If it continued, Lovell and his crew would miss the earth altogether. They called for a midcourse correction. Ordinarily, this would not have been a major event. But to conserve power, the LEM's guidance system had been turned off after the P.C.+2 burn, and a working guidance system was necessary for a trajectory correction. Instead of turning on the guidance system and trying to get an alignment from scratch, which would still be just as difficult as Stafford had warned in the first hours of the crisis, Dave Reed dusted off an untried procedure.

The method involved taking a sighting on the earth's terminator—the line dividing the illuminated and darkened portions of the earth. Specifically, the ground told Lovell, he was to align the cross hairs of the A.O.T. so that they were just grazing the "horns" on the crescent earth he was seeing out the window. Lovell remembered that a similar technique had been devised for Apollo 8 as a last-ditch alternative in case of a crisis where everything else had failed. After Apollo 8, the Trench had discarded it, saying to the astronauts, as Lovell recalled their conclusion, "You might as well use prayers if you're going to try this procedure." It was with these memories in mind that Lovell heard his instructions. "I hope the guys in the back room who thought this up knew what they were doing," he told CapCom.

They did. The alignment was accomplished uneventfully and the burn on Wednesday night put Thirteen back into the middle of the corridor.

By Thursday, the temperature in *Odyssey* was 38 degrees, about that of a refrigerator. Until then, the crew had used *Odyssey* as the bedroom of a two-room apartment; now they huddled together continuously in the little LEM, which was marginally warmer. But *Aquarius*, too, was uncomfortable. The walls and windows of the LEM were moist with condensed water. The men were averaging three hours of fretful sleep per day. Also, concerned about having enough water to cool the electronics, all three astronauts had cut their fluid intake to near zero. They deliberately ignored the Surgeon's instructions to drink—since they didn't feel that thirsty anyway (one doesn't, in space), they chose to save water for the equipment. The result was that, without knowing it, all three were becoming dehydrated. Haise had developed a kidney infection and was running a fever.

Fatigue. Cold. Dehydration. As they neared time for entry, the normal performance levels of all three had deteriorated. Now they were going to have to assimilate a hastily written entry checklist, perform a crucial midcourse correction, and then bring the spacecraft home without any of the backup systems that ordinarily provided some margin for error.

On the ground, even as the rest of the tiger team was racing to finish its checklist, Deiterich and Reed were worrying anew over the tracking data: For reasons that continued to baffle them, the trajectory was once again shallowing. Another midcourse correction would be necessary, this one to take place only five hours before the spacecraft hit the earth's atmosphere.

6

At 4 A.M. on Friday, slightly more than eight hours before the scheduled splashdown and seventy-nine hours after the explosion, Gene Kranz and the White Team took over the consoles in the MOCR. The entry checklist had been given to the crew during an earlier shift. Mattingly, who had practiced it in the simulator, had read it up, a distant voice in Swigert's ear. Swigert had written down each step in full—he didn't trust himself to read abbreviations later—and then read them all back to Mattingly for verification. After two hours of this tedious process, Mattingly had switched over to Haise, reading up the somewhat shorter procedures for the LEM.

As Deiterich seated himself at the Retro console, he was still watching a deteriorating trajectory. Earlier, he had thought that the second burn would be a matter of fine-tuning to bring the spacecraft close to the recovery vessels. But as the time for the burn approached, the rate of shallowing kept increasing, and with it the length of the burn that would be required even to keep Thirteen within the corridor. When it came time to execute the burn, the crew's fatigue became apparent. Guido Ken Russell suddenly noticed on his screen that Lovell had punched P40 into the computer instead of P41. The critical difference was between igniting the large descent engine—not their intention—and igniting the small thrusters. Russell told Kranz: "Should be forty-one, I believe, Flight—not forty," and the mistake was averted. Then Lovell, one of the most skilled and most experienced of the astronauts, mistakenly rolled the spacecraft 18 degrees in the wrong direction.* This mistake, too, was noticed and corrected, and the burn itself went off without a hitch. "The crew was deadly tired," Kranz recalled. For him, this period of the flight was one of the most impressive; an "incredible duet" by crew and flight controllers.

Odyssey was visibly wet from condensed moisture when they powered it up. Pushing aside thoughts of how water-soaked wires might react, Swigert began to turn on equipment in the command module. After feeling like a passenger for the last three days, the command module pilot was glad to have a job to do again. He also watched himself critically, aware of the ways in which exhaustion could betray him. When it came to arming the pyrotechnics, he was taking no chances at all. Once they were armed, a flick of a single switch would separate *Odyssey* from the service module; the flick of another switch right beside it would separate *Odyssey* from *Aquarius*. Because the LEM had to do all the preliminary maneuvering for the entry, the service module would go first. The day before, Swigert had put a piece of tape over the LEM's switch. A piece of paper dangled from it, with "NO" written in large red letters.

Separation of the service module came at 138 hours into the mission, four and a half hours before splashdown. From his station in *Odyssey*, Swigert

---

* It would later be found that Lovell, as commander, had limited his fluid intake even more stringently than the other two, and was the most severely dehydrated. He lost fourteen pounds during the six-day voyage.

couldn't see the service module as it drifted away. He could, however, hear excited voices from *Aquarius*. "There's one whole side of the spacecraft missing," Lovell reported to the ground. Wires were dangling, the area around the oxygen tanks was a tangle of ripped metal—"It's really a mess," Haise said. Until then, no one had realized the magnitude of the explosion. Chuck Deiterich listened over the loops and felt a chill. The oxygen tanks were close to the base of the heat shield. The obvious possibility was that the heat shield had been damaged, perhaps even cracked. "I think everybody in the room had the same idea at the same time," Deiterich recalled. "Everybody knew where the oxygen tank was. Nobody said a word about it. There was nothing anybody could do." In any case, John Aaron kept them from brooding—somewhere, something in the C.S.M. was using 2 amps more than his budget called for, and Aaron was implacable, Kranz recalled, "driving the controllers nuts" as he went from console to console until he found the culprit: a backup control system that had inadvertently been switched on.

The good news was that the command module's guidance system had survived the refrigeration. However, Control and Guido were finding its alignment to be a struggle. The crew had transferred a rough alignment from *Aquarius* to *Odyssey*, but it had been a haphazard proposition— Lovell in the LEM, shouting the angles to Swigert in the command module, hoping the spacecraft didn't shift position between the time he shouted and the time Swigert punched the number into the computer. Getting star sightings from the command module to check the alignment had once again proved difficult, and only after a long struggle was it determined that the rough-and-ready alignment was good.

Then it was time to position *Odyssey* for jettisoning the LEM. Looking out the window, Lovell was supposed to use *Aquarius*'s thrusters to pitch and roll and yaw the docked spacecraft so that he could see a particular star, then shift the spacecraft so that he could see another, then another, until at the end of the sequence he would be in the right position for the jettison. In the process, Lovell brought the command module's guidance system, so recently aligned, perilously near the gimbal lock. Once more instructions had to be shouted across the two spacecraft, this time from Swigert to Lovell, to do this, that, and the other thing to keep *Odyssey*'s guidance system safe.

At 10:43 Friday morning, the crew of Apollo 13, together in *Odyssey* for the first time since they had evacuated it three days earlier, jettisoned

Aquarius. "Farewell, *Aquarius*," Lovell said quietly, "and we thank you." In the MOCR, TELMU wished irrationally that there were some way to preserve *Aquarius*'s life, to bring her back home after all she'd done. In the Grumman plant at Bethpage, a LEM engineer, filled with pride in his machine, remembered a time in his youth when he had pulled a struggling swimmer from the water. That had made him feel good—but nothing like this.

Retro's worst moment that morning came after *Aquarius* had been jettisoned, when *Odyssey* was approaching the earth alone. One may visualize what was happening by imagining a line drawn between *Odyssey* and the earth's horizon. *Odyssey* had to be aimed at a point no less than 5.5 degrees and no more than 7.3 degrees below that line. As *Odyssey* approached earth, Deiterich had been forecasting that the actual angle of attack would be 6.51 degrees, safely in the middle of the corridor. Now, minutes before *Odyssey* reached the atmosphere, Dave Reed handed Deiterich updated figures showing that the angle would be 6.2 degrees. "I don't believe you, FIDO!" he said in frustration, knowing that he had to. Once again, as mysteriously as ever, the spacecraft's trajectory was shallowing.

Without much data, Deiterich now had to make a crucial decision. He was due to pass up to the crew the final PAD (Pre-Advisory Data), containing the "lift vector" that would govern the spacecraft's degree of lift while it fought the atmosphere. If between this moment and the time the spacecraft hit the atmosphere the entry trajectory shallowed by as much as an additional one-tenth of a degree, Deiterich must change the lift vector from the one he had planned. The trajectory had already shallowed by three-tenths of a degree that morning. Would it continue at the same rate? If he made the wrong choice he would certainly affect the landing site substantially. He conceivably could cause the spacecraft to miss the entry corridor altogether. The crew and Kranz periodically inquired when the PAD was going to be ready, and Deiterich kept putting them off until the last minute. His last possible trajectory update from Reed indicated that the rate of shallowing had decreased. Deiterich held his breath and decided to stick with the original PAD figures.*

---

* The cause of the shallowing throughout the return from the moon was eventually determined to be water boiling off the LEM's cooling system. (This effect had not had time to reveal itself during the comparatively short hops that the LEM had taken during previous missions.) Once the LEM was jettisoned, the cause of the shallowing ended, which is why the last updates showed a slowing rate and encouraged Deiterich to guess correctly that the additional change would not exceed a tenth of a degree.

*   *   *

The traffic on the loops in the last minutes before the spacecraft reached what was known as the "entry interface" was indistinguishable from what might be expected for a nominal entry after a nominal mission. Long stretches of silence were broken by routine updates on the trajectory and the disposition of the recovery forces.

The last exchange between the crew and the ground began on an unsettling note. "I know all of us here want to thank all you guys down there for the very fine job you did," Swigert said, as if he thought this might be his last chance to say it. Then the atmosphere lightened. "I sure wish I could go to the FIDO party tonight," added Swigert, then a bachelor with a Center-wide reputation as a ladies' man. CapCom Joe Kerwin advised Swigert that the controllers would be glad to call any phone numbers that Swigert might want to pass down.

When a spacecraft enters the earth's atmosphere, ionization of the air around it prevents radio communication for a period of minutes. On the Apollo flights, it was usually something over three minutes, by which time the spacecraft would be at about 100,000 feet, still slowing until its parachutes could be deployed. On Thirteen, four minutes came and went with no communication from the spacecraft.

The MOCR was silent. If the heat shield had been cracked, it made no difference that EECOM had gotten them to the entry interface with plenty of electricity, no difference that the Trench had gotten *Odyssey* into the correct attitude for entry.

Kranz asked the network controller whether he had gotten a signal from the spacecraft's automatic radio beacon.

"Not at this time, Flight."

For an interminable forty-six seconds there was nothing on the loops but the hiss of the open circuit.

"Network, no ARIA [radio beacon] contact yet?" Kranz asked tightly.

"Not at this time, Flight."

After thirteen more seconds, word came from the network controller, in the best MOCR cryptic:

"ARIA 4 is A.O.S., Flight." One of the ground stations had acquired an automatic signal. It didn't necessarily mean that the crew was alive, but it did mean that *Odyssey* was not a cinder. Kranz's "Rog" was husky.

Still no word from the crew. After a few more seconds, Kranz asked CapCom to call them—"Just advise 'em, standing by," he said.

"*Odyssey*, Houston. Standing by, over," said Kerwin.

Four more seconds passed.

"Okay, Joe," Jack Swigert said.

Still there was silence in the MOCR. As Rocco Petrone, sitting in the back row, would explain it later, getting voice contact was encouraging but not good enough. Who knew what had happened to the electronics of the chutes? Who could be sure of anything about this mission? Petrone prayed.

You can still hear the voices of the Apollo controllers, recorded for posterity on the tapes kept at Houston and in the National Archives. You can hear the voices from the back rooms and the fainter voices of the astronauts themselves. What you cannot hear are the background noises of flight control—the microphones on the controllers' headsets were too highly directional to pick up extraneous sounds. Thus, at the point after the first lunar landing on Apollo 11 when Kranz had to call for quiet in the MOCR, the listener must imagine the hubbub for himself. On the first night of Apollo 13's troubles, when Lunney had to call for quiet, the tapes reveal nothing of the commotion that was distracting him.

But three minutes and fifty-three seconds after Jack Swigert's "Okay, Joe," the screen at the right front of the MOCR lit up with a television image of *Odyssey*, its main parachutes safely deployed. That moment, you can still hear. No microphone could have filtered out that pandemonium.

As *Odyssey* swung beneath its triple parachute, Lovell recalled his landing on Apollo 8, when the spacecraft had smacked into the water flat and very hard. "Gentlemen, be prepared for this landing," he informed his crew. "It's going to be rough." But he was wrong. *Odyssey* caught a wave at just the right angle, and Thirteen ended with barely a bump.

30

"We drank the wine at the pace they handed it to us"

Apollo came to mean many things to the people who were part of it, but for the nation as a whole it will always be Jack Kennedy's Apollo. Kennedy's Apollo was not a spacecraft, not an engineering project, not a means of adding to man's scientific knowledge. Kennedy's Apollo was a heap of chips pushed to the center of the table.

Kennedy's Apollo came out of a long and honored tradition of great American boasts—that we could whip the British, cross the Rockies, build taller buildings, grow more corn and make better mousetraps than anyone else. Childish boasts, some would say, for there was never anything subtle about them: "Anything anyone else can do, America can do better." But there was always an added clause that gave them weight and dignity: "If you don't believe it, just watch us." Kennedy's commitment was the quintessence of this tradition, right down to the gratuitous deadline, "before the decade is out."

Kennedy's Apollo came to an end with the landing of Apollo 13. It hadn't been planned that way, but, as it turned out, Thirteen was the pinnacle of the spirit behind Kennedy's commitment. It was a spectacularly American response to crisis—unorganized (in a way), with people in a hundred different places across the country racing to do their part; youngsters taking charge; improvisation in minutes and invention in hours; courage; indomitability; and then, afterward, a good deal of shrugging it off—"Hey, it was no big deal, we knew after the first half hour we could do it."

Apollo 13 was also the last moment when the nation was transfixed by the adventure. When the initial commitment was made in 1961, it seemed that landing a man on the moon would be just a beginning, a foray by scouts to be followed by outposts and then by settlers. It was this kind of thinking in the early 1960s that enabled Stanley Kubrick to begin a movie about a voyage to Jupiter via a huge space station and lunar colony and plausibly entitle it *2001*.

By the time the movie was released in 1968, the title was already implausible, for the space program's grip on the public imagination had begun to fade even before the first moon landing. Whether this was inevitable or an unlucky juxtaposition of Apollo with Vietnam and domestic upheaval will never be known. But what had been imagined as a natural process of growth in manned space travel had by 1970 come to be seen as a technological exercise that wasn't worth the effort.

In the political arena, the opposition to manned space flight was not just a matter of indifference, but of growing hostility. Editorials recited how many hospital beds could be provided or teachers' salaries paid with the price of a moon shot. NASA reacted with paeans to the many practical spin-offs from Apollo technology—paeans which were accurate, but defensive. Few at that time or subsequently were willing to say passionately that the expansion of mankind into the solar system was an end in itself that justified the cost, whether or not it led to better computers and pacemakers and flame-resistant pajamas back here on earth. Unimpressed by the claims of spin-offs, NASA's critics conceded only that Apollo had shown what the nation could accomplish if it really tried. A new all-purpose political truism entered the language: "If this nation can put a man on the moon, then it should be able to . . ." Cure cancer. Stop crime. End poverty. All it would take, many seemed to think at the time, was the same kind of money and commitment that the United States had lavished on Apollo.

NASA itself had already begun to change by the time Thirteen flew, a point that many of its veterans have made. The truly fun parts of the manned space program, so many said, were Mercury and Gemini and the planning years of Apollo, when the centers were still independent and feisty, collaborating when it suited them, sometimes going their own inefficiently separate ways—but also electric with enthusiasm and imagination, prodigiously inventive. Then, for the last half of the 1960s, NASA seemed to be getting the best of both worlds—superb management without bureaucratic paralysis. By 1970, however, it was apparent that the bureaucracy was going to become more and more dominant, and it was acting more and more like bureaucracies elsewhere. Maybe organizations, like people, can't stay young forever, the veterans conceded. But it sure was fun while it lasted.

The more tangible result of the diminished public interest in the space program and the growing political hostility was a cut in the number of moon landings. When Apollo 13 flew, seven more Saturn Vs were ready or under construction—six more for lunar missions, Apollo 14 through Apollo 19, and one to launch Skylab, a large orbiting manned facility. Before Apollo 14 flew the following January, the last two of those six lunar flights had been cut. The Saturn Vs that were to have launched them were built, complete with engines and pumps and instrumentation units, but they were never used. Instead, they became what must be among the world's most expensive museum exhibits, resting at the Johnson Space Center in Houston and the Marshall Space Flight Center in Huntsville.* The budget for manned space flight, which had been falling since 1966, fell to less than a third of its peak (in purchasing power), where it remained throughout the 1970s and 1980s.

1

After Thirteen, Apollo moved quietly into the voyages of scientific discovery. Apollo 14, which flew at the end of January 1971, was a transitional flight. A few changes had been made to the service module as a result of the Thirteen accident, most notably a third oxygen tank located far from the other two, but the LEM was essentially unchanged, permitting Al Shepard and Ed Mitchell to spend only a little more time in surface exploration than had Conrad and Bean on Twelve—nine hours and seventeen minutes compared to seven hours and forty-five minutes. But the Fra Mauro site, a highlands area, was geologically more interesting than the previous landing sites chosen primarily out of safety considerations.

---

* The Saturn V at Kennedy Space Center is a test article.

The true science missions were Apollo 15, 16, and 17, known within NASA as the "J" Missions.* Each used a modified LEM designed to support an extended stay on the lunar surface and carried in its descent stage the "Lunar Rover," a battery-powered vehicle that let the astronauts leave the vicinity of the LEM and drive for miles across the surface. The result was three missions which were unrecognizably different from the tentative first landing of Apollo 11. Armstrong and Aldrin had spent two hours and forty minutes on a single E.V.A., never moving more than a few hundred feet from the LEM. Apollo 15's lunar astronauts, Dave Scott and Jim Irwin, spent nineteen hours outside the LEM and traversed seventeen miles exploring the terrain around a 15,000-foot mountain. On Apollo 16, John Young and Charlie Duke descended to the lunar highlands and remained on the surface for three days. On Apollo 17, Gene Cernan and Dr. Harrison (Jack) Schmitt spent even longer in the Taurus Littrow area, traversing almost twenty-two miles in more than twenty-two hours of E.V.A.s.

The scientific work of these expeditions involved much more than astronauts picking up rocks, or even exploring and observing. Each flight carried two packages of experiments, either one of which dwarfed the scientific instrumentation carried in the earlier flights. One was the ALSEP (Apollo Lunar Science Experiment Package), an expanded version of the ALSEPs carried on Apollos 12 to 14. Containing seismometers, magnometers, heat-flow probes, solar wind collectors, and the like, weighing about 1,200 pounds, the ALSEP was carried down to the lunar surface in the descent stage of the LEM and deployed during E.V.A.s by the crew. The other was called the Scientific Instrument Module (SIM), stored in a bay in the service module that was larger in volume than the command module itself. The SIM-bay's remote sensing equipment—spectrometers, cameras, laser altimeters—studied the surface of the moon from lunar orbit. After years of paying lip service to the scientific value of the manned space program, NASA was doing serious science on the moon.

One of the most intriguing developments in this process was the scientific contribution of the astronauts. The test pilots among them—which meant all but a handful—were by temperament just about as different from the pure research scientist as it is possible to get. But if their assignment was to bring home scientific knowledge, then that's what they were going to do. With only rare exceptions, they burrowed

---

* "G" had been the first landing. The flights using the original LEM—Twelve, Thirteen, and Fourteen—were all "H" Missions. There were no "I" Missions.

earnestly into their geology training—by the time he flew Apollo 17, Gene Cernan had completed 125 hours of classroom instruction and 300 hours of geological field trips—and some became quite competent amateur geologists. It was striking to watch the best of the astronauts on field trips, remarked Bevan French, a NASA geologist: They had become not only skilled observers, but also highly objective ones, without the specialist's tunnel vision. "They were damn sharp," French said, and they saw things that a specialist might miss.

If the public was bored by the later Apollo flights, the scientists weren't. Some of them had been speculating and arguing about the moon and its geological history for decades. Then the first surface missions had brought home lunar samples—labs all over the world were still settling in for years of work on those. Now, full-scale lunar experiments, planned in a mood of unreality, were actually being put into operation on the moon, and sending a rich stream of electronic data back to the earth.

Some of these data were ambiguous, adding glorious new complications to the existing scientific controversies. When newspaper reporters asked Isadore ("Izzy") Adler, one of the astrophysicists working a J Mission, what the moon was really made of, now that they had made all these new analyses of its chemistry, Adler answered, "Silver." What are you talking about? they asked. It must be silver, replied Ader: "Everyone who sees our results sees his own reflection."

But occasionally there was the moment when everything fell into place and a new truth was established. One of Adler's co-experimenters, Goddard physicist Jack Trombka, described such a moment that occurred during Apollo 15. CMP Al Worden was running their remote-sensing experiment from the command module orbiting the moon. The experiment's purpose was to determine the chemical composition of the lunar surface by analyzing how it radiates solar-excited X-rays back into space. It was two o'clock in the morning, and Trombka was plotting the data that had begun pouring in from his instruments in the SIM-bay. He began by plotting the concentrations of aluminum on the lunar surface. Glancing up from his work for a moment, he saw a plot just like the one he was working on, already finished and hanging in the front of the room. Astonished, Trombka cried out, "Who got our aluminum results before we did?"

In fact, what Trombka mistook for his own results were the results from another remote-sensing experiment, the laser altimeter, which had been measuring the altitudes of lunar surface features. That plot matched the one he'd been drawing because—as had become suddenly clear—the composition of a lunar feature depended on its altitude. In particular, the higher the altitude of a feature, the higher its aluminum content. Suddenly, a central argument about the moon's history was resolved. After more than a century of debate about whether the moon had always been dead and cold or had a history of heating and cooling, Trombka knew the answer: Lighter elements flow to the top of a landscape only if they are molten; therefore, the moon must once have been hot. From a scientific perspective, this discovery (which was amplified by the analysis of the heavier elements, which were concentrated in the lunar lowlands) had a major impact on astronomers' understanding of the whole solar system. For Jack Trombka, working in a discipline where progress usually comes in bits and pieces, it had been a unique moment of "Eureka!"

Just as the astronauts set out to become competent amateur geologists, the Flight Operations people set out to become sponsors of science. John Hodge, once Blue Flight, the second man to become a flight director after Chris Kraft himself, headed the planning for the J Missions. Apollo Program Manager Rocco Petrone became as enthusiastic about the esoterica of lunar science as he had ever been about giant rockets, constantly dropping into the science back rooms and watching over the scientists' shoulders as they examined the incoming data.

During the J Missions, the irrepressible INCO, Ed Fendell, became Captain Video—the man who controlled the television camera during the E.V.A.s. At first Fendell couldn't imagine having to work with scientists, having found it impossible to communciate with the ones he'd met. But the geologists he encountered during Apollo—people like Lee Silver from Cal Tech, Gene Shoemaker from the U.S. Geological Service, Bill Muehlberger from the University of Texas—were exceptions. They didn't talk like intellectuals and they weren't stuck up. They just assumed that Fendell didn't know anything about rocks, and walked him through what he needed to know. Fendell went with them to Warm Springs, Nevada, to look at the lunar-like impact crater there. Later, with Fendell in the MOCR and the geologists in their back room, they ran integrated sims of the E.V.A.s, twelve and fourteen hours at a time, with the astronauts in full kit, deploying the equipment and maneuvering around a mockup of the lunar landscape.

During the missions, their work became a complex exercise in scientific curiosity at long range. When the astronauts first got the television camera set up, Fendell started with a wide-angle pan, working his way around the horizon by increments. In the back room, the geologists took a Polaroid picture of each segment, then assembled them into a mosaic. Studying the patchwork panorama, the scientists told Dick Koos—once the most devious of the SimSups, now the lunar surface experiments officer—which local features they wanted the astronauts to concentrate on. Koos worked with Flight and CapCom and maps of the lunar geography near the landing site to get the astronauts where they were supposed to be, doing what they were supposed to do.

Sitting with them in the lunar surface back room, Koos acted generally as the scientists' guide to and intermediary with the world of Flight Operations. The scientists badly needed such guides and intermediaries—and interpreters as well, it sometimes seemed—for to them Building 30 was a strange new world. Some of them came to the Control Center from academic environments where experiments could be redone if something went wrong. Some were accustomed to geologic field trips where specimens could be taken back to the camp at night and pondered for days or weeks if necessary. At the Control Center, they found themselves indoctrinated into the discipline of "real time," where every step in every procedure and virtually every movement that the astronauts made had to be mapped out long in advance. They learned that they had to anticipate all the things that might go wrong with their experiments and work out detailed fixes and fallback positions. Then they discovered the extraordinary nature of "simulations," which were not just practice sessions but make-believe that for some of them got mixed up with reality. One scientist remembered a colleague who was "livid" when his equipment kept being subjected to simulated failures. The man finally wrote a memo to Chris Kraft complaining that the Flight Operations people weren't running this mission properly if his experiment was breaking down so often.

The flight controllers were particularly exotic to the scientists. "This one fellow, we liked him lots, but we hated him, too," Jack Trombka recalled. He was the one in charge, and he had a phenomenal memory. During the simulations, when something would go wrong with Trombka's equipment and Trombka would report to the front room that he didn't know what to do, this guy would be on the intercom—they called them "loops"—saying something like, "Well, on such-and-such a date, at such-and-such a time, you gave us such-and-such a remedy for this problem. Go do it! Now!" He caught every mistake, seemed never to sleep; and so one day during a simulation it came as a great shock when Trombka was watching the television monitor showing Mission Control and saw this fellow yawn. Up and down the hall from the other science rooms there came the sound of clapping and yells of "He's human!" Trombka couldn't remember his name, but he had a very short crewcut and during missions he always wore a white vest . . .

Trombka's daughter, then a high school senior, went down to Houston for Apollo 16 and helped punch data. She recalled that her father and the other scientists sometimes joked about the controllers, who seemed to them like something out of the Wild West. But she also noticed that after a few weeks these scholarly men began inserting "Rog" and "We're go on that" into their conversations. Another culture had joined Shea's original three, and it was having the time of its life. Maybe nobody in the outside world was paying much attention, but the J Missions were magical for the people involved. Through them, planetary science was transformed.*

2

The last Apollo flight lifted off from Pad 39A at half past midnight on December 7, 1972. It was the only Saturn V ever launched at night, and many of those who witnessed it said it was the most beautiful launch, an explosion of light that brought a false dawn to the Florida sky, a column of fire that as it rose became visible from as far away as North Carolina to the north and Cuba to the south.

Apollo 17 was commanded by Gene Cernan, veteran of Gemini IX and Apollo 10, the only man besides Jim Lovell and John Young to go twice on a lunar journey. His CMP was Ron Evans and his LMP was Jack Schmitt, a geologist and the only scientist to be on an Apollo crew. The command module was named *America*; the lunar module, *Challenger*. On the early afternoon of December 11, Cernan and Schmitt landed *Challenger* in the valley of Taurus Littrow. Seventy-five hours later, at 4:54:37 P.M., Houston time, December 14, 1972, they left the lunar

---

* Aside from settling the major question of whether the moon has a history of heating and cooling, the Apollo missions showed in general that extraterrestrial exploration and science could be fruitful human activities. As NASA astronomer John Wood summarized it, the astronauts neither sank into the dust nor contracted terrible diseases on the moon; moreover, they demonstrated that other bodies in the solar system "can be studied by the same geological sciences that we use on the earth . . . This was a major philosophical change in point of view," Wood said, "and I think the most important thing that came out of the Apollo Program."

surface and rendezvoused with *America* for an uneventful last journey home.

As Apollo 17 approached the earth, the people of Apollo observed the end of the program in characteristically different ways. Max Faget and Caldwell Johnson barely noticed. Both of them were still at M.S.C., both of them still doing what they had always done best, working in tandem, designing flying machines. Faget was still as blunt and opinionated as ever, Johnson still as ornery. By the end of 1972, they had been living in the world of the space shuttle for three years already. Apollo was part of their past.

Apollo wasn't history for Don Arabian over in Building 45. He was still running the MER for the final shift of Seventeen. But in these final hours of the first lunar era, he was supervising a huge gumbo party. Over the past few flights, as the crises and anomalies became fewer, life in the MER had become too tame for his taste. To liven things up, he had started a lunch competition. Each day during a flight a table was set up in Arabian's office and the designated engineer (who may never have so much as heated a can of beans) prepared, or had sent in, the menu of his choice. As time went on, the competition grew intense. Some opted for the outré—one lunch had consisted primarily of enormous banana splits. Sid Jones had arranged for music and candlelight.

Now, for the landing of Seventeen and the last lunch, Arabian had taken charge and decreed that they would have gumbo—not just for the dozen who usually were invited to the lunches, but for everyone in the MER, plus assorted invitees from Building 30, in all probably 200 people. Fulton Plauche, a Louisianian from the cryo and fuel-cell post in the MER, was named chef. The Crew Services Division contributed two gigantic pots. Jack Kinzler's Tech Services Division made up some wooden stirring paddles. Plauche began with a chicken-and-sausage gumbo, but people in the MER also contributed deer sausage, ducks, rabbits, squirrels, and anonymous game that became the subject of grisly rumor. By the time Apollo 17's command module was approaching the entry interface, a line of people waiting to get their rice and gumbo and ice cream snaked around the MER.*

---

* The MER was still staffed with engineers at the time, though they remembered having a hard time concentrating on their work. It should be noted that, by this point in the flight, an hour before entry, almost any problem that might arise aboard the spacecraft would have to be handled by the Control Center anyway—the MER was for longer-term problems.

*   *   *

Owen Maynard was at Raytheon's plant in Sudbury, Massachusetts. Maynard had left NASA in hopes that he could do something on the outside to push the space program forward, to get some exciting new Apollo going. Joe Shea, now head of engineering for Raytheon, had heard that Maynard was available and recruited his old ASPO colleague to join him, putting him to work on a breathtaking project—designing a solar-powered satellite, a monster with wings twenty-five miles wide that would generate up to 50 gigawatts of electricity on earth.

Maynard didn't watch the splashdown. From the time of Alan Shepard's first Mercury launch, he had never watched any of the flights live. Maynard didn't like to admit it to anyone, but to watch them live was an agony of worry and suspense for him—he got very emotional about these things. Some of the others at the Sudbury office crowded around a television, but not Maynard. He kept to his office and his drafting table.

Bob Gilruth was in the viewing room behind the MOCR. Early in the year he had stepped down as M.S.C.'s director to take a job as "Director of Key Personnel Development" for NASA—a pleasant way to end a historic careeer. Gilruth shared Maynard's apprehension during missions, however, and he was glad to see Apollo come to an end without having lost an astronaut in flight. He had never been enthusiastic about the H and J Missions anyway, thinking that the risks of lunar exploration were awfully high. He still hoped that NASA would concentrate on a space station, a project that had been his personal favorite since the Space Task Group began.

Scott Simpkinson, still M.S.C.'s gypsy, had most recently spent three months in Huntsville sitting on the contractor selection board for the space shuttle's solid-fuel boosters. Simpkinson was about to spend a year at Langley helping to prepare the unmanned Viking mission to Mars. But today he was taking his last shift as manager of SPAN, where he had been for all of the Apollo missions. The end of Apollo didn't bother him, either. Simpkinson had never been sorry to see anything end. He figured there was always something better going on somewhere else.

In the MOCR, as the clock moved past one o'clock and the spacecraft *America* reached the entry interface, Flight Director Neil Hutchinson was presiding over a nominal landing. The room was crowded, as was usual for a landing, and the viewing room was full, but the atmosphere was comparatively relaxed. They had done this many times now.

Rocco Petrone watched from his post in the fourth row. Petrone's star was still rising—he had recently learned that in January he would move to Huntsville and replace the retiring Eberhard Rees as Marshall's center director.*

At this moment, however, Petrone was not looking ahead. He couldn't even concentrate fully on the present landing. For Petrone, it was a time for remembering. It seemed a very long time since he had been an Army major sitting with Al Zeiler in a car in the rain outside the Cape Canaveral cafeteria, trying to make out John Kennedy's words through the static. The years in between had been unforgettable ones, and Petrone felt a wave of nostalgia as he watched the television image of *America* descending under its parachutes. No matter what happened to the space program in the future, it could never be like this again. As was his habit, Petrone thought in terms of history: Any number of Pilgrims might follow, but there could be only one Columbus.

At 1:24:59 Houston time on the afternoon of Sunday, December 20, 1972, Apollo 17 splashed down in the Pacific four miles from the carrier *Ticonderoga*. Counting from Apollo 8's launch on December 21, 1968, the first age of lunar exploration had lasted exactly four years.

After the splashdown, the geologists in their back room watched the recovery operation for a while, some of them lighting up cigars in the tradition of the flight controllers in the MOCR. They eventually drifted away, exchanging goodbyes with Dick Koos, who stayed behind to finish up some post-flight paperwork. They were just getting good at this lunar surface business, Koos thought. It had taken time, because you never knew exactly what the terrain was going to be like or what the astronauts would discover; lots of improvisation was required. But by Apollo 17, Koos had become quite adept at determining where the astronauts were and coordinating with his scientists where and how they should proceed next. It was too bad that they had to stop now, just as they were getting the hang of it.

Koos walked out into the high corridor of the Control Center, his

* Rees had become director of Marshall in 1970, when von Braun had moved to Washington as NASA's deputy associate administrator for planning. Von Braun's last two years with the agency—he retired in 1972—were frustrating ones. He had hoped to set a brave new goal for Nixon's space program, perhaps a manned trip to Mars; but Nixon's enthusiasm for an aggressive post-Apollo space program had faded along with that of almost everyone else.

footsteps echoing in the quiet. On impulse, he stopped at the door of the MOCR and walked in. Everyone had left for the splashdown party, leaving their consoles littered with books, logs, checklists, headsets, and ashtrays filled with cigarette butts and half-smoked cigars. Some of them had left behind the little American flags that everyone waved when the spacecraft splashed down.

Koos climbed the steps to the flight director's console and looked out over the deserted room. The big screens at the front of the room were black. The controllers' C.R.T.s showed no flickering columns of numbers, no messages about the health and welfare of men and machines in deep space. They were just small, turned-off television screens. The banks of caution lights were dark.

There were no murmuring voices on the headsets. There was no sound at all anywhere in the room. It was as perfectly silent as the plains and valleys 240,000 miles away where six descent stages still sat. Strange to think of them still up there, Koos reflected. Even the footprints. Strange to think that men might return to those places to look at footprints made by Neil Armstrong or Pete Conrad or Gene Cernan.

It was stranger yet to think that this was the end of it all. From the balky Mercury trainer that Koos couldn't get to work in the spring of 1961 to this—and it had all happened in less than the time it was taking his daughter, born early in Apollo, to reach adolescence. It had come and gone too fast. John Aaron, that philosopher of the MOCR, would say later that it was like gulping good wine. It would have been nice to have had more time to savor it, he said, but "we drank the wine at the pace they handed it to us."

Today they had drunk to the bottom of the glass. There would be Skylab and eventually the shuttle, but Dick Koos, like Rocco Petrone, understood that an Apollo came only once, and he felt immeasurably sad. Koos walked slowly down the steps and out into the corridor, latching the door carefully behind him.

Epilogue:
Twenty Years after the First Landing

> I will admit that I walk out every now and then in the
> morning to get the paper when the moon happens to be up,
> and say, "By God, we were there."
>
> —Rod Loe

Looking back twenty years after the first landing, many of the people of Apollo agree with John Aaron's sentiment—it was a little like gulping good wine. Few had the time to step outside of events and assess what was going on until later. "It was funny," Rod Rose reminisces. "Just a little while ago, when I retired, people asked me to 'look back,' and all that. And it wasn't until then that it came home to me—'My God, I've been involved in the first quarter of a century of manned space flight!' " Another veteran notes that when his children come to visit and look through his memorabilia, they ask him why he doesn't have autographs of the astronauts and presidents and celebrities who came to watch him at work. "You were right there!" they say. But he was always so busy, he tells them, he just didn't pay any attention. "We knew we were doing something, but nobody really felt the impact of what was really going on." The job was too consuming for that —"almost a crusade," said another veteran, nothing like any other job he ever had.

Some are unhappy that the rest of the country has never truly comprehended the historic nature of our first journey to another world.

Most people think of Apollo as just another episode in the tumultuous sixties, secondary to the Vietnam War and America's social upheaval, not as something that will still figure large in the history books when Vietnam and John Kennedy and Lyndon Johnson are consigned to footnotes. But the failure to understand goes beyond that, argues Rocco Petrone. "We've had a lot of reporting of how big the rocket is, how much noise it makes, pictures of guys on the moon. But what was the real meaning of Apollo? What did it symbolize? What were we after?" For a few short years, Apollo was almost like a Renaissance, Petrone thinks, but nobody wants to confront that kind of possibility now.

The dark side of the experience was the sacrifices it required of the Apollo families. Rod Rose again: "If I had to tip my hat to the unspoken heroes of the Apollo Program, it would be to the wives and families of the engineers. The wives were widows. They had none of the glory, none of the kudos—they were the ones left behind with the kids, the bills." Many of the men of Apollo look back at their youthful selves and are appalled at what they took for granted. The same phrases keep recurring—"My wife was a saint," or "I don't know why my wife didn't leave me," or "I didn't see my kids grow up." "Boy, if I hadn't had the kind of wife I had," said tough old Tom O'Malley, "and some of the rest of us who were in the same boat didn't have the wives they had, [Apollo] would have never come off."

In many cases, families didn't survive the stress—during the mid-1960s, the towns around K.S.C. had the highest divorce rate in the country. At the program's height, one Cape engineer counted up the divorces just among the people he knew: seventy-eight of them in the preceding six months. Houston wasn't far behind. Gran Paules pulls out a picture of a party at his house after Apollo 8 landed. Within a week of the party, he says, four of the couples there had revealed they were getting divorced. And yet in 1989 it is striking how many of the people who made the biggest contributions to the program are still married to the same person who, a quarter century ago, came with them to the wilds of Merritt Island or Clear Lake. The consensus is that this is no coincidence. Under the heat of Apollo your marriage either melted or annealed.

The same may have been true of individuals. Apollo no doubt took in its share of miscreants and time-servers and self-aggrandizers, but the program had a way of getting rid of them through self-selection. Among the people who remained, Apollo seems to have counted an extraordinarily high proportion of gentlemen—"gentlemen" in the uncomplicated sense of being straight shooters and hard workers and not show-offs.

Perhaps this can be explained by the nature of the challenge they were ready to take on and by the unworldly rewards with which they had to satisfy themselves. For whatever reasons, the remark Joe Shea made about Houston when he arrived—that he'd never seen a place like it, where no one was jockeying for position, everyone was just trying to do his best—seems to summarize the program's character in the other centers as well. Moreover, the people of Apollo seem to have retained those qualities since. Bill Tindall's insistence he didn't do anything himself is an only slightly exaggerated version of the generosity of spirit that is common among the Apollo veterans. Not everyone was a hero and most certainly not everyone was a saint, but, twenty years after the first landing, the people of Apollo have as a group aged remarkably well. There is nostalgia among them, but no post-Apollo stress syndrome.

Inevitably, some have departed. Wernher von Braun died in 1977, Kurt Debus in 1983, and George Low in 1984. Other less famous characters in the story are also gone. Dan Klute died of a heart attack a few months after the combustion instability problem on the F-1 engine was resolved, worn out, many of his friends believe, by the unremitting strain of that effort. NASA later named a lunar crater in his honor. Others who are not here to celebrate the twentieth anniversary of the first landing include Tecwyn Roberts, the first FIDO; John Disher, who was part of the very first planning for a lunar landing; and George Bliss, of the EECOM back room during Apollo 13.

But many of the famous names of Apollo—famous within NASA anyway—are very much alive. James Webb lives in retirement in Washington, D.C. He still likes to talk, and still does so with all the energy and political insight that he applied to the administration of NASA. Robert Gilruth lives near where it all began, in a little town on the water forty-five miles north of Langley, with a hydrofoil sailboat of his own design tied to the dock. Abe Silverstein divides his time between Cleveland, close to Lewis Research Center, which he directed after leaving NASA headquarters, and his retirement home in Florida. Sam Phillips retired to the seaside south of Los Angeles; his successor, Rocco Petrone, is only a few miles away, newly retired from Rockwell. George Mueller is up the coast in Santa Barbara. Chris Kraft is still in Houston, Eberhard Rees still in Huntsville. Walt Williams lives in Tarzana, California, but it is easier to locate him in Washington, the Cape, Huntsville, or Houston—he still consults for NASA. John Houbolt is in Williamsburg, and still keeps an office at Langley, as does the mentor to so many in the Space Task Group, William Hewitt Phillips.

Robert Seamans has for many years been on the faculty of M.I.T. Jerome Wiesner, with whom he tangled in the early days of the Kennedy administration, is president-emeritus of M.I.T., still working in his office in the Jerome B. Wiesner Building. In 1989 they had a new colleague, Joe Shea, who is taking some time off from his senior vice-presidency at Raytheon to teach a course there.

For the many figures in Apollo who were young men when the program finished, the twentieth anniversary finds them still in the middle of their careers. It is bemusing to walk away from a conversation with Glynn Lunney about the dawn of manned space flight and realize that, to his colleagues at Rockwell, Glynn Lunney is—as he was in Chris Kraft's fledgling Flight Operations group—a guy on the way up.

Some of the kids of Apollo are running the manned space program. The director at M.S.C.—now the Lyndon B. Johnson Space Center, J.S.C.—is Aaron Cohen. One of his predecessors was Gerry Griffin. John Aaron runs J.S.C.'s space station planning. Go to the J.S.C. cafeteria and, if you know where to look, you are likely to see Steve Bales, Chuck Deiterich, Bob Holkan, Ed Pavelka, Sy Liebergot, Rod Loe, Jay Greene, Jack Garman, and Emil Schiesser. All still work on the J.S.C. campus, either for NASA or for NASA contractors. Gene Kranz is still at J.S.C., directing the Mission Operations Division. The main difference between Kranz now and the Kranz of twenty years ago is the advent of the automobile tape deck: He can play his Sousa marches on the way to work. He is too senior to be a flight director any more, but if you're trying to find him when a shuttle is flying, you might want to look first in the MOCR.

Though they don't work at J.S.C. any more, many others remain in the Houston area—Scott Simpkinson and his co-conspirator in shaving down the Big Joe heat shield, Jack Kinzler, live within a few miles of each other. Gerry Griffin, Jerry Bostick, Dick Koos, Rod Rose, Cliff Charlesworth, John Llewellyn, and Ed Fendell live in and around the Clear Lake housing developments that were so raw and new in 1962 and now are green and shady and established.

Max Faget and Caldwell Johnson are in Houston, too, past retirement age but hardly retired. They still work together in a private company that Faget founded, Space Industries, Inc., collaborating now into their fifth decade. They have most recently designed a man-tended space station that many NASA engineers will say—off the record—is a masterpiece of economy and elegance, and may yet be the first space station deployed. Johnson still builds model airplanes in his spare time.

In Huntsville, Karl Heimburg, who ran von Braun's testing lab, is only one of many of the Germans—and now Alabamians—who remain near Marshall in their retirement. Bob Wolf, who waited out the crippled S-II stage on Apollo 6, is still at Marshall. Jerry Thomson heads the Advanced Propulsion Systems Office. Thomson's Rocketdyne colleague during the struggle with combustion instability, Paul Castenholz, has retired from the rocket business to live in Colorado Springs.

One prominent Marshall veteran is not in Huntsville. In 1983, the Justice Department, acting on information that had been in the U.S. government's hands for almost forty years, told Arthur Rudolph, program manager of the Saturn V, that he must surrender his U.S. citizenship and leave the country or face charges that he had been involved in the forced-labor camp at Nordhausen where the V-2 was manufactured. Rudolph, then seventy-six, denied the charges but returned to Germany. His colleagues during Apollo continue to hold him in high esteem, saddened and sometimes angered by this denouement of his career at Marshall.

At the Cape, those who have retired have almost all stayed nearby— Merritt Preston, Tom O'Malley, Ed Fannin, Grady Corn, Sam Bedding-field, Walt Kapryan, and Don Buchanan may be found scattered in Titusville, Melbourne, and Cocoa Beach. Joe Bobik tried to retire, but after the *Challenger* accident he was lured back to work for Lockheed, to oversee quality assurance for the shuttle's orbiter. Ike Rigell has been no more successful in pulling himself away from rockets; he is an executive with United States Boosters Inc., with an office a half-dozen miles from Pad 39.

Don Arabian may be found just south of K.S.C., still insisting he has no sentiment about space flight but nonetheless living in Cape Canaveral. Finally free to do everything himself, he lives in a home that looks like a showcase for every kind of craftsmanship, which he built by himself from a concrete shell. Beside the house, a thirty-two-foot oceangoing sailboat is nearing completion, also entirely his own work. Arabian has never sailed, but he intends to take it around the world. Alone, of course. He still has Bill Tindall's band saw which he borrowed fifteen years ago, but he promises to return it when he is finished. Tindall himself is in Washington, D.C., where he is now as enthusiastic about the mysteries of civilian air traffic control as he used to be about lunar flight.

The X-ray experimenters Izzy Adler and Jack Trombka are also near Washington. Adler continues to teach physics at the University of Maryland and Trombka continues to work at the Goddard Space Flight

Center on gamma-ray studies, using high-altitude balloons these days instead of lunar spacecraft. Trombka also teaches an occasional evening course on astrophysics for laymen at the Smithsonian Institution. One evening after class in the fall of 1983, he captivated two of his students with stories about Apollo—of what it was like to be in the Control Center, and especially of the strange and wonderful people who ran the place. Someone, he observed upon parting, really ought to write a book about them.

Apollo *as History*

Writing definitive history is a solemn undertaking and *Apollo* was not. Our objective has been to tell stories—true stories, but stories rather than analysis—about how an epic triumph was achieved.

The reader should be aware that in this process perspective has often been foreshortened. We devote pages to the semi-comic fiasco of M.R.-1 at the Cape and mention great managerial controversies at headquarters in passing. The design and development of the command module are described over several chapters; the design and development of the ingenious lunar module get a few pages. Brainerd Holmes, who made major contributions to Apollo as head of O.M.S.F. from 1961 to 1962, is mentioned briefly; his successor, George Mueller, is the subject of a long profile. These and many similar discrepancies do not reflect the relative importance of the topics, but the idiosyncrasies of weaving together a narrative.

Much has been omitted altogether. Most conspicuously missing in *Apollo*, or at least underrepresented, are the astronauts. The astronauts' story is fascinating and they in fact were much more than pilots, with vital roles in designing the spacecraft and planning the missions. But most of their story has been well and fairly told elsewhere, along with much of the story of the training and support they received from the Flight Crew Operations Division.

The contractors got the shortest shrift in *Apollo*. The program's strategic decisions and flight operations were dominated by people in NASA, but most of the detailed design of the hardware and all of the fabrication of the hardware were done by contractors. One book tells the story of Grumman: *Chariots for Apollo: The Making of the Lunar Module*, by Charles B. Pellegrino and Joshua Stoff (New York: Atheneum, 1985), from which we drew heavily in our own discussion of the lunar module. With that exception, only the official NASA histories give a hint of the challenges that had to be overcome in actually building the Saturn stages, the command and service module, and the ground support facilities at the Cape.

The recovery operations have been ignored. We remember John Stonesifer, who directed the Recovery Systems Branch, wistfully telling us that somehow none of the historians had ever paid any attention to what happened after the spacecraft splashed down. He described for us the complex and delicate task the recovery forces accomplished. And eventually, with apologies, we too left his deserving story on the cutting room floor.

There is a book at least the length of this one to be written about the people of Marshall and the building of the Saturn V, perhaps the greatest technological triumph of the program. The F-1 story barely scratches the surface, and the development of the hydrogen-fueled S-II stage, which was as difficult and important, is barely mentioned. We say nothing at all about the extraordinary manufacturing processes that had to be developed to produce the Saturn V.

The work of spacecraft operations at the Cape got brutally short treatment. How could one write about Apollo without talking about people like George Page, Paul Donnelly, Ted Sasseen, John Williams, or Ernie Reyes? We left them out nonetheless, and can only hope that someone else mines that rich lode of material. The omission that caused us the greatest personal regret involves the inspectors at the Cape. Joe Bobik, chief inspector for the spacecraft, was one of the first people we met during the research. His stories, which so intrigued us that we eventually compiled long accounts not only from him but from many of his inspectors, are reduced to a few sentences.

The historical framework for the account in this book—technical information, dates, times, the weather, job titles, and the like—comes from the published archive of material about the space program. Five books in particular became dog-eared with use: for the early days of manned space flight, *This New Ocean: A History of Project Mercury*, by Loyd S. Swenson, Jr., James M. Grimwood, and Charles C. Alexander (Washington, D.C.: NASA, 1966); for Apollo through the first moon landing, *Chariots for Apollo: A History of Manned Lunar Spacecraft*, by Courtney G. Brooks, James M. Grimwood, and Loyd S. Swenson, Jr. (Washington, D.C.: NASA, 1979); for the development of the Saturn V, *Stages to*

Saturn, by Roger E. Bilstein (Washington, D.C.: NASA, 1980); for the construction and operation of Kennedy Space Center, *Moonport: A History of Apollo Launch Facilities and Operations*, by Charles D. Benson and William Barnaby Faherty (Washington, D.C.: NASA, 1978); and as our treasured encyclopedic source, *The History of Manned Space Flight*, by David Baker (New York: Crown Publishers, 1982).

Valuable as these have been, however, they contain little about the personal stories we have tried to recount. The autobiographies of the astronauts were useful—Michael Collins's *Carrying the Fire: An Astronaut's Journey* (New York: Farrar, Straus, and Giroux, 1974) is a classic of its kind. Not surprisingly, however, they have focused on the training of astronauts and their activities during the flights. Neither do the official histories, which tend to focus on administrative decision-making, say much about people like Caldwell Johnson, Scott Simpkinson, Bill Tindall, and Don Arabian. Only a few published accounts have featured anyone except the astronauts or top NASA managers. From among the handful that have, we gratefully acknowledge our use of the Pellegrino and Stoff book on Grumman and the superb account of Apollo 13 by Henry S. F. Cooper, Jr., *13: The Flight That Failed* (New York: Dial Press, 1973).

The great bulk of the narrative in *Apollo*, therefore, has been developed through interviews. The NASA history offices in Washington and Houston contain hundreds of such interviews, compiled in the course of preparing the official histories, plus a scattering of interviews conducted by independent outsiders who were kind enough to send copies to NASA.

Another invaluable source of information was Robert Sherrod's archive. In 1968, Sherrod, a well-known war correspondent and journalist, began research for a major history of the Apollo Program. In the process, he gained access to the major figures of the program while the program was ongoing. His collection of interviews and a voluminous file of index cards are kept at the NASA history offices in Washington. He gave us full access to them and to his handwritten notebooks with their day-by-day chronicle of events.

Even this vast collection of interviews, valuable as they were, only occasionally contained the relaxed, unofficial recountings of what it was like to be part of the Apollo Program that we sought. Further, they tended to omit people who were not senior executives. To obtain material about Apollo not yet available elsewhere, we conducted interviews with 157 people over a period of more than three years. In three instances, people we approached declined to be interviewed; otherwise, everyone we approached cooperated—in most cases, with an openness and generosity beyond any reasonable expectation.

This raises a question of detachment, for it will be obvious to readers of *Apollo* that we found most of the Apollo people to be both admirable and likable. Having acknowledged that, however, we should also say that we have not suppressed skulduggery. Finishing the book, we are reminded of what public affairs officer Dick Young told us on our first visit to the Cape. "I used to have a lot of fun at NASA's expense back in my newspaper days," he said. "When I took a job with NASA, I thought, 'Well, they've been hiding these things from me all these years, and I'm going to get out and find out what they are.' Hey, I'm still looking."

That we found this to be true of the Apollo Program seems to surprise some people. But what, after all, should one expect? A small group of men were told in 1961 to put a man on the moon by the end of the decade. They worked indefatigably for eight years, recruiting thousands of fellow workers along the way. They surmounted obstacles, overcame setbacks, and achieved their goal. Should we be surprised that the people who were able to do this tend to have been talented rather than dull, honest rather than venal, cooperative rather than selfish?

Within the framework of our objective, these were the ground rules:

For the stories we did choose to tell, we strove to present competing viewpoints. In one case, the account of President Kennedy's decision to make the lunar commitment, we drew what may be a controversial portrait of a president who was not an advocate of manned space flight and who made the decision to go to the moon only reluctantly. We should state explicitly that we were unable to find any counterbalancing evidence. Kennedy became deeply interested in manned space flight, but only (as far as we could determine) after he was led to his decision by Gagarin's flight and the Bay of Pigs.

Statements of historical and technical fact are as pristine as we could make them. Errors doubtless remain—there is simply too much on too many topics to hope otherwise—but we have tried to check everything against both the archival record and the knowledge of the people involved. Events that could easily have been perceived differently by different people we report through the eyes of specific observers. When we describe an event without assigning an observer to it, then it may be assumed that we have multiple accounts saying the same thing. We present a few stories that were reported to us as fact, but which we suspect have been embellished over the years. Variations of the phrase "so the story is told" indicate our reservations.

All of our interviews were tape-recorded, and all comments within quotation marks are taken verbatim from transcriptions of them. Reports of conversations have been more difficult to handle. Conversations cannot be remembered exactly even a day later, let alone twenty-plus years later. Any dialogue within quotation marks is as reported by a person who either said the words or heard them said; we have attempted to make the context indicate which. The conversations over the loops in the Control Center were transcribed directly from tape recordings preserved at Houston and in the National Archives.

The core of the book came from interviews by one or both of the authors with the following people:

For the story of Langley and the early years of the Space Task Group: Chris Critzos, Constance Critzos, Charles Donlan, Maxime Faget, Robert Gilruth, Paul Havenstein, Jack Heberlig, John Houbolt, Carl Huss, Caldwell Johnson, Jack Kinzler, Richard Koos, Eugene Kranz, Glynn Lunney, Charles Mathews, Henry Pearson, William Hewitt Phillips, Paul Purser, H. Kurt Strass, and Walter Williams. For the Space Task Group from the point of view of the "Canadians": John Hodge, Owen Maynard, Tecwyn Roberts, and Rodney Rose (of whom only Maynard was actually Canadian).

For the story of headquarters in Washington: John Disher, George Mueller, Samuel Phillips, Robert Seamans, Joseph Shea, Abe Silverstein, James Webb, Jerome Wiesner, DeMarquis Wyatt.

For the story of Marshall: Joseph Bethay, Walter Haeussermann, Karl Heimburg, Fletcher Kurtz, Thomas (Jack) Lee, Alexander McCool, George McDonough, James Mizell, George Smith, William Sneed, Walter Wiesman, and Robert Wolf. Details about the trials of the F-1 engine came from Leland Belew, Paul Castenholz, Saverio (Sonny) Morea, and Jerry Thomson.

For the story of launch vehicle operations and ground support equipment at the Cape: Donald Buchanan, Forrest Burns, Jewel (Jay) Campbell, Ray Clark, Graydon Corn, Edward Fannin, Terry Greenfield, John Humphrey, Walter Kapryan, Albert Martin, Rocco Petrone, Andrew Pickett, Isom (Ike) Rigell, Glover Robinson, Orval (Buddy) Sparkman, and Chester Wasileski.

For spacecraft operations at the Cape: Samuel Beddingfield, Joseph

Bobik, Clarence (Skip) Chauvin, Martin Cioffoletti, Charles Clary, Paul Donnelly, Larry Lettow, Bryce Lowry, Thomas O'Malley, George Page, G. Merritt Preston, Raul (Ernie) Reyes, George (Ted) Sasseen, John Tribe, Charles Welly, Donald Whiting, Gary Woods, and Edward Zirnfus.

For the story of flight operations and mission planning in Houston: many of the same people who told us about Langley, plus John Aaron, George Abbey, Arnold Aldrich, Steven Bales, Michelle Brekke, Jerry Bostick, Clifford Charlesworth, Henry (Pete) Clements, Michael Collins, John Cox, Charles Deiterich, Lyn Dunseith, Robert Farquhar, Edward Fendell, John Garman, Jay Greene, Gerald Griffin, Claiborne Hicks, Robert Holkan, Richard Hoover, Seymour Liebergot, Edward Lineberry, Charles Llewellyn, John Llewellyn, Rodney Loe, Jack Lousma, John Mayer, Harold Miller, Granville Paules, Edward Pavelka, Donald Puddy, David Reed, Emil Schiesser, Carl Shelly, Howard (Bill) Tindall, Manfred (Dutch) von Ehrenfried, and Kenneth Young.

For the story of the development and operation of the spacecraft: others already listed, plus Donald Arabian, Aaron Cohen, James Cooper, Robert Fricke, Billie Gibson, George Jeffs, Sidney Jones, Kenneth Kleinknecht, Richard Kohrs, Thomas Markley, Joseph Mechelay, Owen Morris, Fulton Plauche, Scott Simpkinson, and Judith Wyatt.

For the story of science during the Apollo missions: Isadore Adler, Michael Duke, Bevan French, Wilmot (Bill) Hess, David McKay, William Phinney, Barbara Trombka Blaustein, Jacob Trombka, and John Wood.

For other aspects of M.S.C.: James Elms, Emily Ertle, Mary R. Low, Warren North, Jack Riley, Jack Sleith, John Stonesifer, Terrance White, and Raymond Zedeker.

For the story of M.I.T. and the guidance system: others already listed, plus John Miller, Norman Sears, and George Silver.

In addition to the interviews conducted by the authors, the following interviews have been used for direct and indirect quotations in the text:

Interviews conducted by NASA historians: John Leland Atwood, 7/16/70; Richard Battin, 4/29/66; William Bergen, 6/21/71; John Bird, 6/20/66; John H. Boynton, 4/27/70; Clinton Brown, 4/23/69; Kurt Debus, 5/18/64; Paul Dembling, 8/2/73; John Disher, 1/27/67; Charles Donlan, 6/20/66; Stanley Faber, 4/22/70; Maxime Faget, 12/15/69 and 8/20/73; Charles Frick, 6/26/68; Robert Gilruth, 8/24/73; John Healey, 7/16/70; Bastian Hello, 12/20/68; John Houbolt, 12/5/66; George Jeffs, 1/26/70;

÷ÿÿÿÿÿ

Caldwell Johnson, Jr., 12/9/66; Walter Kapryan, 5/25/67; Alan Kehlet, 1/26/70; Christopher Kraft, Jr., 8/20/73; Eugene Kranz, 4/28/67; Thomas Markley, 1/17/68; Owen Maynard, 1/9/70; Riley McCafferty, 11/15/69 and 1/28/71; George E. Mueller, 10/4/66 and 6/27/67; Dale Myers, 5/12/69; Warren North, 5/1/70; John Paup, 6/7/66; Rocco Petrone, 5/19/70, 5/21/70, 9/17/70, and 5/25/72; Samuel Phillips, 7/22/70 and 9/25/70; William Rector III, 1/27/70; Rodney Rose, 5/6/70; Robert Seamans, Jr., 5/8/68, 5/20/68, and 6/3/68; Joseph Shea, 5/6/70 and 1/12/72; Joseph Thibodeaux, 8/23/73; Wernher von Braun, 11/30/71; Walter Williams, 1/27/70.

Interviews conducted by Robert Sherrod: George Abbey, 6/29/74 and 7/26/72; Neil Armstrong, 9/23/71; John Leland Atwood, 6/24/69; Charles Berry, 12/15/69; Ben Cate, 7/18/72; Walt Cunningham, 3/4/71; John Disher, 9/24/70 and 4/15/71; James Elms, 12/4/69 and 10/24/73; Llewellyn Evans, 6/11/69 and 10/5/70; Robert Gilruth, 3/12/71 and 11/14/72; Gerald Griffin, 3/8/74 and 4/11/74; Paul Haney, 1/6/70, 1/7/70, and 6/17/71; Thomas Kelly, 12/13/72; Christopher Kraft, Jr., 7/23/72 and 7/27/72; James Lovell, 6/28/74; George Low, 11/7/69, 12/30/69, 1/17/70, 2/14/70, 7/14/70, 8/12/70, 6/21/72, 6/28/72, 7/5/72, 9/7/72, 1/16/74, and 2/12/74; Charles Mathews, 2/17/70; George Mueller, 11/19/69, 4/21/71, 8/19/71, and 3/20/73; Dale Myers, 3/31/70; Thomas Paine, 1/23/69, 5/13/69, 10/13/69, 8/14/70, and 10/7/71; Rocco Petrone, 9/25/70, 7/30/71, 7/29/74, 9/26/74, and 10/8/74; Samuel Phillips, 7/2/71; Robert Piland, 12/3/71 and 1/7/72; Eberhard Rees, 2/6/71; Robert Seamans, Jr., 6/24/69; Joseph Shea, 5/6/71, 5/16/71, and 3/10/73; Sigurd Sjoberg, 7/28/72; Wernher von Braun, 11/19/69 and 8/25/70; James Webb, 8/2/68, 11/15/68, 6/8/69, 6/16/69, 9/17/69, and 4/28/71; Walter Williams, 6/29/68 and 2/10/71.

Other interviews: George Low, for the John F. Kennedy Oral History, 5/1/64; James Webb, by John Logsdon, 12/15/67; Harrison Storms, by the British Broadcasting Company, 5/14/79.

Notes

These notes identify the sources for direct quotations and descriptions of events obtained from published sources, memoranda, letters, or other written material. All other quotations and descriptions of events were drawn from the interviews listed in "*Apollo* as History."

Prologue

Pages 13–14   Stories from the front page of *The Washington Post*, 25 May 1961: "Mississippi Jails 27 Riders," "No Troops Asked, Says Johnson," and "President to Address Congress," by Chalmers M. Roberts.

Page 15   "I believe that this nation should commit itself." John F. Kennedy, quoted in John M. Logsdon, *The Decision to Go to the Moon: Project Apollo and the National Interest* (Chicago: University of Chicago Press, 1976), p. 128.

Page 17   "Robert Gilruth . . . was . . . aghast." Robert R. Gilruth, "Experts Were Stunned by Scope of Mission," *New York Times*, Moon Special Supplement, 17 July 1969; and Robert Gilruth, NASA interview, 21 March 1968, quoted in Courtney G. Brooks, James M. Grimwood, and Loyd S. Swenson, Jr., *Chariots for Apollo: A History of Manned Lunar Spacecraft* (Washington, D.C.: NASA, 1979), p. 31.

1. "That famous Space Task Group is akin to the *Mayflower*"

Page 21   Chapter title: Authors' interview with Clifford Charlesworth.

Page 24   "some advantages of tight, totalitarian control." *Dallas News* quoted in Walter A. McDougall, . . . *the Heavens and the Earth: A Political History of the Space Age* (New York: Basic Books, 1985), p. 143. Other examples of the world's reaction to Sputnik are taken from *Heavens*, pp. 142–44.

Page 24   "some fat Roman lolling in the baths." George R. Price, "Arguing the Case for Being Panicky," *Life*, 18 November 1957, p. 126.

Pages 24–25   Choice of the N.A.C.A. as the basis for NASA: McDougall, *Heavens*, chapter 7. Eisenhower's message to Congress proposing NASA is given in "President Details NACA Space Mission," *Aviation Week*, 14 April 1958, pp. 51–53.

Pages 25–26   Description of the founding of Langley is taken from James R. Hansen, *Engineer in Charge: A History of the Langley Aeronautical Laboratory 1917–1958* (Washington, D.C.: NASA, 1987), pp. 9–22.

Page 26   Description of John Victory and Langley's administrative system is taken from Hansen, *Engineer in Charge*, pp. 24, 28–36.

2. "I could picture the astronauts looking down at it with binoculars"

Page 37   Chapter title: Authors' interview with Maxime Faget.

Pages 38, 42–43   Glennan's meeting to discuss lunar landing: The official history of the Apollo 38, 42—43 program, Courtney G. Brooks, James M. Grimwood, and Loyd S. Swenson, Jr., *Chariots for Apollo: A History of Manned Lunar Spacecraft* (Washington, D.C.: NASA, 1979), dates this meeting from mid-1960 (p. 62). In an interview with the authors, Faget was sure it was in the spring of 1959, prior to any of the formal discussions of a lunar landing that began at the time of the Goett Committee (which met for the first time on 25 May 1959). Faget's memory of the approximate date is consistent with the naïveté of the discussion. By 1960, Faget and von Braun were already thinking about a lunar landing in much more sophisticated ways.

Pages 43–45   Account of the Goett Committee meetings: Brooks et al., *Chariots*, p. 8.

3. "Those days were out of the Dark Ages"

Page 46   Chapter title: Authors' interview with Joseph Bobik.

Pages 50–51   The German team's early days in Huntsville: Frederick I. Ordway III and Mitchell R. Sharpe, *The Rocket Team: From the V-2 to the Saturn Moon Rocket* (New York: Crowell, 1979); plus authors' interviews.

Page 54   Eisenhower's space policy in 1960: Courtney G. Brooks, James M. Grimwood, and Loyd S. Swenson, Jr., *Chariots for Apollo: A History of Manned Lunar Spacecraft* (Washington, D.C.: NASA, 1979), pp. 13–14.

Page 54   Naming the rockets: Erik Bergaust, *Wernher von Braun* (Washington, D.C.: National Space Institute, 1976), p. 406.

Page 55   Participants in the lunch where Silverstein floated "Apollo" as a name: Ivan D. Ertel and Mary Louise Morse, *The Apollo Spacecraft: A Chronology* (Washington, D.C.: NASA, 1978), vol. 1, p. 36. The account of the lunch comes from authors' interviews with Charles Donlan and Abe Silverstein.

Pages 55–56   The failure of M.A.-1: Loyd S. Swenson, Jr., James M. Grimwood, and Charles C. Alexander, *This New Ocean: A History of Project Mercury* (Washington, D.C.: NASA, 1966), p. 275; plus authors' interviews.

Page 56   The July 1960 Washington conference: Brooks et al., *Chariots*, p. 15; plus authors' interviews.

Pages 57–58   "Subject: Manned Lunar Landing Program." A copy of Low's memorandum with Silverstein's scrawled "OK" was provided courtesy of Mary R. Low.

4. "He would rather not have done it"

Page 59   Chapter title: Authors' interview with Jerome Wiesner.

Page 60   Space as a campaign issue: Walter A. McDougall, . . . *the Heavens and the Earth: A Political History of the Space Age* (New York: Basic Books, 1985), pp. 221–26.

Page 61   ". . . could not be convinced that all rockets were not a waste of money." NASA interview with Charles Draper, quoted in McDougall, *Heavens*, p. 302.

Page 61   Space as Kennedy's weak spot: Hugh Sidey, *John F. Kennedy, President* (New York: Atheneum Press, 1964), p. 118; plus authors' interviews with Sidey and Jerome Wiesner.

Pages 63–64   The cause of the M.R.-1 fiasco: Low memorandum to Glennan, quoted in David Baker, *The History of Manned Space Flight* (New York: Crown Publishers, 1982), p. 62. Low's memo indicates that the plug was designed with one prong shorter than the other. We have used Terry Greenfield's account of the technician filing off one prong (authors' interview) because he was the man on the scene and more directly associated with the investigation.

Page 65   Eisenhower's plan to drop the National Aeronautics and Space Council: McDougall, *Heavens*, p. 309.

Page 66   The reversal of Kennedy's tentative decision to drop the Space Council: John M. Logsdon, *The Decision to Go to the Moon: Project Apollo and the National Interest* (Chicago: University of Chicago Press, 1976), p. 67.

Page 67   "We should stop advertising Mercury." Wiesner Report, quoted in Loyd S. Swenson, Jr., James M. Grimwood, and Charles C. Alexander, *This New Ocean: A History of Project Mercury* (Washington, D.C.: NASA, 1966), pp. 304–8.

Page 68ff.   Quotations from the Disher notebooks courtesy of unpublished materials provided to the authors by John Disher.

Page 69   Outcome of the Glennan review of the lunar program: Logsdon, *Decision*, p. 61.

Page 69   The number of people who turned down the NASA job: Seamans remembered that it was only nine. Johnson told Eugene Emme, the NASA historian, that the number was seventeen (contained in Emme's exit interviews with Robert Seamans, 8 May 1968).

Pages 70–71   The account of Webb's hiring is drawn from John Logsdon's interview with James Webb on 15 December 1967.

Page 74   Kennedy put funds for Apollo spacecraft on indefinite hold: Memorandum from James Webb to President Kennedy, 23 March 1961; attached memorandum from Robert Seamans to James Webb, 23 March 1961; and Robert Seamans' notations on results of the meeting with Kennedy. Memoranda provided courtesy of Robert Seamans.

5. "We're going to the moon"

Page 75 Chapter title: Statement of Ted Sorensen, recounted in authors' interview with Hugh Sidey.

Page 75 "hour of euphoria." Arthur M. Schlesinger, Jr., *A Thousand Days: John F. Kennedy in the White House* (Boston: Houghton Mifflin, 1965). The phrase is the title of chapter nine.

Page 75 "He is not only the handsomest." James MacGregor Burns, "John Kennedy and His Spectators," *New Republic*, 3 April 1961, p. 7.

Page 75 "the magic touch." Interview with Theodore Sorensen on ABC News, "Our World": "One Day: April 12, 1961."

Page 76 "fifty feet high," and following account of Jackie Kennedy's party for the women reporters: "Exposure," *Time*, 21 April 1961, p. 16.

Page 76 "Give me the news in the morning." Kennedy quoted in Hugh Sidey, "How the News Hit Washington—With Some Reactions Overseas," *Life*, 21 April 1961, p. 26.

Page 77 "In these matters, what people believe." *Washington Post*, 13 April 1961, p. A18.

Page 77 "the American people must be 'properly alerted.' " From hearings of the House Space Committee, quoted in John M. Logsdon, *The Decision to Go to the Moon: Project Apollo and the National Interest* (Chicago: University of Chicago Press, 1976), p. 103.

Page 77 "see in all this the supposed superiority of the Communist way of life." This and following reactions from abroad are from United States Information Agency studies of the reaction to Vostok I, quoted in Walter A. McDougall, . . . *the Heavens and the Earth: A Political History of the Space Age* (New York: Basic Books, 1985), pp. 246–47.

Page 77 "young, active, and vigorous leader." *New York Times*, 16 April 1961, Sec. 4, p. 3.

Page 77 "it is a fact that it's going to take some time." This and following quotation from the Kennedy press conference are taken from the transcript in *The New York Times*, 13 April 1961, p. 18.

Page 78 "today belonged to the Russians." Excerpt from Edwin Newman's commentary shown on ABC News, "Our World."

Page 78 Meeting in Sorensen's White House office: Theodore C. Sorensen, *The New York Times*, Moon Special Supplement, 17 July 1969.

Page 79 "Kennedy turned back to the men around him." Hugh Sidey, *John F. Kennedy, President* (New York: Atheneum Press, 1964), pp. 122–23.

Page 80 Description of the White House during the Bay of Pigs crisis: Schlesinger, *Thousand Days*, pp. 276–84.

Page 80 "In accordance with our conversation." Memorandum from Kennedy to Lyndon Johnson, quoted in Logsdon, *Decision*, pp. 109–10.

Page 81 Participants at the meeting called by Johnson: Logsdon, *Decision*, p. 114.

Page 83   "a use of technological means for political ends." Wiesner quoted in Logsdon, *Decision*, p. 118.

6. "The flight article has got to dominate"

Page 87   Chapter title: Authors' interview with Rocco Petrone.

Page 88   "a fair-sized nuclear weapon might one day explode." Fact book provided by Kennedy Space Center, 1987 edition, p. 74. While the figure of 1,000,000 pounds (500 tons) of T.N.T. is theoretically correct and they had to plan for the worst case, Rocco Petrone pointed out that an actual explosion of the Saturn V was more likely to be on the order of 300–400 tons of T.N.T.

Page 89   Sites under consideration: Charles D. Benson and William Barnaby Faherty, *Moonport: A History of Apollo Launch Facilities and Operations* (Washington, D.C.: NASA, 1978), p. 91.

Page 90   Projected launches per year: Benson and Faherty, *Moonport*, pp. 67–68.

Page 90   Use of mobile launch facilities at Peenemünde: Benson and Faherty, *Moonport*, p. 74.

Page 91   Meeting of Seamans and Debus regarding mobile launch system: Benson and Faherty, *Moonport*, p. 77.

Pages 93–99   Specific figures regarding the size and construction of the V.A.B., the launcher, and the crawler were taken from Benson and Faherty, *Moonport*, chaps. 4 and 6, and from materials provided courtesy of Donald Buchanan.

Pages 96–98   Account of the decision on the barge: Benson and Faherty, *Moonport*, p. 118ff., plus authors' interviews.

7. "We had more harebrained schemes than you could shake a stick at"

Page 100   Chapter title: Authors' interview with Caldwell Johnson.

Page 108   The first article in the *Collier's* series was "What Are We Waiting For?" *Collier's*, 22 March 1952, p. 23ff.

8. "Somewhat as a voice in the wilderness . . ."

Page 113   Chapter title: Letter from John Houbolt to Robert Seamans, 15 November 1961, provided courtesy of John Houbolt.

Page 113   The earliest thinking about lunar-orbit rendezvous: John M. Logsdon, "Selecting the Way to the Moon: The Choice of the Lunar Orbital Rendezvous Mode," *Aerospace Historian* (June 1971), pp. 63–70.

Page 114   For estimates of weight savings using L.O.R.: Courtney G. Brooks, James M. Grimwood, and Loyd S. Swenson, Jr., *Chariots for Apollo: A History of Manned Lunar Spacecraft* (Washington, D.C.: NASA, 1979), p. 66.

Page 115n   "Herein lies contention." The article that prompted the subsequent controversy over John Houbolt's role was D. Sheridan, "How an Idea

No One Wanted Grew Up to Be the LEM,'' *Life*, 14 March 1969, pp. 20–27.

Page 116   "Almost spontaneously . . . it became clear.'' John C. Houbolt, "Lunar Rendezvous,'' *International Science and Technology* (February 1963), p. 63.

Page 119   "Somewhat as a voice in the wilderness.'' Letter from John Houbolt to Robert Seamans, 15 November 1961, provided courtesy of John Houbolt.

Page 120   "Nevertheless, I feel that Houbolt's message.'' Copy of memorandum from George Low to Brainerd Holmes, 5 December 1961, reprinted with permission of the George M. Low Papers, Institute Archives, Rensselaer Polytechnic Institute.

9. "What sonofabitch thinks it isn't the right thing to do?"

Page 124   Chapter title: Words attributed to John Paup in authors' interview with Owen Maynard, confirmed in authors' interview with Caldwell Johnson.

Pages 125–26   "rendezvous schemes may be used as a crutch.'' Letter from Robert Gilruth to Nicholas Golovin, 12 September 1961, quoted in Barton C. Hacker and James M. Grimwood, *On the Shoulders of Titans: A History of Project Gemini* (Washington, D.C.: NASA, 1977), p. 61.

Page 130   "in a climate 'permitting year-round, ice-free water transportation.' '' This and the other site-selection criteria are given in Courtney G. Brooks, James M. Grimwood, and Loyd S. Swenson, Jr., *Chariots for Apollo: A History of Manned Lunar Spacecraft* (Washington, D.C.: NASA, 1979), Appendix A.

Page 130   a "politically arranged gift.'' Brooks et al., *Chariots*, p. 53.

Page 135   "Most of the M.S.C. people seem enthusiastic about L.O.R.'' Memorandum from Joseph Shea to Brainerd Holmes, quoted in John M. Logsdon, "Selecting the Way to the Moon: The Choice of the Lunar Orbital Rendezvous Mode,'' *Aerospace Historian* (June 1971), p. 68.

Page 135   Changing weight estimates for the L.O.R. mode: Brooks et al., *Chariots*, p. 78.

Page 139   "A drastic separation of these two functions.'' Brooks et al., *Chariots*, p. 82.

10. "It aged me, I'm sure"

Page 144   Chapter title: Authors' interview with Jerry Thomson.

Page 145   Doubt within the President's Science Advisory Committee (PSAC) that the F-1 could be made to work: Roger E. Bilstein, *Stages to Saturn: A Technological History of the Apollo/Saturn Launch Vehicles* (Washington, D.C.: NASA, 1980), p. 113.

Pages 146–47   Description of the workings of the F-1 engine: The initial draft was drawn from Bilstein, *Stages to Saturn*, pp. 104–25, and revised on the basis of critiques by Jerry Thomson and Paul Castenholz (who are not responsible for technical simplifications retained by the authors for purposes of clarity).

Page 152   "Lunar program in crisis." *New York Times*, 13 July 1963, p. 9.

Page 152   Holmes as a "restless, dynamic worker." "Reaching for the Moon," *Time*, 10 August 1962, p. 52ff.

Page 152   "An American Tragedy." *Missiles and Rockets*, 17 June 1963, p. 54.

Page 153   "It is probably too much to say." Richard Austin Smith, "Now It's an Agonizing Reappraisal of the Moon Race," *Fortune* (November 1963), pp. 124–29.

Page 153   "lunar landing cannot likely be attained." Copy of the Disher and Tischler briefing for George Mueller, provided courtesy of John Disher.

11. "It sounded reckless"

Page 155   Chapter title: Wernher von Braun, "Saturn the Giant," in *Apollo Expeditions to the Moon*, edited by Edgar Cortright (Washington, D.C.: NASA, 1975), p. 50.

Page 160   "an official schedule reflecting the philosophy outlined here." Quoted in Roger E. Bilstein, *Stages to Saturn: A Technological History of the Apollo/Saturn Launch Vehicles* (Washington, D.C.: NASA, 1980), p. 349.

Page 161   [a sign which] "raises a serious question." *New York Times*, 21 November 1963, p. 24.

Pages 161–62   Arthur Rudolph's encounter with George Mueller: Bilstein, *Stages to Saturn*, p. 351.

Page 162   Von Braun overriding the objections of his senior staff on the all-up decision: Bilstein, *Stages to Saturn*, note 7 to Chapter 12.

Page 162   "just as Dr. Mueller could not guarantee." Bilstein, *Stages to Saturn*, p. 350.

Page 162   "George Mueller visited Marshall." This and the paragraph's other quotations of von Braun are from "Saturn the Giant," *Apollo Expeditions*, p. 50.

Page 164   "I think you know that I would be enthusiastic." Memorandum from Joseph Shea to George Mueller, 17 September 1963, provided courtesy of Joseph Shea.

12. "Hey, it isn't that complicated"

Page 166   Chapter title: Authors' interview with Joseph Shea.

Page 168   "The Martin Company is considered the outstanding source." Quoted in Courtney G. Brooks, James M. Grimwood, and Loyd S. Swenson, Jr., *Chariots for Apollo: A History of Manned Lunar Spacecraft* (Washington, D.C.: NASA, 1979), p. 43.

Pages 168–69   The rumors and coincidences surrounding the C.S.M. contract: John Noble Wilford, *We Reach the Moon: The* New York Times *Story of Man's Greatest Adventure* (New York: Bantam, 1969), pp. 106–9.

Page 175*n*   Tracking down the origins of "the better is the enemy of the good":

Robert Sherrod, unpublished manuscript, provided courtesy of Robert Sherrod.

13. "We want you to go fix it"

Page 179   Chapter title: Words attributed to Gus Grissom in unpublished manuscript by Joseph Shea.

Page 182n   Changes in the Block I schedule of manned flights: Courtney G. Brooks, James M. Grimwood, and Loyd S. Swenson, Jr., *Chariots for Apollo: A History of Manned Lunar Spacecraft* (Washington, D.C.: NASA, 1979), p. 130.

Page 183   "I am definitely not satisfied." Quoted in Roger E. Bilstein, *Stages to Saturn: A Technological History of the Apollo/Saturn Launch Vehicles* (Washington, D.C.: NASA, 1980), p. 226.

Page 183   "I believe NASA has to resort." Quoted in Bilstein, *Stages to Saturn*, p. 227.

Pages 184–85   "This is not a meeting to bring up." This and subsequent quotes from the CARR are from a partial transcript of the CARR, provided courtesy of Robert Sherrod.

Page 185   "Specific NASA direction." "CARR Action Responses—CSM 012," 26 September 1966, Internal Letter, North American Aviation, J.S.C. History Office.

Page 185   "I do not think it technically prudent." Letter from Hilliard Paige to Joseph Shea, 30 September 1966, J.S.C. History Office.

Page 185   "our usual press of business." Memorandum from William Bland to Joseph Shea, 23 November 1966, J.S.C. History Office.

Page 186   "our inherent hazards from fire." Letter from Joseph Shea to Hilliard Paige, 5 December 1966, J.S.C. History Office.

14. "Did he say 'fire'?"

Page 189   Chapter title: Authors' interview with John Tribe.

Pages 189–91   The description of the probable source of the fire is taken from the *Final Report of the Apollo 204 Review Board* as summarized in David Baker, *The History of Manned Space Flight* (New York: Crown Publishers, 1982), pp. 279–83, plus authors' interviews.

Page 191ff.   The description of the crew's transmission is taken from Part V.c of the *Report of the Apollo 204 Review Board*, "Analyses of Crew Voice Transmission During the Fire," pp. 5.8–5.9, plus authors' interviews with Martin Cioffoletti and Donald Arabian, who listened to the tape repeatedly as part of the investigation.

Pages 191–98   The accounts of Gary Propst, Donald Babbitt, James Gleaves, Clarence Chauvin, Deke Slayton, Stephen Clemmons, and L. D. Reece are taken from statements given in testimony before the Apollo 204 Review Board as published in Appendix B to the *Final Report of the Apollo 204 Review Board*.

Page 197   *New York Times'*s report of the spacecraft and crew's condition: This was a scandal at the time, with NASA angrily denying the report (which in fact had been grossly inaccurate). References to the article indicate that the *Times*'s article appeared on Sunday. But a review of the *Times* coverage of the fire on Sunday, January 29, and the rest of the week as well, using the microfilmed *New York Times* held by the Library of Congress, failed to uncover the offending passage—it apparently ran only in the paper's early editions (the final edition is used for archival purposes). But it clearly existed. Here is *Time* magazine's summary of it: "Quoting an unidentified 'official source,' the *New York Times* said that the three had suffered horribly as the fire spread: that they shrieked repeatedly, pleading for help; that they died scrambling frantically at the sealed hatch cover of the capsule, leaving shreds of flesh on the metal; and that their bodies were incinerated until little more than bones remained." "Inquest on Apollo," *Time,* 10 February 1967, p. 9.

Page 197   Condition of the crew after the fire: Taken from the Report of Panel 11, Medical Analysis Panel, Appendix D-11 to the *Final Report of the Apollo 204 Review Board,* pp. D11-7 to D11-8, plus unpublished materials provided courtesy of Joseph Shea.

Page 201   "a spine-tingling affair." Claudia Alta Johnson, *A White House Diary* (New York: Holt, Rinehart, and Winston, 1970), p. 481.

Page 202   Gilruth catching a plane to the Cape: This is Gilruth's recollection from the authors' interview with him. It is a minor point, but the plane may have been North American's or McDonnell's—during his research, Robert Sherrod tried valiantly to pin this point down but without success.

15. "The Crucible"

Page 206   Chapter title: From Joe Shea's original title for the Goddard Lecture, "The Crucible of Development."

Pages 206–18 The account of Shea's activities: From the week after the fire until he left Houston for the headquarters job, Shea kept a daily technical diary, supplemented with personal observations. A few years after the fire, Shea also wrote up a detailed account of the first weekend after the fire. Much of the account in this chapter is drawn from these materials, provided courtesy of Joseph Shea.

Page 211   "What is done is done." From "Talk to Apollo Contractors," 3 February 1967, provided courtesy of Joseph Shea.

Page 212   "I may be over optimistic." From Shea's technical diary.

Page 215   "a more thorough inspection was required." Quoted in Ivan D. Ertel and Roland W. Newkirk with Courtney G. Brooks, *The Apollo Spacecraft: A Chronology* (Washington, D.C.: NASA, 1978), vol. 4, pp. 80–81.

Page 215   1,407 errors: Charles D. Benson and William Barnaby Faherty, *Moonport: A History of Apollo Launch Facilities and Operations* (Washington, D.C.: NASA, 1978), p. 411.

Page 215   "The more we probed." Frank Borman and Robert J. Serling, *Countdown: An Autobiography* (New York: William Morrow, 1988), p. 175.

Page 215   "getting drunk seemed like a good idea." Borman and Serling, *Countdown*, p. 173. Other sources for this account are Shea's technical diary and Robert Sherrod's interview with Paul Haney, 7 January 1970.
Page 217   Description of the Goddard Lecture is drawn from the typescript of the speech, provided courtesy of Joseph Shea.
Page 220   "The dry technical prose of the report." "Incompetence and Negligence" (editorial), *New York Times*, 11 April 1967.

16. "You've got to start biting somewhere"

Page 226   Chapter title: NASA history interview with Bastian Hello, 20 December 1968.
Page 227   Recruiting George Low for ASPO: For Low's very carefully worded account of this incident, see George Low, "The Spaceships," in *Apollo Expeditions to the Moon*, edited by Edgar Cortright (Washington, D.C.: NASA, 1975), pp. 59–60.
Page 229   "I hope you don't think I'm nuts." Memorandum from Julian Scheer to George Low, provided courtesy of Judith Wyatt.
Page 230   "My general impression." George Low, Apollo Notes to Dr. Gilruth, 18 May 1967. Reprinted with permission of the George M. Low Papers, Institute Archives, Rensselaer Polytechnic Institute.
Page 230   "North American was positively schizophrenic." Frank Borman and Robert J. Serling, *Countdown: An Autobiography* (New York: William Morrow, 1988), p. 182.
Page 230   "No copies to *anyone* except RRG." George Low, Apollo Notes to Dr. Gilruth, 26 September 1967. Reprinted with permission of the George M. Low Papers, Institute Archives, Rensselaer Polytechnic Institute.
Page 235   Grumman "needs a Healey." George Low, Apollo Notes to Dr. Gilruth, 5 October 1967. Reprinted with permission of the George M. Low Papers, Institute Archives, Rensselaer Polytechnic Institute.
Page 235   "fewer discrepancies than on any spacecraft." Courtney G. Brooks, James M. Grimwood, and Loyd S. Swenson, Jr., *Chariots for Apollo: A History of Manned Lunar Spacecraft* (Washington, D.C.: NASA, 1979), p. 253.

17. "And then on launch day it worked"

Page 237   Chapter title: Authors' interview with Isom (Ike) Rigell.
Page 238   The numbering of the Apollo flights: Courtney G. Brooks, James M. Grimwood, and Loyd S. Swenson, Jr., *Chariots for Apollo: A History of Manned Lunar Spacecraft* (Washington, D.C.: NASA, 1979), pp. 231–32.
Page 239   Cracking pressure of eggs as the standard for the crane operators: Charles D. Benson and William Barnaby Faherty, *Moonport: A History of Apollo Launch Facilities and Operations* (Washington, D.C.: NASA, 1978), p. 412. Details of the joining of the stages and the lists of tests also come from *Moonport*, pp. 412–13. Description of the problems with the

C.D.D.T. are taken from *Moonport*, pp. 427–28, and Roger E. Bilstein, *Stages to Saturn: A Technological History of the Apollo/Saturn Launch Vehicles* (Washington, D.C.: NASA, 1980), pp. 352–55, plus authors' interviews.

Page 243   Arthur Rudolph's history with von Braun: Frederick I. Ordway III and Mitchell R. Sharpe, *The Rocket Team: From the V-2 to the Saturn Moon Rocket* (New York: Crowell, 1979), pp. 22–23.

Pages 244–48 Sequence of operations in the countdown: Roger E. Bilstein, *Stages to Saturn: A Technological History of the Apollo/Saturn Launch Vehicles* (Washington, D.C.: NASA, 1980), Appendix B: Saturn V Prelaunch-Launch Sequence.

Pages 245–46 Sequence of operations in engine start: Bilstein, *Stages to Saturn*, p. 111.

Pages 245–46 Mike Collins's reflections on watching the launch of 501: Michael Collins, *Carrying the Fire: An Astronaut's Journey* (New York: Farrar, Straus, and Giroux, 1974), p. 286.

18. "We're going to put a guy in that thing and light it"

Page 253   Chapter title: Authors' interview with Walter Williams.

Page 267   "It's almost impossible to describe." Unpublished manuscript by Walter Williams, NASA History Office, Washington.

19. "There will always be people who want to work in that room"

Page 270   Chapter title: Authors' interview with Michelle Brekke (the first woman flight director).

20. "The Flight Director may take any necessary action"

Page 282   Chapter title: From Rule 1–6, from *Flight Mission Rules, Apollo 11*: "The Flight Director may, after analysis of the flight, choose to take any necessary action required for the successful completion of the mission." Provided courtesy of Eugene Kranz. The statement of the rule remained the same throughout the Apollo Program.

21. "There was no mercy in those days"

Page 291   Chapter title: Authors' interview with Edward Fendell.

Page 292   "Specifically, his job is to determine the operational rules." George M. Low, "Announcement: Chief of Apollo Data Priority Coordination, Apollo Spacecraft Program Office," 3 August 1967, provided courtesy of Howard W. Tindall, Jr.

Page 296   spacecraft velocity as "teensy weensy." From Howard W. Tindall, Jr., "Transearth midcourse correction philosophy—a major operational breakthrough!" 3 June 1968, provided courtesy of Howard W. Tindall, Jr. The discussion of the "unbelievable" proposal is in "LM rendezvous radar is essential," 1 August 1968.

Page 297   "The present LM weight and descent trajectory." From Howard W.
                  Tindall, Jr., "LM DPS low level light fixing," 25 November 1968,
                  provided courtesy of Howard W. Tindall, Jr.

22. "You've lost the engines?"

Page 308   Chapter title: Clifford Charlesworth, transcribed from the flight director's
                  loop for Apollo 6.
                  Details regarding launch times and timing of events during flights, for
                  this and subsequent chapters: David Baker, *The History of Manned Space
                  Flight* (New York: Crown Publishers, 1982), especially Tables 7–15 at
                  the back of the book.
Page 313   Fixes for pogo in A.S.-502: Roger E. Bilstein, *Stages to Saturn: A
                  Technological History of the Apollo/Saturn Launch Vehicles* (Washing-
                  ton, D.C.: NASA, 1980), p. 363.
Pages 313–14   Fix for the engine failures in A.S.-502: Samuel C. Phillips, "The
                  Shakedown Cruises," in *Apollo Expeditions to the Moon*, edited by
                  Edgar Cortright (Washington, D.C.: NASA, 1975), p. 168; plus authors'
                  interview with Jerry Thomson.
Page 316   "Chris Kraft and I agreed." George Low, Apollo Notes to Dr. Gilruth,
                  24 May 1968. Reprinted with permission of the George M. Low Papers,
                  Institute Archives, Rensselaer Polytechnic Institute.
Page 316   Low's first thoughts about a circumlunar flight in 1968: On 19 August
                  1968, George Low dictated a thirteen-page memorandum for his private
                  files called "Special Notes for August 9, 1968, and Subsequent." He
                  began by discussing the period of June–July 1968, describing the delays
                  in preparing the first lunar module and stating, "In this time period also
                  the possibility of a circumlunar or lunar orbit mission during 1968, using
                  AS 503 and CSM 103, first occurred to me as a contingency mission to
                  take a major step forward in the Apollo Program" (p. 1). "Special
                  Notes" were provided courtesy of Judith Wyatt. Copies are also filed in
                  the J.S.C. History Office and the George M. Low Papers at Rensselaer.
Pages 318–19   The August 9 meeting at Marshall: See Samuel Phillips account in "The
                  Shakedown Cruises," *Apollo Expeditions*, p. 171, and "Minutes of
                  Meeting Held at MSFC, 9 August 1968, at Dr. Gilruth's Request,"
                  J.S.C. History Office.
Page 318   "It doesn't matter to the launch vehicle." Attributed to von Braun by
                  George Low, "Special Notes for August 9 and Subsequent," p. 8.

23. "It was darn scary"

Page 320   Chapter title: Authors' interview with Martin Cioffoletti.
Page 322   "We'll have a hell of a time." Samuel Phillips, "The Shakedown
                  Cruises," in *Apollo Expeditions to the Moon*, edited by Edgar Cortright
                  (Washington, D.C.: NASA, 1980), p. 172.
Page 322   "Mueller was 'skeptical and cool.' " Phillips, p. 172.
Page 324   "there is no technical reason not to fly Apollo 8." From an unpublished

manuscript by Robert Sherrod, who was present at the meeting, provided courtesy of Robert Sherrod.

Page 324   "There are grave risks." Letter from George Mueller to Robert Gilruth, 4 November 1968.

Page 324   Account of meetings on November 10–11: George Low, "Special notes for November 10 and 11, 1968," 14 November 1968.

Page 328   "Suddenly the familiar map of the earth." Ibid.

24. "We . . . we're go on that, Flight"

Page 335   Chapter title: Steven Bales, transcribed from the flight director's loop for Apollo 11.

Page 336   "Suddenly we were in a very, very free-form world." Will Bischoff, quoted in Charles R. Pellegrino and Joshua Stoff, *Chariots for Apollo: The Making of the Lunar Module* (New York: Atheneum, 1985), p. 33. To avoid confusion with Brooks et al., *Chariots for Apollo*, the short citation for the Pellegrino and Stoff book will be *Lunar Module*.

Page 336   Stories about volatile fuel and pressures in the tanks: Pellegrino and Stoff, *Lunar Module*, pp. 41–44ff.

Page 338*n*   The reason for choosing 47,000 feet as the cutoff point for Apollo 10: John Noble Wilford, *We Reach the Moon: The* New York Times *Story of Man's Greatest Adventure* (New York: Bantam, 1969), p. 233.

Page 346   "The Apollo 12 crew was in the simulator." This detail and others regarding the simulations preceding Apollo 11 were provided courtesy of Eugene Kranz, personal communication, after he had reviewed the simulation logs for that period.

25. "Well, let's light this sumbitch and it better work"

Page 356   Chapter title: Authors' interview with Glynn Lunney.

Page 364   "There's no telling what it will do." Charles R. Pellegrino and Joshua Stoff, *Chariots for Apollo: The Making of the Lunar Module* (New York: Atheneum, 1985), p. 167. Their account of the little-known problem with the blocked fuel line, pp. 166–68, is by far the most complete ever published, and is the main source for our description. Additional technical information was obtained from Donald Arabian et al., *Mission Evaluation Report: Apollo 11*. Such a report was prepared by the Test Division for each of the flights, and was provided to the authors courtesy of Robert Fricke.

Page 365   Neil Armstrong breaking the arming switch: Pellegrino and Stoff, *Lunar Module*, pp. 168–69. Additional technical information from Arabian et al., *Mission Evaluation Report: Apollo 11*.

Page 369   Mike Collins awaiting *Eagle*'s liftoff: Michael Collins, *Carrying the Fire: An Astronaut's Journey* (New York: Farrar, Straus, and Giroux, 1974), p. 418.

Page 370   Display on the MOCR's screens after splashdown: David Baker, *The*

History of Manned Space Flight (New York: Crown Publishers, 1982), p. 358.

26. "I think we need to do a little more all-weather testing"

Page 371   Chapter title: Charles (Pete) Conrad, transcribed from the air-to-ground loop for Apollo 12.
Page 379   Details about the lightning strike: Donald Arabian et al., *Mission Evaluation Report: Apollo 12*; plus authors' interviews.

27. "You really need to understand that the C.S.M. is dying"

Page 387   Chapter title: Authors' interview with John Aaron.
Page 387   Safety record of manned space flight: Baker, *History*, Table 27, p. 540.
Page 388   Swigert's position as a last-minute replacement: Henry S. F. Cooper, Jr., *13: The Flight That Failed* (New York: Dial Press, 1973), pp. 12–13.
Page 397   "If somebody had thrown that at us in the simulator." Quoted in Cooper, *Flight That Failed*, p. 41.
Page 404   For unusually clear illustrations of the systems involved in the accident, see Lovell, "Houston, We've Had a Problem," in *Apollo Expeditions*, pp. 252–53.
Pages 404–6   Account of the history of O_2 Tank 2 is drawn from David Baker, *The History of Manned Space Flight* (New York: Crown Publishers, 1982), pp. 374 and 387; plus authors' interviews with Seymour Liebergot and Donald Arabian.
Page 406   The times of the critical events are taken from the chronology in James A. Lovell, "Houston, We've Had a Problem," in *Apollo Expeditions to the Moon*, edited by Edgar Cortright (Washington, D.C.: NASA, 1980), pp. 250–51.

28. "You don't have time to worry"

Page 408   Chapter title: Authors' interview with Glynn Lunney.
Page 419   "It's up to you now." Quoted in LIFE *in Space*, edited by Robert Grant Mason (New York: Time-Life Books, 1983), p. 182.
Page 420   The scene at Grumman: Charles A. Pellegrino and Joshua Stoff, *Chariots for Apollo: The Making of the Lunar Module* (New York: Atheneum, 1985), pp. 189–90.
Pages 420–21   National and worldwide reaction: David Baker, *The History of Manned Space Flight* (New York: Crown Publishers, 1982), pp. 378, 382; "Apollo's Return: Triumph Over Failure," *Time*, 27 April 1970, pp. 12–14; and "Apollo 13: Three Who Came Back," *Newsweek*, 27 April 1960, pp. 21–27.
Pages 421–22   Deiterich's first meeting in Room 210: Henry S. F. Cooper, Jr. *13: The Flight That Failed* (New York: Dial Press, 1973), pp. 68–70; plus authors' interviews.

Page 422   Analogy between burn and coming about in a sailing ship: Cooper, *Flight That Failed*, p. 71.

29. "I hope the guys who thought this up knew what they were doing"

Page 424   Chapter title: James Lovell on the air-to-ground loop during Apollo 13, quoted in Henry S. F. Cooper, Jr., *13: The Flight That Failed* (New York: Dial Press, 1973), p. 130. The full quotation is given later in the chapter: "I hope the guys in the back room who thought this up knew what they were doing" (p. 440).

Page 424   "Nothing had been built yet." Tom Kelly, quoted in Charles R. Pellegrino and Joshua Stoff, *Chariots for Apollo: The Making of the Lunar Module* (New York: Atheneum, 1985), p. 190.

Page 426n   "They can throw anything at us." Dave Reed, quoted in Cooper, *Flight That Failed*, p. 88

Pages 426–27 Data on consumables: Pellegrino and Stoff, *Lunar Module*, p. 194; David Baker, *The History of Manned Space Flight* (New York: Crown Publishers, 1982), pp. 378–79.

Pages 427–28 Account of the Abort Sensor Assembly: Pellegrino and Stoff, *Lunar Module,* pp. 194–95.

Page 428   John Strakosch's memory of the crisis: Pellegrino and Stoff, *Lunar Module*, p. 196.

Pages 429–31 Account of the meeting in the viewing room: Our account draws from Cooper, *Flight That Failed*, pp. 82–85, but we differ with him on one important point. Cooper ascribes the decision to Gilruth, writing, "He was the senior man present, and when NASA people met to make a decision there was no voting; rather, after discussion the top man made the decision" (p. 84). In an interview with the authors, Kranz described the decision as having been made earlier, at the meeting of the flight directors with Kraft. Lunney, who attended the viewing room meeting, saw it as a process in which, as customary with operational decisions, senior management made sure that the flight controllers had thought through their decision, then ratified it. In any case, the senior man in the chain of command attending the meeting that morning was not M.S.C. director Gilruth, but Rocco Petrone, head of the Apollo Program (and, if it came to it, Thomas Paine, administrator of NASA). The account of the meeting also draws from authors' interviews with David Reed and Charles Deiterich.

Page 432   "the biggest heart-stopper." This and other quotations from Lovell regarding the alignment are from James A. Lovell, "Houston, We've Had a Problem," in *Apollo Expeditions to the Moon*, edited by Edgar Cortright (Washington, D.C.: NASA, 1980), pp. 247–65.

Page 438   Writing the instructions for constructing the lithium hydroxide box: Cooper, *Flight That Failed*, pp. 127–28; plus authors' interviews.

Page 440   "by Saturday or Sunday." Cooper, *Flight That Failed*, p. 143.
Page 440   "I hope the guys." Lovell, quoted in Cooper, *Flight That Failed*, p. 130.

Page 443   "I think everybody in the room." Deiterich, quoted in Cooper, *Flight That Failed*, p. 166.

Page 444   TELMU's wish to bring Aquarius home: Cooper, *Flight That Failed*, p. 185.

Page 444   Grumman engineer's memory of saving a swimmer: Pellegrino and Stoff, *Lunar Module*, p. 209.

Page 444   "I don't believe you, FIDO!" Deiterich, quoted in Cooper, *Flight That Failed*, p. 188.

30. "We drank the wine at the pace they handed it to us"

Page 447   Chapter title: Authors' interview with John Aaron.

Epilogue: Twenty Years after the First Landing

Page 463   Arthur Rudolph's departure: "German-Born Nazi Expert Quits U.S. to Avoid a War Crimes Suit," *New York Times*, 18 October 1984, p. 1. Rudolph left the United States in November 1983 under an agreement with the Department of Justice and renounced his citizenship on 25 May 1984, though the news did not become public until October 1984. Apparently Rudolph was not accused of committing atrocities, but of being aware of them and failing to acknowledge that in his application for U.S. citizenship. Rudolph maintained his innocence, but said that because of his age and financial situation he would not undertake the prolonged litigation that the government suit would have entailed.

Index

About the Authors

Catherine Bly Cox was educated at William and Mary, Oxford, and Yale. Charles Murray was educated at Harvard and the Massachusetts Institute of Technology. He is the author of the best-selling *Losing Ground* and *In Pursuit: of Happiness and Good Government*. They share a long-standing fascination with the exploration of space, and live with their children in Washington, D.C.

629.454 Murray, Charles A.
MUR

Apollo, the race
to the moon

$24.45

DATE			

11/89

© THE BAKER & TAYLOR CO.